This book belongs to:

Fil♡

FORENSIC EXAMINATION OF FIBRES

ELLIS HORWOOD SERIES IN FORENSIC SCIENCE
Series Editor: JAMES ROBERTSON, Head of Forensic Services Division,
Australian Federal Police, Canberra, Australian Capital Territory

Aitkin and Stoney — **THE USE OF STATISTICS IN FORENSIC SCIENCE**
Cole and Caddy — **THE ANALYSIS OF DRUGS OF ABUSE: An Instruction Manual**
Ellen — **THE SCIENTIFIC EXAMINATION OF DOCUMENTS**
Pounder — **SUDDEN NATURAL DEATH: Forensic Pathology and Medico-Legal Investigation**
Robertson, Ross and Burgoyne — **DNA IN FORENSIC SCIENCE: Theory, Techniques and Applications**
Robertson — **FORENSIC EXAMINATION OF FIBRES**
Tebbett — **GAS CHROMATOGRAPHY IN FORENSIC SCIENCE**

FORENSIC EXAMINATION OF FIBRES

Editor:
JAMES ROBERTSON
Head of Forensic Services Division,
Australian Federal Police,
Canberra, Australia

ELLIS HORWOOD
NEW YORK LONDON TORONTO SYDNEY TOKYO SINGAPORE

First published in 1992 by
ELLIS HORWOOD LIMITED
Market Cross House, Cooper Street,
Chichester, West Sussex, PO19 1EB, England

A division of
Simon & Schuster International Group
A Paramount Communications Company

Printed and bound in Great Britain
by Redwood Press, Melksham

British Library Cataloguing in Publication Data

A Catalogue Record for this book is available from the British Library

ISBN 0–13–325309–0

Library of Congress Cataloging-in-Publication Data

Available from the publisher

Table of contents

Dedicated to the memory

of Christopher Kidd and

Joy Grieve

Preface

It is often said that everyone has one book in them. Having co-edited a previous volume in this series it is probably fair to question the mental health of doing it a second time! As it says on one of my more amusing coffee mugs, '*if at first you don't succeed, try again and then give up, no sense being a damned fool*'.

My second foray into the editorial field has been less harrowing not least because of the excellent efforts of my authors. I wish to thank all of them for the support they have given me. I hope the whole product reflects the individual quality of their contributions. I thank my secretary Sue Martinsen for all of her invaluable assistance and good humour. My own chapter has benefited from the efforts of my wife, Margaret, who gasps in despair at my sometimes less than perfect grammar and spelling.

Finally, I wish to explain the dedication to this volume. My entry into the field of fibres entailed research into factors affecting fibre transfer and persistence. Much of this work was done by Chris Kidd. Chris had a wonderful sense of humour. He was a big person in every way, and his loss to cancer left a huge gap in the lives of all who knew him. If there is a world beyond the mere physical I know Chris will get a kick out of reading this. During the production of this book, cancer claimed a second victim in Joy Grieve. Again, to all who had the pleasure to know Joy her loss is tragic.

Dedicating this book to Chris and Joy is one small way of remembering them.

About the authors

Glenn R. Carroll
Glen joined the Royal Canadian Mounted Police (RCMP) in 1970 upon graduation from Carleton University, Ottawa. He has served in a number of RCMP laboratories and has testified in Courts in Canada and Bermuda. Since 1985 he has been based in Ottawa, taking part in research, development, and training, becoming acting Chief Scientist—Hair and Fibre from 1988. He is treasurer of the Canadian Society of Forensic Science and has published and presented papers on fibre research.

John Challinor

John graduated from Manchester University, England with a BSc (Hons) in Chemistry. After a spell in industry John emigrated to Australia and joined the Government Chemistry laboratories in Perth in 1972. Since then he has specialized in the examination of physical evidence with a special interest in the application of pyrolysis techniques. He is President of the Western Australia branch of the Australia and New Zealand Forensic Science Society and has published a number of papers on the application of pyrolysis techniques to forensic problems.

Shantha David

Shantha graduated MSc from the University of Waikato in 1979 and gained her PhD from the University of British Columbia in 1986. She completed two postdoctoral appointments in Rome and in California before taking up an appointment as a lecturer in textile chemistry at the University of New South Wales. Her special interests include fibre identification techniques and the dyeing and finishing of textile fibres.

James Dunlop

James graduated with a BSc degree in Botany in 1975 from the University of Glasgow, and the following year was awarded his MSc in Forensic Science by the University of Strathclyde. After spending nine years at Strathclyde Police Laboratory in Glasgow, Scotland, he specialized in fibre examinations at the US Army Criminal Investigation Laboratory in Frankfurt, Germany, from 1986–1991. He has recently been appointed Senior Biologist at the Police Forensic Science Laboratory in Dundee, Scotland.

Michael Grieve

Michael graduated from the University of Durham with a BSc in Zoology in 1964, and gained his initial experience in fibre examination at the Metropolitan Police Forensic Science Laboratory, London. After 15 years of continued specialization in this area of forensic science at the United States Army Criminal Investigation Laboratory—Europe, he has recently accepted an appointment at the Bundeskriminalamt in Wiesbaden, Germany. He has been the author of many publications concerning forensic examination and comparison of textile fibres.

K. Paul Kirkbride

Paul Kirkbride was awarded a PhD by the University of Adelaide after completing a research project dealing with synthetic and mechanistic organic chemistry. He then became a post-doctoral research fellow and was involved in projects dealing with organic synthesis and nuclear magnetic resonance spectroscopy of carbocations at low temperature and nitrogen nuclei. Paul is now a Senior Forensic Scientist with State Forensic Science of South Australia. He investigates cases requiring organic microanalysis, fires and matters related to illicit drugs, particularly their manufacture.

Michael Pailthorpe
Michael is an Associate Professor in the Department of Textile Technology at the University of New South Wales. He gained a BSc (Hons) degree in 1966 and a PhD degree in 1970, both from the University of New South Wales. In 1985 he was admitted as a Fellow of the Textile Institute (UK) and is a Chartered Textile Technologist. Michael is the author or co-author of over eighty-five patents, journal papers, reviews, etc., and is very active in the supervision of sponsored textile chemistry type research projects. Michael has been a consultant to the textile industry for the past twenty years. Many of these consultancies have been in the forensic science area.

James Robertson
James graduated BSc (Hons) in 1972 and PhD in 1975, both from the University of Glasgow. He lectured in forensic science at the University of Strathclyde from 1976 to 1985, then moved to Australia where he worked for nearly five years at State Forensic Science, Adelaide before taking up his present appointment in 1989 as Head of Forensic Services for the Australian Federal Police. He has published a number of papers across a wide range of topics, but fibres and hairs have always been his major interest.

Ian Tebbett

Ian graduated BPharm (Hons) in 1980 from the University of London and PhD in 1984 from the University of Strathclyde. He lectured at the University of Strathclyde from 1984–1988 before moving to the University of Illinois at Chicago where he is an Assistant Professor and Director of Forensic Toxicology. He has published extensively on various analytical aspects of forensic science.

Peter White

Peter is a lecturer at the University of Strathclyde, having moved there in 1990 from the Metropolitan Police Forensic Science Laboratory (MPFSL) where he completed 15 years' service ending up as a senior scientific officer. For much of his time in the MPFSL Peter worked on the development of HPLC methods, and he has published extensively in this area. He is a C. Chem. FRSC, and recently graduated PhD from Brunel University.

1

Classification of textile fibres: production, structure, and properties

Shantha K. David, MSc, PhD and
Michael T. Pailthorpe, BSc, PhD
Department of Textile Technology, School of Fibre Science and Technology, University of New South Wales, PO Box 1, Kensington, New South Wales, 2033, Australia

1.1 INTRODUCTION

From ancient times textiles have been employed by man for protection from the elements, modesty, and adornment. Carpets, tents, sails, ropes, and cordages can be traced back over three thousand years. More recently textiles have found wider applications and may be designated as architectural textiles, industrial textiles, geotextiles, etc.

The rapid growth in the demand for textile fibres, together with the more demanding applications for the fibres, has led to the invention and production of an ever increasing range of man-made fibres. At the present time, total world-wide production of man-made fibres is on a par with the production of natural fibres.

It would be worthwhile at this stage to define some of the words that are employed in common parlance to describe textile fibres. 'Natural fibres' are fibres that occur in nature: wool, cotton, asbestos, etc. 'Man-made fibres' is the term applied to those fibres that have been manufactured by man from either naturally occurring fibre-forming polymers (for example viscose) or synthetic fibre-forming polymers (for example polyester). 'Synthetic-fibres' are man-made fibres spun from synthesized fibre-forming polymers. 'Regenerated fibres' are man-made fibres that have been produced from naturally occurring polymers by a technique that includes the regeneration of the original polymer structure.

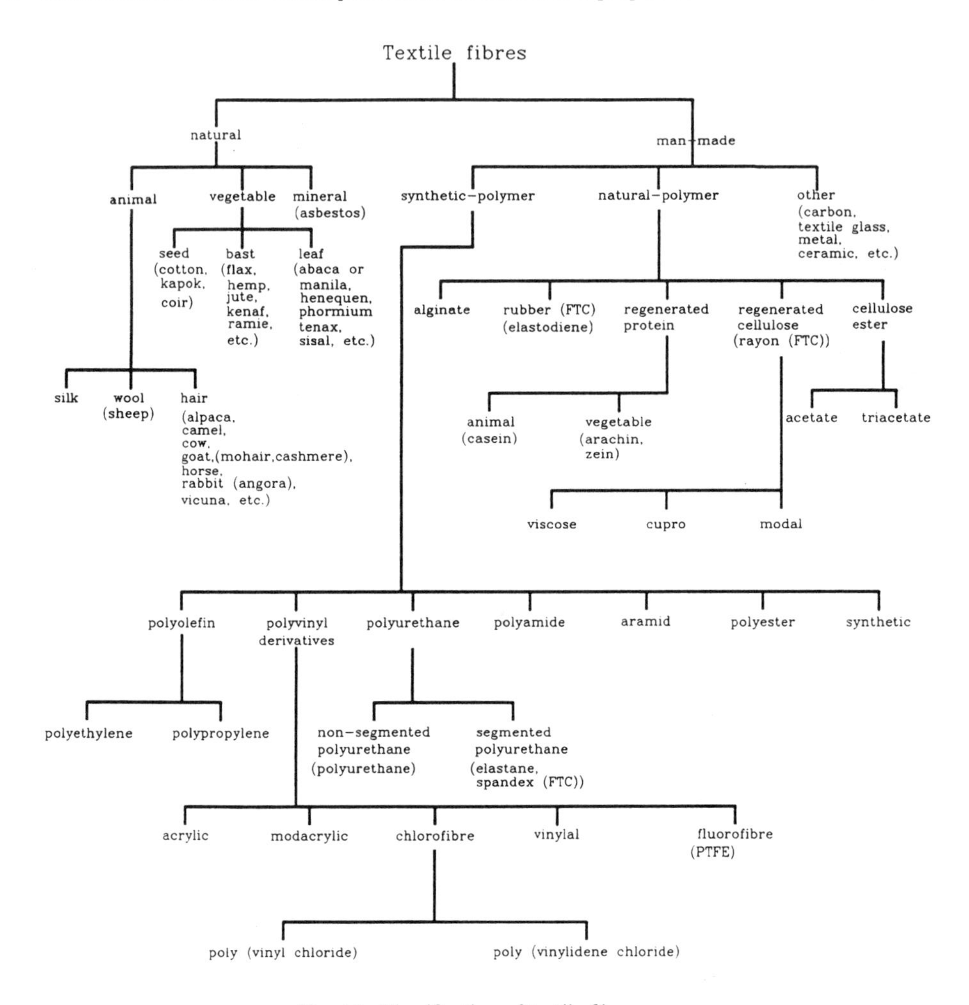

Fig. 1.1. Classification of textile fibres.

A scheme for the classification of textile fibres is given in Fig. 1.1. The classification begins by dividing fibres into two basic groups, natural fibres and man-made fibres.

The natural fibres may then be subdivided into three classifications, viz. animal (protein), vegetable (cellulose), and mineral (asbestos). Animal, or protein, fibres may be further subdivided into three groups, depending on the protein composition and/or utilization:

(i) silk (fibroin),
(ii) wool (keratin), and
(iii) hair fibres (also keratin).

The vegetable, or cellulosic, fibres are also subdivided into three groups, depending on which part of the plant is the source of the fibre. Thus the subdivisions become:

(i) seed fibres (cotton, kapok, and coir),
(ii) bast (stem) fibres (flex, hemp, etc.), and
(iii) leaf fibres (manila, sisal, etc.).

The naturally occurring mineral fibres are asbestos.

Man-made fibres are subdivided into three groups, namely:

(i) synthetic-polymer fibres,
(ii) natural-polymer fibres (casein, viscose, acetate, etc.), and,
(iii) other fibres (carbon, glass, metal, etc).

As can be seen From Fig. 1.1, each of these subdivisions may be further subdivided on the basis of polymer type. The various polymers employed will be described in section 1.4 *et seq.* of this chapter.

1.2 FIBRE-FORMING POLYMERS

Irrespective of the fibre type or the chemical composition of the polymer, fibre-forming polymers must have some, or all, of the following characteristics:

(a) linear molecular chains which possess some degree of extension or orientation to the fibre axis, thereby giving a structure which is much stronger longitudinally than transversely;
(b) a high molecular weight thereby imparting both a high melting point and low solubility in most solvents;
(c) streamlined molecular chains thereby allowing for close packing of the polymer chains;
(d) the molecular chains should be flexible and hence impart extensibility to the fibres.

1.3 NATURAL FIBRES

1.3.1 Animal (protein) fibres

Proteins have a wide occurrence in nature and are probably the most important substances in living matter. The word 'protein' is derived from the Greek word 'proteios', meaning 'first', and was used by Mulder in 1839 to describe the nitrogen-containing substances that are found in all plant and animal tissues that are involved in vital phenomena. Proteins are essentially chain-like molecules formed by the union of α-amino acids, which are joined together by the peptide linkage. An α-amino acid has the general formula:

$$H_2N-\overset{\displaystyle H}{\underset{\displaystyle R}{\overset{|}{\underset{|}{C}}}}-COOH \qquad [1.1]$$

where both the acidic and basic groups are attached to the same α-carbon atom. These α-amino acids link together, with the elimination of water molecules, to give a peptide which retains one terminal amino group and one terminal carboxylic acid group. Hydrolysis of the polypeptide units results in the reformation of the amino acid units.

The difference between proteins arises from the differences between the side groups (R) pendant to the main chain. Over twenty amino acids with different side groups are known, but, by virtue of their different sizes, shapes, and chemical functions, they have apparently sufficed for all the proteins of life. This limited number of side groups is sufficient to give rise to an astronomical number of polypeptides and proteins because of the various arrangements that are possible. The α-amino acids found in silk, wool, and casein are given in Table 1.1.

Table 1.1. Side-groups in protein fibres

Amino acid	g amino acid per 100 g protein		
	Silk Fibroin	Wool Keratin	Casein
INERT			
Glycine	43.8	6.5	1.9
Alanine	26.4	4.1	3.5
Valine	3.2	5.5	6.1
Leucine	0.8	9.7	10.6
Isoleucine	1.4	0.0	5.3
Phenylalanine	1.5	1.6	6.5
ACIDIC			
Aspartic acid	3.0	7.3	6.7
Glutamic acid	2.0	16.0	22.0
BASIC			
Lysine	0.9	2.5	8.3
Arginine	1.1	8.6	3.9
Histidine	0.5	0.7	3.3
HYDROXYL			
Serine	12.6	9.5	5.9
Threonine	1.5	6.6	4.5
Tyrosine	10.6	6.1	6.3
MISCELLANEOUS			
Proline	1.5	7.2	10.5
Cystine	0.0	11.8	0.4
Cysteine	0.0	0.1	0.0
Methionine	0.0	0.4	3.5
Tryptophan	0.0	0.7	1.4

The end of chain amino and carboxylic acid groups, together with those contained in the di-acidic amino acids (aspartic and glutamic acids) and di-basic amino acids (lysine and arginine), form 'Zwitter Ion' pairs as follows:

$$\text{—protein—NH}_3^+ \qquad {}^-\text{OOC—protein—}$$

which attract each other by Coulomb's law and form the so-called 'salt links' found in protein fibres. These salt links are affected by the presence of acids and bases. In the presence of dilute acids ($1 < pH < 5$) the carboxylate anion is titrated back to the un-ionized carboxylic acid group thereby leaving the protonated amino groups to act as dye sites. Such dyes are known as acid dyes.

The end of chain amino groups, and especially those in the side chains of the di-basic amino acid lysine, act as nucleophiles for both nucleophilic substitution and nucleophilic addition reactions with reactive dyes.

Raw silk, whether cultivated silk or wild silk, contains about 75% fibre and 25% of a globular protein (sericin). The sericin is usually left on the silk filaments to protect them from mechanical damage during processing. The silk yarn or fabric is 'de-gummed' to remove the sericin; hence de-gummed silk is essentially pure fibroin. In the past it was common practice to restore the weight lost in de-gumming by a process known as 'tin-weighting'. This multi-step process deposits insoluble tin silicates within the filaments to restore the weight to par. Thus tin-weighted silk would contain about 75% fibroin and 25% tin silicates. Tin weighting has become unpopular because of the cost of the process.

Raw wool, after shearing, contains substantial quantities of impurities: grease, swint (sweat), dirt, vegetable matter (VM), etc. A typical yield might be 65% clean wool after scouring (washing) and VM removal. After scouring, the scoured wool might contain about 0.5% residual grease which is left on the wool to protect the wool during carding, combing, etc. The scoured wool is essentially pure keratin.

Raw animal fibres may also contain impurities similar to those found on wool; however, the yield is usually higher after scouring. Clean animal fibres are essentially pure keratin.

1.3.1.1 Silk

Silk, which is composed of the protein fibroin, has a markedly different amino acid composition from that of wool and other animal fibres, all of which are made from the protein keratin. The amino acid cystine, for example, is not present is silk but comprises some 11–12% of wool. Cystine provides the disulphide crosslinks which hold the polymer chains together in wool and other animal fibres. Since disulphide crosslinks are not present in silk, it will dissolve in powerful hydrogen bond breaking solvents such as cuprammonium hydroxide, whereas wool and animal fibres will not dissolve.

As can be seen in Table 1.1, silk fibroin consists essentially of four amino acids, glycine (31%), alanine (26%), serine (14%), and tyrosine (10%). The three amino acids glycine (gly), alanine (ala), and serine (ser) are joined in the sequence:

$$(\text{—gly.ala.gly.ala.ser—})_N$$

and contribute to the crystalline regions where the protein chains are fully extended in the β-sheet structure with maximized hydrogen bonding. There are virtually no covalent crosslinks or salt linkages present to contribute to the stabilization of the structure of silk.

Silk is obtained from a class of insects called *Lepidoptra* (scaled winged insects). The fibre is produced in filament form by the larvae of caterpillars when the cocoons are formed. The principal species cultivated for the commercial production of silk is the *Bombyx mori* or mulberry silkworm.

The use of the Bombyx silkworm for the production of silk started some 4000–5000 years ago in China. The methods of producing silk yarns were kept a secret in China for many centuries, but the knowledge eventually spread to Japan and, later, via the Middle East, to Europe. At the present time, the important silk-producing countries are Japan, China, Italy, and France.

Silk production is divided into two sections:

(i) sericulture, which deals with the production of cocoons from the eggs, and,

(ii) silk reeling which deals with the conversion of the cocoon into a thread.

In the raw state Bombyx silk filaments consist of two fibres of fibroin embedded in the sericin. The width of the filaments is uneven, and the surface shows many irregularities such as fissures and folds. Each raw filament is roughly elliptical with the two triangular fibres having their bases adjacent to one another. After de-gumming the fibroin fibres are transparent, uniform in width (9–12 μm) with smooth and structureless surfaces. Tussah silk (wild silk) fibres are darker in colour, considerably coarser (average 28 μm), and less uniform in width, with pronounced longitudinal striations. Anaphe silk often has cross-striations at intervals along the fibre.

In cross-section, the brins of cultivated Bombyx silk are roughly triangular, the corners of the triangle being rounded. Those of Tussah silk are wedge-shaped, and those of Anaphe silk are roughly triangular, but the apex of the triangle is elongated and bent. Thus, the bave of Anaphe silk is crescent-shaped in cross-section being formed by two curved triangles, joined along their bases.

The breaking strength of silk fibres is about 4–4.5 g/denier dry and 3.5–4.0 g/denier wet, with corresponding elongations of 20–25% and 25–30%. Silk has a reversible extension of up to 4%, beyond which recovery is rather slow with some permanent set. Silk almost equals nylon in resistance to abrasion and toughness (the ability to absorb work).

Raw silk has a standard regain of 11%, but de-gummed silk about 10%, the sericin having a higher water absorbing capacity than the fibroin.

Like all proteins, silk is amphoteric, and adsorbs acids and alkalis from dilute solutions. It has an iso-electric point of 3.6, the pH at which the fibre is electrically neutral. Silk is not readily attacked by warm dilute acids but dissolves with decomposition in strong acids. Silk is not as resistant to acids as wool, but is more resistant to alkalis, though low concentrations at high temperatures will cause some tendering. Silk is very resistant to all organic solvents, but is soluble in cuprammonium hydroxide and cupriethylene diamine, the latter solvent being used for fluidity tests. Silk is less resistant to oxidizing agents and exposure to sunlight than cellulosic or synthetic fibres, but more resistant to biological attack than other natural fibres. It has a large capacity for adsorbing dyes, and can be dyed with a wide range of dyestuffs including acid, basic, and metal complex dyes.

The outstanding properties of silk are its strength, toughness, high regain, excellent soft handle, resistance to creasing, good draping properties, and luxurious appearance. Silk, therefore, has a wide variety of uses in the apparel and drapery fields. Its high cost, however, has restricted its use mainly to top quality apparel goods such as ladies' frocks, underwear, stockings, and handkerchiefs.

Since silk is a continuous filament fibre, with a smooth surface, fibre transfer during contact is unlikely.

1.3.1.2 Wool

Wool is just one member of a group of proteins called keratin. Other members of this group include hair, feathers, beaks, claws, hooves, horn, and even certain types of skin tumour.

Wool is produced in the fibre follicle in the skin of the sheep. The cells of the wool fibre begin growing at the base of the follicle, which is bulbous in shape. The cells complete their growth immediately above the bulb where the process of keratinization occurs. The keratinization process is completed before the fibre emerges above the surface of the skin of the sheep. The keratinization process involves the oxidation of thiols to form disulphide bonds which stabilize the fibre structure.

The fineness, quality, and properties of wool depend on the sheep breed from which the wool was shorn. Merino wool is sought after for its fineness, softness, strength, elasticity, and crimp. Merino wool has superior spinning properties and is used for spinning the finest woollen and worsted yarns. Medium wools, produced by breeds such as Leicester, Cheviot, Corriedale, and Polwarth, are used in the manufacture of woollens, knitting yarns, hosiery, blankets, etc.

The keratin of wool, like the keratins from all mammals, is of the α-type so called because unstretched wool keratin gives the characteristic α–X-ray diffraction pattern. Stretched wool fibres, on the other hand, give a quite different X-ray diffraction pattern—the β-pattern.

Because of the multitude of variations possible in diet, breed, health, climate, etc., wool fibres show a great variation in both their physical and chemical properties. For example, one expects considerable variation in physical properties such as fibre diameter, length, and crimp as well as the chemical constitution of the fibres. The properties of the wool fibre, at any one time, are found to vary from tip to root. For example, for a given extension the root end of the fibre always stress relaxes more than the tip end, and the thiol content of the fibre decreases from the root to the tip. When fibres have been stored for a long time, however, the properties become uniform from root to tip as the thiols become oxidized. These differences in mechanical properties from the root to tip of single wool fibres are eliminated by the dyeing and bleaching processes.

1.3.1.2.1 Morphological structure

Wool fibres are composed of two types of cell, namely the cuticle cells and the cortical cells. As shown in Fig. 1.2, the cuticle cells form an outer sheath which encases the inner cortical cells.

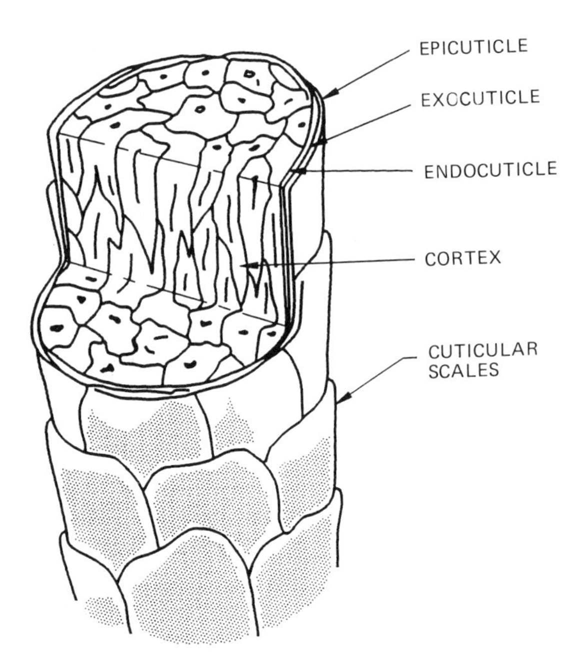

Fig. 1.2. Schematic of a wool fibre.

Whilst the cells of wool are mainly keratin, the nonkeratinous proteins of wool must also be considered to be important in the dyeing of wool. When the keratin–nonkeratin ratio is increased, by partly removing the nonkeratinous proteins, the diffusion coefficients for acid dyes become smaller. Diffusion of dyes in wool with a low nonkeratinous protein content becomes increasingly difficult as the nonkeratinous protein content decreases.

The cuticle cells comprise about 10% of the mass of the whole fibre and overlap each other with the exposed edges pointing toward the tip of the fibre. The structure of the cuticle can be further subdivided into three regions, namely, an enzyme-resistant exocuticle, an enzyme digestible endocuticle, and a thin outer hydrophobic epicuticle. These cuticle cells are separated from the underlying cortical cells by a so called 'intercellular cement' which acts like a 'glue' and cements the cells together.

The epicuticle of wool is strongly hydrophobic in character and forms a resistant barrier to the penetration of dyes. The epicuticle is readily damaged, however, by weathering, and mechanical or chemical processes. When the epicuticle has been damaged, dyes can penetrate the fibre more readily, especially at low temperatures, because the epicuticle membrane is missing. This feature has often been utilized in staining tests used to assess certain types of wool damage. Chemical treatments, such as chlorination, cause extensive damage to the epicuticle, and this process has often been used to increase the ease with which wool can be dyed.

The remaining 90% of the wool fibre is made up of cortical cells which comprise the cortex of the fibre. The cortex of wool fibres has a bilateral structure and can be further subdivided into two parts, the orthocortex and the paracortex. The orthocortex has a more open structure and is more accessible to dyes and more reactive chemically than the paracortex.

Wool which has been thoroughly cleaned by scouring and solvent extraction is essentially pure keratin. Acid hydrolysis of wool yields eighteen amino acids, the relative amounts of which vary considerably from sheep breed to sheep breed and even within the same breed of sheep.

Wool contains both di-basic and di-acidic amino acids which appear within the structure as basic and acidic side chains. Since wool has both acidic and basic groups, it is amphoteric in character. The basic amino acid residues in wool are arginine, lysine, and histidine which collectively total about 900 μM/g and hence far outweigh the contribution from the N-terminal residues. These basic groups are considered to be the predominant dye-sites for the attraction of acid dyes to wool. Furthermore, the side chains of lysine and histidine are the sites for the formation of covalent bonds between reactive dyes and wool.

Wool, like other hair fibres, contains a substantial quantity of the amino acid cystine. Cystine residues in wool play a very important role in the stabilization of the fibre structure owing to the crosslinking action of their disulphide bonds. The disulphide bonds are responsible for the relatively good wet strength of wool and particularly for its low lateral swelling. While the amino acid cystine accounts for the majority of the crosslinks in wool, there are several other types of crosslink present, including the isopeptide crosslink. The isopeptide group links the ε-amino group of lysine to a γ-carboxyl group of glutamic acid or to a β-carboxyl group of aspartic acid.

Since wool is a staple fibre, with a rough 'scaly' fibre surface, wool fibres may transfer readily during contact, especially from loosely constructed fabrics.

1.3.1.3 Hair fibres

The hair fibres, sometimes referred to as 'specialty' hair fibres, can be divided roughly into three groups:

(i) fibres from the goat family (mohair, cashmere),
(ii) fibres from the camel family (camel hair, alpaca, vicuna), and,
(iii) fibres from other fur-bearing animals, in particular the rabbit (angora).

Mohair is the fibre from the angora goat *Capra hircus aegagrus*. Mohair fibres have a length of 20–30 cm for a full year's growth. The fibre diameters range from 10–70 μm. Kid mohair averages 25 μm, in diameter while adult mohair averages 35 μm. Mohair has similar physical and chemical properties to wool.

Cashmere was the name originally given to hair from the Asiatic goat *Capra hircus laniger*. Hair produced from selectively bred feral goat populations of Australia, New Zealand, and Scotland is also called Cashmere. Cashmere fibres are from 5–10 cm in length with diameters of 14–16 μm. Cashmere is chemically identical to wool, but because of its fineness and better wetting properties, is more susceptible to chemical damage, especially with respect to alkalis. Cashmere is highly regarded for producing garments which are comfortable and have a soft handle. The main outlets are highclass ladies dress goods and knitwear.

Camel fibres are the hair from the camel *Camelus bactrianus* or dromedary. The finer inner camel hair has found use in men's high grade overcoating, while the longer outer hair is used mainly in beltings and interlinings.

Alpaca is the hair from the fleece of the alpaca *Lama pacos* which inhabits South America. Alpaca fibres have a diameter of 24–26 μm, with a distinctive scale structure and medullation. They are similar to mohair and have similar uses.

Vicuna is the undercoat hair of the vicuna, the rarest and smallest of the llama family. Each animal yields about 500 g of fibre with a diameter of 13–14 μm. The fibre is as valuable as cashmere and has similar uses.

Rabbit hair is obtained from the pelts of the angora rabbit. The hair is shorn from the pelts and separated by blowing. The fine hair is used to make felts for the hat trade, while the long guard hair is spun into yarn. The best hair is 6–8 cm long and about 13 μm in diameter. Rabbit hair is often blended with wool or nylon.

Other animal fibres may be encountered in forensic examinations. Such fibres would include hairs from domestic pets and farm animals: cat, dog, cow, horse, etc. and hairs from humans, for example head hairs, body hairs, pubic hairs.

1.3.2 Vegetable fibres

1.3.2.1 Introduction

The vegetable fibres are divided into three groups, depending on the section of the plant from which they are harvested—seed, bast (stem), and leaf. Depending on the source of the vegetable fibre, the chemical constituents of the fibre can vary considerably. The principal constituents are cellulose, hemi-cellulose, pectins, lignins, water solubles, and fats and waxes.

1.3.2.1.1 Cellulose

The term 'cellulose', in the strict scientific sense, applies only to the plant cell materials consisting of macromolecules of at least several hundred to several thousand anhydroglucose units. It is the carbohydrate part of the cell wall of plants, formed out of only glucose molecules condensed and linked together by means of 1,4-β-glucosidic bonds. The repeat unit is shown in Fig. 1.3a.

The chain molecules in natural cellulose are not of the same length. This is revealed for different samples of cellulose which exhibit no detectable chemical difference but which have different alkali solubilities and viscosities. The degree of polymerization (DP) of unopened cotton has been reported at 15 300, decreasing rapidly to 8100 on exposure to the atmosphere. Bast fibres have an average DP of about 9900, while wood celluloses vary between 7500 and 10 500. Chemical damage to cotton can be assessed by a fluidity measurement on cotton dissolved in cuprammonium hydroxide.

1.3.2.1.2. Hemi-cellulose

In any kind of plant cell, such as the flax fibre, the hemi-cellulose is usually of quite constant chemical composition, and although several polysaccharides may be present, one molecular structure usually predominates. The hemi-cellulose in flax is, however, different from that in jute, and that in hemp and in wood fibres is different again,

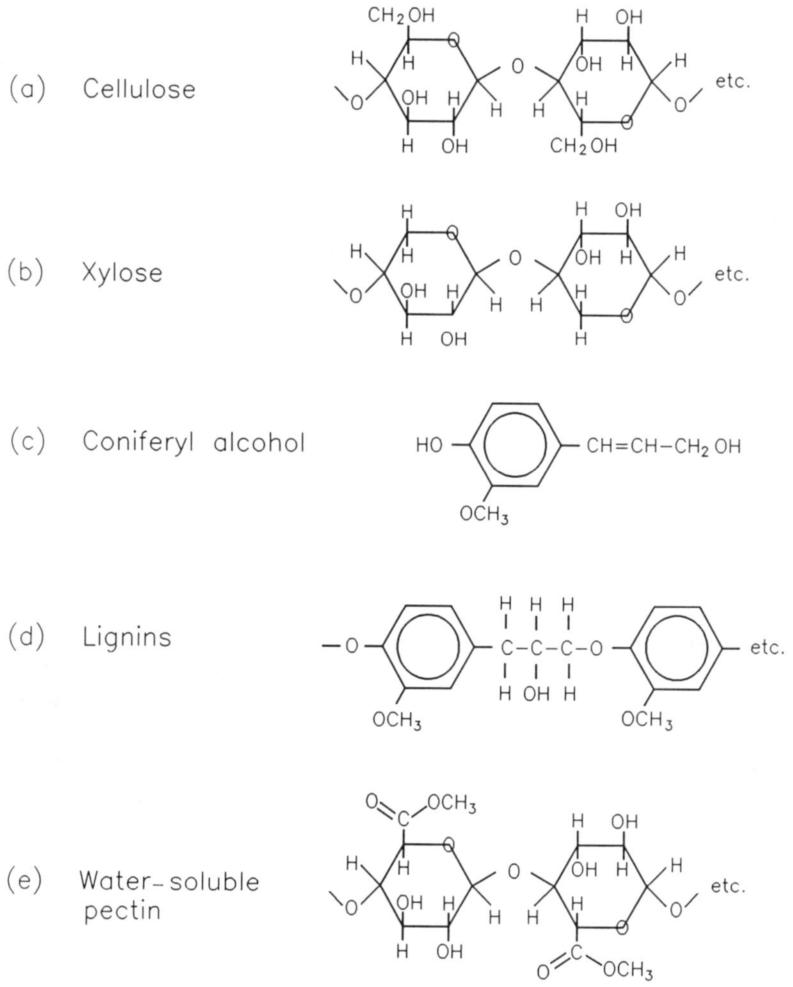

(a) Cellulose

(b) Xylose

(c) Coniferyl alcohol

(d) Lignins

(e) Water-soluble pectin

Fig. 1.3. Vegetable fibre components.

and so on, although most hemi-celluloses show many similarities of structure. A typical hemi-cellulose is xylose, which is chemically similar to cellulose but does not contain the projecting CH_2OH groups (see Fig. 1.3b).

Hemi-cellulose has a much shorter chain length than cellulose (average DP of xylose is about 120) and is soluble in dilute alkali solution.

1.3.2.1.3 Lignin

Lignin is the name given to the group of substances which are deposited in plant cell walls, particularly woody tissue, and which are based on the phenyl propane skeleton. Coniferyl alcohol (Fig. 1.3c) is built from this skeleton and is found in immature cells of coniferous wood in which liquefication is proceeding, and it is probable that this is the precursor of the lignin molecule.

A distinctive property of this group of molecules is the very large number of ways in which they can become linked together in the presence of the enzymes found in wood sap. This undoubtedly accounts for the many forms in which the phenyl propane unit is found in studies of lignin breakdown products, and also explains the overall complexity of the lignin structure (see Fig. 1.3d).

Most lignin studies have been devoted to wood lignin, and very little is known about the lignin in bast fibres, although it is unlikely that appreciable differences exist. The lignin in plant cell walls can be fairly readily, and almost completely, removed by chlorination procedures (bleaching) whereby a soluble chloro–lignin complex is formed.

It is believed that the presence of lignin in cellulosic material may increase the susceptibility of the cellulose to degradation in the presence of sunlight. For example, the high lignin content of jute is associated with the colour changes that occur when jute is exposed to light.

1.3.2.1.4 Pectins

Pectins are mixtures of non-crystalline carbohydrates of high molecular weight occurring in the cell walls of plants and vegetables. They are not homogeneous and exist in a number of different forms.

In growing plant cells, pectin is a jelly-like substance, soluble in water, and it is very suitable for maintaining the growing cells in close proximity to each other, while still allowing small mutual displacements. The structure of a water soluble pectin is given in Fig. 1.3e.

In mature cells, however, more rigid bonding is required between cells, and the pectin in such tissues is in the form of a complex with calcium (calcium pectate), which is a cement, insoluble in water.

1.3.2.2 Seed fibres

The seed fibres are cotton, kapok, and coir.

1.3.2.2.1 Cotton

Cotton is a fibre attached to the seed of several species of the genus *Gossypium*. The cotton plant is a shrub which grows to a height of 1.2 to 2 m. The plant is indigenous to nearly all subtropical countries, though it thrives best in warm humid climates where the soil is sandy, near the sea, lakes, or large rivers.

The plants are raised from seed each year, and require 5 to 7 months of warm to hot weather with ample moisture for optimum development. About 3–4 months after planting, the plant attains its full height and blossom pods form, which expand until they burst, displaying the flower. After about 24 hours the flower falls off, leaving a boll (or pod) containing the seeds and immature fibres. As the fibres mature they cause the boll to expand until, about 2 months after flowering, it bursts into sections.

The cotton fibres begin to grow at the time of flowering as an elongation of a single cell from the epidermis, or outer layer, of the seed. The diameter is immediately established, and the cylindrical cell continues to grow in length for about 4 weeks. At this stage the fibre is a greatly elongated cell about 12 to 36 mm in length, and it consists of protoplasmic material (lumen) bounded by the primary wall with the cuticle on its exterior surface. For the next 25–30 days, the maturing period, thickening of the fibre occurs by the deposition of cellulose on the interior surface of the primary wall forming a secondary wall. This growth ceases a few days before the boll splits open. When the boll opens the moisture inside evaporates and the fibres lose their tubular form. As drying proceeds, the walls of the fibre shrink and collapse, the lumen becomes smaller and flatter, and the fibre develops convolutions (twists). There are from 110 (fine cottons) to 60 (coarse Asian cottons) convolutions per centimetre in cotton fibres. These convolutions improve the flexibility of the fibres and hence the spinning properties of the cotton. The convolutions are an important morphological feature used in the microscopic identification of cotton.

As a result of attack by disease or pests, or of unfavourable climatic conditions, the growth of the fibre in the boll may be terminated before the fibre is fully developed. This is the origin of immature or 'dead' fibres. If the growth ceases after the fibre has attained more or less its full length, but when little or no thickening has taken place, the immature fibre consists of a hollow tube formed by little more than the thin primary wall and the cuticle. When the boll opens, and the immature fibres collapse, they assume a ribbon-like appearance, and, because of the absence of the secondary wall, they are devoid of convolutions. Dead fibres have little rigidity, easily crease and crumple, and can be difficult to remove in carding, drawing, and combing. Dead fibres also cause unlevel dyeing problems.

Cotton fibres can be dyed with a wide range of dyestuffs, including direct, azoic, vat, sulphur, reactive, and metal complex dyes.

Since cotton is a staple fibre, with a convoluted surface structure, cotton fibres may transfer readily during contact. Since a very high proportion of cotton is marketed in 'white' products, colourless cotton fibres have little, if any, value as evidence, However, 'white' cotton fibres would have been bleached both chemically and 'optically', so that the fluorescent whitening agent (FWA) employed may provide some evidence of common source.

1.3.2.2.2 Kapok

These seed fibres are obtained from the pods of the kapok tree, *Ceiba petrandra.* The pods are picked from the tree and broken open by hand. The seed and the attached fibre are then dried in the sun and, when dry, the fibre is removed by hand.

Kapok fibres have an average length of 18 mm and a diameter of 20–30 μm. Kapok fibres have a low density. The fibre is oval in cross-section with a wide lumen and a very thin wall. Kapok fibres are composed of about 65% cellulose with the balance being mainly lignins and hemi-celluloses.

Kapok fibres are unsuitable for spinning into yarn and are principally used as a filling in such products as lifebouys, belts, mattresses, and pillows. Before World War II, world production of kapok exceeded 40 million kg per annum. However, production since the war has dropped dramatically owing to destruction of plantations and the development of alternative fibre fillings, for example Dacron (polyester).

1.3.2.2.3 Coir

Coir can also be classified as a fruit fibre. It is the fibrous mass contained between the outer husk of the coconut and the shell of the inner kernel. The fibres can be recovered either by hand or by an industrial process. In the normal mill process, the husks are immersed in water for about a week, dried, and passed through a crusher fitted with spikes. The spikes tear away the non-fibrous material from the long coarse fibres. The fibres are then combed and spun into hanks.

The best quality coir, which is produced from green or unripe nuts, is used in the manufacture of ropes, twines, fishing nets, and matting. Bristle fibre, which is produced from mature nuts, is a coarser fibre and is used in the manufacture of brushes, brooms, and door mats.

Coir fibres are between 15 and 36 cm in length, with diameters ranging between 0.1 and 1.5 mm. The virgin coir fibres comprise about 40% cellulose, 40% lignin, 2% pectin, and 18% hemi-celluloses. The process of retting removes much of the pectins and hemi-celluloses so that the commercial fibre consists essentially of cellulose and lignin. Coir has a high tensile strength which is retained after prolonged immersion in water. Coir also has a good resistance to microbiological attack.

1.3.2.3. Bast fibres

Bast fibres are obtained from the stems of *dicotyledonous* plants, the fibre being located in bundles in the plant stalk, under the outer bark and forming an inner bark around the woody central portion of the stalk. The fibres are firmly held together and to the core of the stem by non-cellulosic materials, in particular pectin. Unlike cotton, bast fibres are not homogeneous, but consist of many different substances. Cellulose is the chief component, comprising about 65–75% of the fibres. The other components consist mainly of hemi-cellulose, pectins, lignins, and some other water soluble materials. These are, in fact, the principal components of the cell walls of all vegetable matter.

The principle bast fibres of commerce are flax (linen), ramie, hemp, kenaf, and jute. It is important to note, however, that man-made fibres have taken a very large share of the traditional markets for bast fibres.

1.3.2.3.1 Flax

Flax fibres are obtained from the stalk of the plant *Linum usitatissimum*. This plant is one of the earliest known to civilization, having been cultivated for over 6000 years. The flax plant is an annual which grows in any temperate climate and in a variety of soils. It is grown for either the fibre (linen) or for the seed (linseed) which is a valuable source of oil and stock feed. The varieties grown for fibre have straight slender stalks to a height of 1 to 1.4 m and are planted close together to prevent branching

The seeds are sown in late spring, and, after 10–12 days, the plants begin to appear. After 3 months the plant starts to ripen from the root up, with the green leaves changing to brownish-yellow and finally dropping off. The plant is harvested when the stems start to change colour at the root, before the fibres become dry and harsh. Flax can be harvested by either pulling or cutting. The bundles of harvested plants are allowed to dry for one to two days before being deseeded (rippling). The dried flax plants at this stage are known as flax straw.

The fibres are recovered from the stem of the plant by a process known as retting. The flax fibres are bound to the other parts of the plant and to each other by pectins which are not soluble in water. The object of retting is to decompose these substances by fermentation, so that the fibres may be removed from the stem and, at the same time, to rot the woody portion of the plant, so that it will break up easily, thus facilitating its removal.

Water retting is the most common technique employed and can be carried out in rivers, ponds, or tanks. Bacteria, present in the straw, water, and tank walls, breed and feed on the pectins, reducing them to water soluble sugars. When the retting process is judged to be complete, the straw is dried to stop the fermentation action. Chemical retting can be achieved in boiling alkali, in which case all of the pectins are removed, the fibre bundles break up into ultimates, and the material becomes 'cottonized'. The flax fibres are recovered from the straw by a mechanical process known as scutching.

In the flax plant the fibres are arranged in bundles of 12 to 40 fibres. These bundles, which are up to 100 cm in length, run the full length of the stem and are joined to each other at intervals. The individual, or ultimate, fibres in the bundles average about 3 mm in length and 15–20 μm in diameter. Flax fibres are mainly polygonal in cross-section, caused by the manner in which they are packed together in bundles. Each ultimate fibre is pointed at both ends, and there is a small, but well defined, lumen running lengthwise but disappearing near the ends. The fibres have a smooth surface except at intervals where they are ringed with transverse nodes. These nodes, which are useful in identifying the fibre, help to bind the fibres together, and their regular and frequent occurrence is important for the formation of fine, strong yarns.

The flax fibre has a much higher content of non-cellulosic material than cotton. Flax contains about 75% cellulose, 15% hemi-cellulose, 2.5% pectins, and 2% lignins. The quality and spinning properties of flax are very dependent on the 1–1.5% wax present. This gives the fibres their high lustre as well as imparting suppleness.

Owing to its compact structure, the flax fibre is stronger than other natural fibres and, for the same reason, is much less pliable and elastic than other common natural fibres. Flax has about the same regain as cotton but absorbs and desorbs moisture much more rapidly, making it very suitable for towels and drying cloths. The smooth fibre surface makes it very easy to launder, hence linen found great use in the manufacture of table cloths and bandages.

Linen is more difficult to dye than cotton. However, it bleaches to a full white which is enhanced by its natural lustre. This characteristic is much prized in household linens, so only a small portion is dyed.

1.3.2.3.2 Ramie

Ramie, a member of the hemp family, has a dozen or so varieties, but only one type, the *Boehermeria nivea*, is a commercial source of fibre. The name 'ramie' comes from the Malayan word 'rami'.

Unlike other bast fibres such as flax, jute, and hemp, which are grown as annuals, ramie is grown as a perennial. Propagation is by seed, cutting, or more commonly by rhizomes (root stocks). The heavy root stock sends up numerous stems which are harvested every 2–3 months for a period of 5–6 years. Harvesting is carried out when the stems turn brown near the root.

The extraction of ramie fibres is much more difficult than that of flax, jute, and hemp fibres, and this difficulty has restricted its use. Decortification machines consists of a series of crushing rollers and beaters. The gums, waxes, and pectins remaining on the fibres after decortification make the fibres weak and brittle, therefore before spinning, the fibres must be de-gummed. De-gumming is usually effected by an alkali boil.

The ultimate ramie fibres vary in length from 2.5–30 cm with diameters ranging from 40–75 μm. The cells, which are elongated in cross-section, have thick walls, with a well defined lumen. The fibre surface is characterized by small node-like ridges and striations. The orientation of the molecules in ramie is very regular, and it is the most crystalline of the natural cellulosic fibres.

Natural ramie consists essentially of about 75% cellulose, 16% hemi-cellulose, 2% pectins, 1% lignins, and 6% water solubles and waxes. After decortification the cellulose content has risen to about 85%, while the fully de-gummed fibre contains 96–98% cellulose on a dry basis.

Ramie fibres are very white with a silk-like lustre and hence make attractive fabrics. Ramie is unaffected by exposure to sunlight and has a very high resistance to bacteria, fungi, and mildew. Ramie fabrics and yarns, like linen, are highly absorbent and dry quickly. Fabrics are easily laundered and show only minor strength loss after repeated washings. Durable and attractive sheets, table cloths, and towels can be made from ramie.

1.3.2.3.3 Hemp

True hemp is the bast fibre produced from the stalk of the plant *Cannabis sativa.* The term 'hemp', however, has been applied indiscriminately to a variety of fibres used for cordages, some obtained from stalks but mostly from leaves, such as Manila hemp and New Zealand hemp.

The plant is grown from seed as an annual in a temperate climate. Stalks are ready for harvesting when the lower leaves turn yellow three to four months after sowing. For good quality fibres the stalks should not exceed 1.8 m in height. The plants are pulled, retted, and scutched in the same manner as the flax plant.

Fibre strands vary from 1.0–2.0 m in length with the ultimate cells ranging from 0.5–5.0 cm in length and 15–50 μm in diameter. These ultimates are similar in appearance to flax except that the fibre surface has longitudinal fractures (no nodes) and the cell ends are blunt and irregularly shaped.

The composition of dry hemp fibres is about 75% cellulose, 17% hemi-cellulose, 1% pectins, 3.6% lignins, 2.7% water solubles, and 0.8% fats and waxes. Hemp fibres are grey–green to brown in colour with the better quality fibres having a lustre. Their strength, elongation, etc are very similar to flax, and, in some instances, good quality hemp can be used as a substitute for flax. The main commercial uses of hemp are in the manufacture of ropes, cordages, string, and twine.

1.3.2.3.4 Jute

Jute is the bast fibre obtained from the plants *Corchorus capsularis* and *Corchorus clitorius*. It has been used on the Indian sub-continent from time immemorial both as a vegetable (leaves) and as a source of textile fibres (stalk). The term 'jute' has been traced back to the Indian 'jhat'.

The jute plants are annuals which flourish in alluvial soils with damp, tropical climates. The seeds are sown by broadcasting in February in the lowlands and up to June in the higher lying country. The crop matures in about 3–4 months, by which time the plants are about 3–4 m high with stems 1.2–1.9 cm thick. The leaves are mainly near the top of the plant, leaving the stalk fairly free from leaf. The stems are harvested by cutting shortly after flowering. The harvested stems are retted and the fibres separated by hand.

The jute fibre strands comprise bundles of spindle shaped single cells or ultimates. These ultimates vary in length from 1–6 mm, with an average of 3.1 mm, being much shorter than cotton fibres. The diameters vary from 15–25 μm. They have a polygonal cross-section with a large lumen and thick cell walls. The surface of the cells is mainly smooth with only occasional markings.

The main components of raw dry jute fibres are 71% cellulose, 13% lignin, 13% hemi-cellulose, 0.2% pectin, 2.3% water solubles, and 0.5% fats and waxes. The presence of the hemi-cellulose makes jute more sensitive to alkalis and acids than pure cellulose. Unbleached jute is extremely light sensitive and turns yellow or brown on prolonged exposure with a loss in tensile strength. The sensitivity to light appears to be connected with the high lignin content.

Jute is a fairly lustrous fibre with moderate strength. However, it is inextensible and brittle. The coarse nature of jute limits the count to which jute yarns can be spun. Jute finds application in packaging for foods, cotton, etc, as backing for linoleum and carpets, and in the manufacture of ropes and cordages.

Kenaf. The fibre kenaf is obtained from the plant *Hibiscus cannabinus* and is similar to jute in many of its properties. Hence kenaf is used as a substitute for jute.

1.3.2.4 Leaf fibres

Leaf or 'hard' fibres are the textile fibres obtained from the leaves of certain tropical and subtropical plants. The leaf fibres consist of bundles of individual cells with overlapping ends to form continuous strands along the length of the leaf. The term 'hard' fibre is a misnomer in so far as the textile fibres encompassed by the term are not harder than the 'soft' bast fibres, though they are thicker and stiffer. The main commercial fibres in this group are sisal, henequin, abaca, and New Zealand hemp.

As for the bast fibres, man-made fibres have taken a large share of the traditional markets for leaf fibres.

1.3.2.4.1 Sisal

There are many different species of the genus *Agave* which have been used for fibre production, but by far the most important commercial type is *Agava sisalana*, from which sisal is produced. The name sisal is taken from the name of the port in Yucatan from which the first sisal fibres were exported.

Sisal is grown in tropical areas. Propagation is by suckers or bulbils which are raised in nursery beds. When 1 year old the plants are transplanted in the field and are ready for initial harvesting 3 years after planting. The life of the plant for fibre production is 7–8 years. The fibres are removed from the leaf by a decortification process comprising crushing followed by scraping away the cellular tissue. The fibres are finally washed.

The individual strands vary from 1–2 m in length and are white to yellowish in colour. The fibre bundles consist of ultimates which are from 3–7 mm in length and have an average diameter of 24 μm. The fibres have a broad lumen and the fibre ends are broad and blunt and sometimes forked. The fibres are polygonal in cross-section, sometimes with rounded edges. The main use of sisal is for the manufacture of ropes, cordages, twines, etc.

Henequin. Henequin is another important *Agave* leaf fibre from the plant *Agave forcroydes*. Like sisal, henequin is a native of the Mexican state of Yucatan. It is grown in a similar manner to sisal but in more arid, less fertile areas. The physical and chemical properties of henequin are almost identical to sisal.

1.3.2.4.2 Abaca (Manila Hemp)

Abaca fibre is produced from the leaf of the plant *Musa textilis*, a member of the banana family indigenous to the Philippines. The name 'Manila hemp', which is often used, is a misnomer, as Abaca is not a true hemp, and, although a native to the Philippines, it is not grown near Manila. Abaca was widely sought for the manufacture of better quality ropes and cordages.

The abaca is a perennial plant which thrives in a moist topical climate with high humidity. The leaves are 40–50 cm wide and 1.5–2.3 m long, and are ready for harvesting when the blossom first appears. The fibres are recovered by stripping the outer fibre layer from the inner fleshly layer by means of knives. The fibres are then freed of adhering pulpy material by drawing the fibres over serrated knives. Prompt drying is essential to preserve the strength and lustre of the fibres.

The fibre strands vary in length from 1–3 m and from 0.05–0.30 mm in diameter. The ultimate fibres are from 3–12 mm in length and from 16–32 μm in diameter. The cross-section is irregularly round or oval in shape, and the lumen is very large and generally empty. The fibre cells have thin smooth walls and sharp or pointed ends.

1.3.2.4.3 Phormium tenax (New Zealand hemp)

Phormium tenax, or New Zealand hemp, is a perennial plant indigenous to New Zealand. It is the only leaf fibre that has been grown in commercial quantities outside the tropics. The use of Phormium has declined over recent years.

The plant grows to a height of about 2 m in four years, and consists of about ten shoots each bearing 5–6 sword-shaped leaves, 5 cm wide and 2.8 m long. The outer leaves are cut and scraped mechanically to extract the fibres. The ultimate fibres vary from 2.5–15 mm in length and from 10–20 μm in diameter. The cells are nearly circular in cross-section with a circular lumen. *Phormium tenax*'s composition differs from other leaf fibres in having a much lower cellulose content but higher hemi-cellulose and lignin content.

1.3.3 Mineral fibres (asbestos)

The name 'asbestos' is of Greek origin and means incombustible. Until relatively recently substantial quantities of asbestos were mined. However, health considerations have since greatly reduced the industry. Asbestos has been replaced wherever possible by other fibres such as glass or Nomex.

The most important form of asbestos is *chrysotile* which accounted for about 90% of total asbestos production. Chrysotile belongs to the serpentine group of minerals and occurs only in serpentine rocks. These are metamorphic rocks, and the development of chrysotile seems to depend on a hydrothermal recrystallization, which is probably initiated at cracks in the rock. The fibres grow at the expense of the adjacent rock. The ideal chemical formula for chrysotile is $Mg_3Si_2O_9H_4$.

In economically workable chrysotile deposits, the fibre content of the rock is between 2% and 10%. The whole rock is mined, and the fibre is separated from the rock and graded for length by a series of crushing, screening, and air flotation processes. The proportion of fibres over 25 mm in length is usually quite low.

Chrysotile is very strong. However, its ultimate fibrils are so fine, about 250 Å in diameter, that any handleable fibre contains millions of these fibres and breakage commonly occurs by a succession of breaks in different fibrils, leading to fraying and thence to complete rupture.

The other commercially important types of asbestos are grouped together under the heading of amphibole asbestos and are crocidolite, amosite, tremolite, actinolite, and anthophyllite. The range of compositions of the amphiboles is very wide indeed, and, in addition to magnesium, can contain sodium, calcium, and iron in their molecular structure.

1.4 MAN-MADE FIBRES

1.4.1 Introduction

This umbrella grouping encompasses fibres that may be derived from naturally occurring polymers (regenerated fibres) or synthesized from simple starting materials (synthetic fibres) such as coal and oil. The term 'man-made fibres' thus refers to the method of fibre production and not to the origin of the polymer material.

1.4.2 Production methods

The basis of man-made fibre formation is the same for both regenerated and synthetic fibres. The polymer in a concentrated, viscous form, either in solution or in a molten state, is forced through the tiny holes of a spinneret, and the emerging filaments are immediately precipitated or cooled to form the solid state fibre. This process is termed 'spinning' or 'extrusion', and may be accomplished in three different ways: *wet spinning*, *dry spinning*, and *melt spinning*. The method of choice depends on the chemical and physical properties of the individual polymer in question.

Both wet and dry spinning processes use concentrated solutions of the polymer. In the wet spinning process, the polymer may be dissolved in either aqueous or organic solvents, and the extruded filaments are immediately precipitated in a coagulation bath. The dry spinning process employs only volatile solvent systems which allow rapid evaporation of the solvent, effecting fibre solidification after extrusion. The melt spinning process is applicable exclusively to thermoplastic, synthetic polymers, which melt without decomposition. The molten state polymer is extruded and the filaments are solidified by simple cooling. This latter method is the most economical, and thus preferable if polymer properties are suited to the process.

The fibre filaments thus formed are usually stretched or 'drawn' mechanically. This operation causes the polymer chains to become more aligned (or oriented) in the direction of the longitudinal axis of the fibre. The increase in molecular chain orientation maximizes the inter-polymer forces of attraction between polymer chains, leading to an increase in polymer crystallinity, and fibre strength. Polymer crystallinity is expressed as the 'degree of crystallinity', which refers to the percentage of the polymer network that is present in a crystalline form, the remainder being in an amorphous state where polymer chains are disordered (or oriented at random). Amorphous and crystalline regions do not occur in any particular order within the polymer system of the fibre, and the extent to which either region predominates within the polymer network largely determines the overall properties of the fibre.

Fibre filaments are twisted to form continuous filament yarns, or are cut into short lengths to form staple fibre which is spun into yarn. Staple fibre lengths are often compatible with those of cotton or wool and may be spun on the same spinning frames. (The 'spinning' of fibre into yarn should not be confused with the 'spinning' of polymers into fibre filaments, discussed above.)

The longitudinal and cross-sectional appearance of man-made fibres is largely affected by the method of fibre extrusion (for example the shape of spinneret holes, or rate of coagulation or cooling). Fibre microscopy related to these determining factors will be discussed in a separate chapter.

1.4.3 Synthetic fibres

1.4.3.1 Condensation and addition polymers

There are two broad classes of synthetic polymers separated by the nature of their starting materials and polymer formation, namely: *condensation* and *addition polymers*.

Condensation polymerization is defined as occurring when monomers bearing two reactive or functional groups from one or more compounds condense by normal chemical reaction to produce a linear polymer. Examples in this class include: polyamides and aramids, polyesters, and polyurethanes. The structures of these various polymer chains are all different and will be discussed separately below. In each case the repeating units are joined by inter-unit functional groups along the polymer backbone (for example, amide, ester, urethane linkages). These linkages are somewhat susceptible to hydrolysis (cleavage) by chemical reagents such as acids and alkalis, which can lead to polymer degradation.

Addition polymerization is defined as occurring when monomers containing double bonds from one or more compounds add together at the double bond to form a polymer confined to an aliphatic carbon chain. The degree of polymerization attained for addition polymers is usually much higher than that for condensation polymers

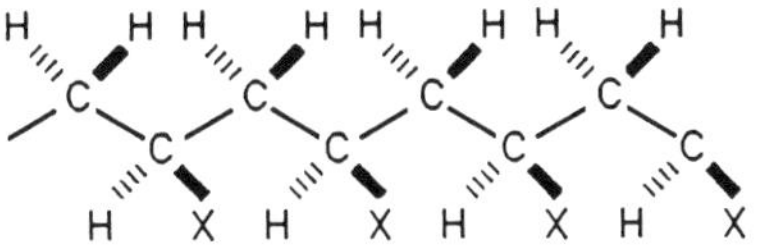

Fig. 1.4. Addition polymer: all polyolefin and polyvinyl derivatives belong to this polymer class.

because of the nature of their respective polymerization reactions. The strong C–C bonds along the main chain of addition polymers offer no sites for easy cleavage by corrosive reagents, thus these polymers are generally more stable that condensation polymers. Polymers in this class include: polyolefins and polyvinyl derivatives. Unless otherwise stated, these polymers may be represented by the straight carbon chain structure, shown in Fig. 1.4, differing only by the nature of the side group, X, attached to every alternative carbon atom. Three different types of solid state stereochemical conformations are possible for the polymer chain, depending on the spacial direction of the X group. These are termed: *atactic* (no stereoregularity; that is the X group may point in any direction); *isotactic* (the X group points in only one spacial direction, as shown); and *syndiotactic* (the X group alternates in spacial direction).

1.4.3.2 Polyamides
(Nylon 6.6: Blue C, ICI-nylon, Perlon T, Ultron)
(Nylon 6: Celon, Dederon, Enkalon, Perlon)

Polyamides may be synthesized via the condensation of diamine ($H_2N—R^1—NH_2$) and dicarboxylic acid ($HOOC—R_1{}^2—COOH$) monomers, or by the polymerization of an ω-amino acid ($H_2N—R—COOH$). In either case a polymer is formed, containing an amide functionality [—CONH—] within the repeating unit: [$—NHR^1—CONHR^2—CO—$] or [—NHRCO—], respectively. The R group may be either aliphatic (straight carbon chain), alicyclic (saturated ring), or aromatic (benzene ring); the last named group, termed 'aramids', will be dealt with in section 1.4.3.3.

Aliphatic polyamides are defined according to the number of carbon atoms in the repeating unit. For instance, if the repeating unit is derived from a diamine and a di-acid, the polymer is designated by two numerals, the first indicating the number of carbon atoms in the diamine and the second the number of carbon atoms in the dicarboxylic acid (for example nylon 6.6, Fig. 1.5). Nylons synthesized from an ω-amino acid are designated by a single numeral indicating the number of carbon atoms in the ω-amino acid (e.g. nylon 6)

Nylon 6.6 and nylon 6 are the most widely manufactured polyamides; their properties are very similar and they may therefore be discussed together. Nylon fibres are melt spun, drawn, and manufactured predominantly as continuous filament yarn (80%) as opposed to staple fibre (20%). The smooth, regular longitudinal appearance results in a highly lustrous and translucent fibre requiring the addition of a delustring

$$\left[-\underset{\displaystyle H}{\underset{|}{N}}-CH_2-CH_2-CH_2-CH_2-CH_2-CH_2-\overset{\displaystyle H}{\overset{|}{N}}-\underset{\displaystyle O}{\underset{\|}{C}}-CH_2-CH_2-CH_2-CH_2-\overset{\displaystyle O}{\overset{\|}{C}}- \right]_n$$

Fig.1.5. Nylon 6.6 (n = 50–80): a condensation polymer of hexamethylene diamine and adipic acid.

agent (for example titanium dioxide) to the spinning dope. Nylon filament or staple is usually subjected to an additional heat treatment after stretching in order to improve the yarn texture and bulk.

Nylon is a linear polymer in which the flexible carbon chain forms a zigzag configuration in between the stiffer amide groups (Fig. 1.5). These polar amide linkages [—CONH—] all along the backbone of the polymer, allow numerous hydrogen bonds to form between adjacent polymer chains. In addition, the regularity of the polymer chains and lack of bulky side groups allow a close approach of adjacent chains (0.3 nm apart), which maximizes the hydrogen bonding interactions which are formed across an interpolymer distance of less than 0.5 nm. This results in a highly oriented and crystalline polymer network (65–85%), with good tensile strength and excellent elastic recovery (Table 1.2). When the fibre is stretched or placed under strain the numerous and strong hydrogen bonds prevent polymer slippage, so that on release of the strain the molecular chains return to their original configuration within the polymer system. Nylon fibres and fabrics also exhibit outstanding resistance to abrasion, again a consequence of the high degree of crystallinity; in fact, nylon is the most abrasion resistant of all common textile fibres.

Despite the predominance of polar amide residues which attract water molecules, the highly crystalline nature of the polymer system limits the penetration of water molecules to the amorphous regions, as evidenced by the rather low moisture regain of nylon. On wetting, a 10–20% loss in tensile strength results as water molecules disrupt to some extent the hydrogen bonding interactions in the amorphous regions, with an accompanying increase in fibre extensibility and loss in elasticity. A particular disadvantage related to the low moisture absorbency of nylon is its propensity to develop static electricity. This limitation may be overcome by the addition of hygroscopic polymers to the nylon spinning dope or by blending nylon fibres with highly conductive fibres capable of dissipating the static charge.

Under controlled conditions of applied heat and stress nylon can be 'heat set'. During the heat setting treatment inter-polymer hydrogen bonds are broken by the application of heat energy, and they reform in the new heat set configuration on cooling, thus stabilizing the set. This thermoplastic character is used to an advantage in introducing bulk and texture into yarns. Fine hosiery yarns as well as coarse carpet yarns are texturized to improve their aesthetic appeal and resilience.

Nylon is relatively stable to alkalis. However, strong acids will readily hydrolyse the amide linkage, weakening the fibre. Nylon fibres are soluble in phenols and concentrated acid, and a useful distinction between nylon 6.6 and nylon 6 is that the latter is soluble in 4.4N HC1, while the former is not. On prolonged exposure to

Table 1.2. Properties of condensation polymers

	Nylon	Aramids	Polyester	Polyurethane
Specific gravity (g/cm^3)	1.14	1.45	1.22–1.38	1.10
Degree of polym. (%)	50–80	100	115–140	unknown
Degree of cryst. (%)	65–85	100	65–85	extremely amorphous
Tenacity (cN/Tex) normal to H. T.	40–90	250	35–85	5–7
Extensibility (%) normal to H.T.	45–150	2.8	37–70	very high
Initial modulus (N/Tex) normal to H.T.	2–4.5	90	8–9	--
Moisture regain (%) at 20°C and 65% RH	4.0–4.5	<0.1%	0.1–0.3	0.4–1.3
Thermal properties	Heat sensitive	Flame-retardant	Heat sensitive	Very heat sensitive
Softening range (°C)	220–230		220–240	110–120
Melting range (°C)	250–260	>500 (decomp.)	250–260	230
Handle	Medium-hard waxy	Stiff	Medium-hard waxy	Medium waxy

sunlight nylon will degrade via oxidation of the amide groups, causing the material to weaken considerably.

Nylon is readily dyed with a wide range of dyestuffs; this versatility, combined with its easy texturizing, high wear resistance, and recovery from deformation provides a fibre that is ideal for the carpet and hosiery industries, the main outlets of nylon production.

By increasing the extent of stretching after extrusion, high fibre tenacities can be attained with reduced breaking extension; these high tenacity yarns have several industrial uses in ropes, belts, parachutes, etc. Bicomponent nylon fibres manufactured by the fusion of two polymers to form a single yarn during the extrusion stage, exhibit special characteristics (for example as crimped fibres or bonding agents in non-woven materials).

Other aliphatic polyamides that are commercially manufactured include nylon 4.6, 6.10, nylon 7, and nylon 11. Some of these fibres have been developed for their exceptional resistance to abrasion and thermal degradation in engineering applications.

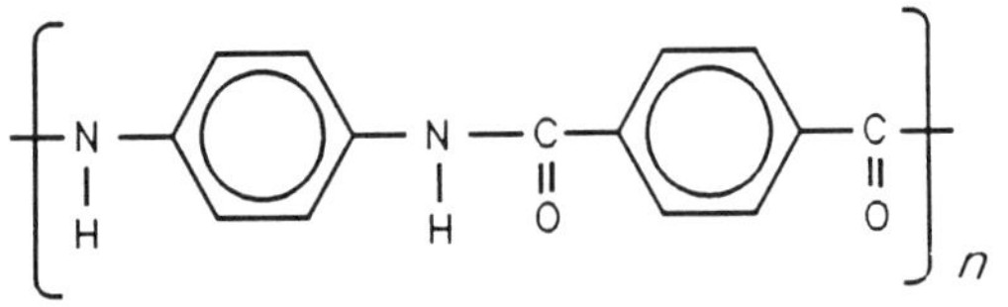

Fig. 1.6. Kevlar ($n = 100$): a condensation polymer of *para*-phenylene diamine and terephthalic acid.

Qiana is a polyamide of undisclosed structure, considered to contain an alicyclic R group and a long 12-membered carbon chain. This fibre is primarily a fashion fabric known for its silk-like handle and higher moisture regain compared to other polyamides; this latter property is effected by the incorporation of additional polar hydroxyl functional groups into the polymer network.

1.4.3.3 Aramids (Nomex, Kevlar, Conex)

Aramids are aromatic polyamides defined as having at least 85% of their amide linkages attached directly to two aromatic rings. Nomex and Kevlar are the most commonly encountered aramids, and the structure of the latter is shown in Fig. 6. The para-substituted phenyl rings of Kevlar form a rigidly linear polymer chain resulting in a rod-like structure and a high degree of inter-chain hydrogen bonding. The molecular chain linearity and hydrogen bonding combine to give a polymer network with almost perfect molecular orientation as the filament emerges from the spinneret. Such fibres need not be stretched to increase polymer chain alignment, and, in solution form, they are referred to as liquid crystals. The structure of Nomex is identical to that of Kevlar, except that the benzene rings are meta-substituted.

Nomex and Kevlar have a high specific gravity compared with aliphatic polyamides as a result of the better alignment of their linear polymer chains. Aramid fibres are distinctive for their very high tensile strength, high modulus (resistance to stretching), and extremely high chemical and heat resistance (up to 300°C). At these very high temperatures the fibres char and decompose rather than melt. Aramid fibres are insoluble in most common solvents and therefore require the use of special solvent systems for extrusion into filaments.

The expense of aramid fibre manufacture limits their applications to those of high performance products. Nomex is principally used for fire/heat retardant purposes such as in heat protective clothing and hot-gas filtration fabrics, while the outstanding performance of Kevlar relates to its strength-to-weight ratio. When compared with conventional engineering materials, Kevlar is five times stronger than steel and more than ten times stronger than aluminium. There are several different grades of Kevlar, denoted by a number following the name, each with a different balance of properties associated with fibre tenacity and modulus. Some of the applications of Kevlar include high performance tyres and conveyor belts, and reinforcement fibres in sporting goods and in the aerospace industry.

1.4.3.4 Polyesters (polyethylene terephthalate: Crimplene, Dacron, Diolen, Fortrel, Grilene, Tergal, Terital, Terylene, Tetoron, Trevira)

Polyesters are synthesized by the condensation of a diol (for example ethylene glycol) and a dicarboxylic acid (for example adipic acid) with formation of an ester linkage [—COO—] along the polymer chain backbone. The selection of an aromatic di-acid and short carbon chain diol as starting monomers affords a fibre with a high melting temperature and a desirable degree of stiffness. Alternative monomers with longer carbon chains form unsuitably low melting point fibres which are also very susceptible to hydrolytic attack by corrosive reagents. Hence the most common polyester manufactured is based on the polymer polyethylene terephthalate (PET, Fig. 1.7)

Polyesters are melt-spun and produced in roughly equal amounts of both filament and staple fibre. The fibres are fine, regular, and translucent, and are usually textured, as is nylon, by a heat treatment.

The PET polymer contains only weakly polar carbonyl groups, thus the polymer chains are largely held together by hydrophobic interactions such as weak van der Waals' forces. The para-substituted benzene rings, however, reinforce the linearity of the polymer, thus maximizing the van der Waals' attractive forces. The resultant polymer network has a very high degree of crystallinity (65–85%) and strength, with no loss of strength on wetting because of the hydrophobic nature of the polymer chains. The rigidity of the para-substituted benzene ring further imparts a stiff handle to polyester textiles, which prevents the polymer system from yielding when under stress. This latter property is particularly important in differentiating polyester from nylon, in that polyester has a much higher modulus, requiring twice the force required by nylon to produce a similar extension. A high fibre modulus results in a crisp handle and good dimensional stability, rendering polyester fabrics extremely resistant to wrinkling. In other respects, polyester is similar to nylon, with both fibres possessing good strength, abrasion resistance, and thermal stability with the ability to hold a heat setting treatment (see Table 1.2.).

The hydrophobic character of polyester, combined with the high degree of crystallinity, results in a fibre with very low moisture regain. The limitations associated with such low moisture absorbency may, however, be overcome by blending with absorbent fibres to increase wear comfort. Treatment with antistatic agents serves to dissipate static charge, and dyeing with carriers swells the fibre, facilitating the penetration of disperse dyes.

The ester linkage in polyester is hydrolysed by alkalis. The fibre is, however,

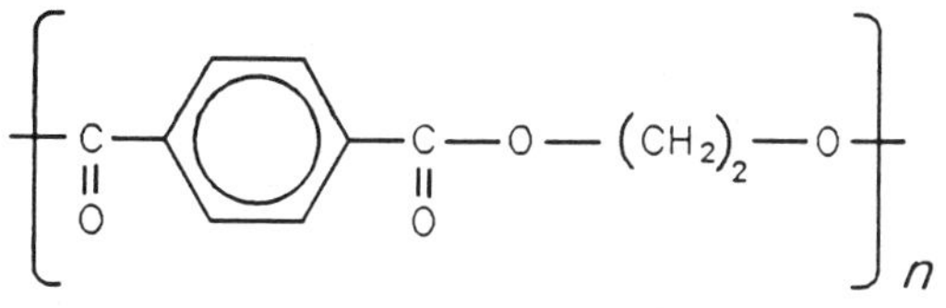

$$\left[-\underset{\substack{\| \\ O}}{C}-C_6H_4-\underset{\substack{\| \\ O}}{C}-O-(CH_2)_2-O- \right]_n$$

Fig. 1.7. Polyester (n = 115–140): a condensation polymer of adipic acid and ethylene glycol.

extremely stable even in concentrated acids, dissolving only in hot sulphuric acid (of concentration $>80\%$). The highly crystalline nature of polyester resists the entry of corrosive reagents, and even alkalis will mostly cause only surface damage. The fibre is also more resistant to sunlight degradation than nylon. Polyester is insoluble in most organic solvents, except chlorophenol and hot *m*-cresol. It is also resistant to dissolution in 90% *o*-phosphoric acid, which will digest most other textile fibres derived from organic polymers.

The higher fibre modulus of polyester, combined with the lower cost of production compared with nylon, has given polyester the lead over nylon in many applications. The easy care and toughness of polyester render it very suitable for staple fibre blending with wool or cotton in a wide variety of apparel end uses. Continuous filament polyester may be produced in a high tenacity form and hence has displaced nylon in various industrial applications such as ropes, conveyor belts, seat belts, and tarpaulins.

Modified polyesters (for example Dacron), produced by copolymerization with an acidic component, are known for their ease of dyeability with basic dyes.

1.4.3.5 Polyurethanes (Dorlaston, Enkaswing, Glospan, Lustreen, Lycra, Sarlane, Spanzelle)

Polyurethane elastomeric fibres, also referred to by the generic term 'elastane', are defined as containing at least 85% by mass of recurrent aliphatic groups joined by urethane linkages [—NH—COO—]. Polyurethane fibres are very complex polymers combining a flexible segment, which provides the high degree of stretch and rapid recovery of an elastomer, with a rigid segment that confers the necessary strength of a fibre.

The flexible segments can be one of two types: a long polyether or polyester chain (the former is shown in Fig. 1.8a), while the rigid segments are composed of a diphenyl methyl group attached to a urethane group. The polymerization reaction involves the initial formation of a pre-polymer (Fig. 1.8b) which is then reacted with a diamine to form additional polyurea linkages. In the final stage of polymerization some

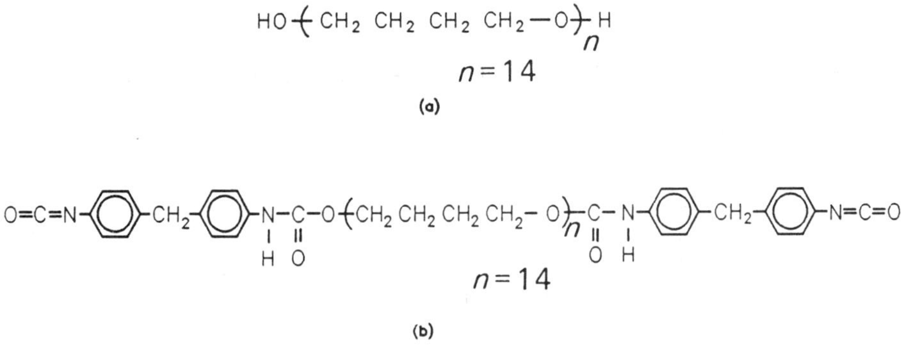

$$HO\!\left(CH_2\,CH_2\,CH_2\,CH_2{-}O\right)_n\!H \qquad n=14$$

(a)

$$O{=}C{=}N{-}C_6H_4{-}CH_2{-}C_6H_4{-}N(H){-}C({=}O){-}O\!\left(CH_2\,CH_2\,CH_2\,CH_2{-}O\right)_n\!C({=}O){-}N(H){-}C_6H_4{-}CH_2{-}C_6H_4{-}N{=}C{=}O \qquad n=14$$

(b)

Fig. 1.8. (a) Long chain polymer segment. (b) Polyurethane pre-polymer.

crosslinking occurs, and the degree of polymerization of the final polymer is not known.

Polyurethane elastomers are usually solution spun and manufactured in the form of continuous monofilaments or coalesced filaments, off-white in colour; the multi-filament yarns show individual filaments that are fused together.

The polymer network of polyurethanes is largely amorphous as a result of the long, flexible polyether of polyester chains which are folded upon themselves. The interconnecting hard segments tend to be more aligned, with hydrogen bonding between the urethane groups of adjacent chains contributing to polymer strength and crystallinity. When the fibre is stretched the flexible segments become extended, allowing up to a maximum of 500–600% elongation; interchain interactions within the crystalline regions, however, are not broken, so that on release of the stress the polymer chains recoil immediately to their original configuration, conferring the distinctive highly extensible, 'snap-back' property of elastomeric fibres. Furthermore, on extension of the elastomer, the molecular chains of the soft segments straighten and crystallize somewhat, reinforcing the fibre and enhancing its breaking strength. Compared to rubber, elastomeric fibres have high tensile strength for all levels of extension; on the fibre scale, however, they are relatively weak fibres (see Table 1.2).

The flexible segments of polyurethanes are completely hydrophobic, hence they do not attract water molecules within the polymer structure. The urethane linkages, while being polar, are compactly aligned in the crystalline regions into which penetration is not possible, so that polyurethane fibres on the whole have very low moisture regain values. Elastomeric textiles are therefore prone to develop static electricity and are very difficult to dye. Particularly as a result of the latter property elastomeric filaments are rarely used on their own, but are more often combined with conventional, readily dyeable yarn such as nylon. During processing, the elastomeric filament is stretched under tension and the inelastic yarn (continuous or staple) is wound around the core of the stretched elastomer. When the resultant yarn retracts, the coils of the covering yarn jam together. Such yarns may be directly woven or knitted into stretch fabrics. The elastomeric fibre itself is always in continuous form and undyed, while the wrapping yarn is dyed, and is consequently easily separable from the elastomeric component. More recently, a bicomponent yarn consisting of a nylon and polyurethane filament extruded together is being developed.

Polyurethanes are thermoplastic, although excessive amounts of heat can disrupt the elastic properties of the fibre. They are generally not susceptible to attack by corrosive reagents, owing to their hydrophobic character, although alkalis will readily attack the ester linkages of the polyester type of polyurethanes (mentioned above).

The single most important benefit of polyurethane elastomers lies in their stretch and recovery properties when incorporated into fabrics for the purpose of comfort and fit (for example swimwear, active wear).

1.4.3.6 Polyolefins (polyethylene: Courlene X3, Drylene) (polypropylene: Aberclare, Deltafil, Fibrite, Gymlene, Herculon, Meraklon, Neofil, Polycrest, Pylen, Reevon, Spunstron, Tritor)

Polyethylene (X=H, Fig. 1.4) and polypropylene (X = CH_3) polymers are polymerized from the very common petroleum based products, ethylene and propylene, respectively. Free radical polymerization of these starting monomers, under conditions of high temperature and pressure, leads to the formation of low density polyethylene (LDPE) and atactic polypropylene, unsuitable as fibrous polymers (in fact atactic PP is a grease at room temperature). The use of a special Ziegler–Natta catalyst system, developed in the 1950s, facilitates the polymerization reaction such that little or no chain branching occurs, allowing dense packing of the adjacent polymer chains with the formation of highly crystalline polymer networks. Polyethylene produced in this high density form (HDPE) is 85% crystalline. In the case of polypropylene, stereochemical control is achieved with the formation of the isotactic polymer, which is >90% crystalline. Further, the catalytic system allows polymerization to proceed at a relatively low pressure (30 atm.) and temperature (100°C), with a very high degree of polymerization attained (*circa* 3000 for polypropylene).

Both polypropylene (PP) and polyethylene (PE) are melt extruded into either monofilaments or sheet film. Film extrusion through a die is 25% cheaper, and the film is slit into tapes which are further handled like a yarn. The properties of PP and PE are very similar, and will be discussed below with respect to the more commonly encountered fibre, namely polypropylene.

Table 1.3. Properties of addition polymers

	Polyolefins	Acrylics	Chlorofibres	Vinylals
Specific gravity (g/cm^3)	0.90–0.95	1.14–1.18	1.38–1.70	1.26–1.30
Degree of polym. (%)	2500–3000	2000	--	1700
Degree of cryst. (%)	85–90	70–80	40	70
Tenacity (cN/Tex)	40–80	20–27	20–40	20–60
Extensibility (%)	10–20	25	100–150	26–90
Initial modulus (N/Tex)	8–10	6.2	1.3–4.4	3–22
Moisture regain (%) at 20°C and 65% RH	<0.1	1.2–2.0	<0.1	4.5–5.0
Thermal properties	Very heat sensitive	Very heat sensitive	Very heat sensitive	Heat sensitive
Softening range (°C)	PP: <155	235	60–100	200
Melting range (°C)	PE: 133 PP: 160	Does not melt	160	220
Handle	Very waxy	Soft, waxy		Soft

The dense packing of polymer chains and the high degree of polymerization tend to maximize the van der Waals' interchain forces, giving a drawn fibre with high tenacity and good elongation (Table 1.3), similar to high tenacity nylon and polyester. In competition with nylon and polyester, the modulus of PP is much less than that of polyester, which has somewhat hindered its exploitation in applications requiring stiffness, while the abrasion resistance of PP, although excellent, is not equal to that of nylon.

At *circa* 100°C the tenacity drops by half, with increased extensibility and lower modulus. Gradual extension (or 'creep') under load also sets in at higher temperatures. The melting points of PE and PP fibres, being 135°C and 165°C, respectively, are rather low for many textile applications, which is a severe limitation on their use. The fibre is also flammable, although self-extinguishing.

The highly crystalline and hydrophobic nature of the polyolefin polymer systems result in fibres that have zero moisture regain, are extremely difficult to dye, and are very resistant to attack by corrosive chemicals. The low moisture absorbency is accompanied by a water transport property, known as 'wicking', which allows moisture to be transferred rapidly between fibres without actually being absorbed by the fibres. Typical applications that utilize this property to advantage include medical products and next-to-skin active wear.

Coloration is usually accomplished by pigmentation before melt spinning; however, several types of dyeable PP polymers are manufactured, namely: disperse dyeable, acid-dyeable, and metal modified polypropylene. They are all obtained by the addition of suitable modifying chemicals to the spinning dope, allowing versatility in coloration at a later stage; the latter type usually contains a metal chelate which is incorporated primarily as a light stabilizer although it also serves as a mordant for certain dyes.

Polyolefins are highly resistant to chemical degradation by acids and alkalis. They are, however, soluble in hot hydrocarbon solvents and cannot be dry cleaned. They are sensitive to sunlight degradation, requiring the addition of ultraviolet stabilizers for a variety of outdoor uses and indoor applications such as carpets and furnishings.

Polypropylene film tape, slit into 2–3 mm widths, is woven and used principally as primary backing for carpets and also in sacks, etc. Wider tapes of 20–50 mm width are twisted under tension to form twine with a fibrillated texture, used in fishing nets and cordage etc. In these end uses, PP has almost completely displaced the natural fibres jute and sisal.

Polypropylene fine fibres are primarily used in carpet face yarns, offering excellent cover, abrasion resistance, recovery, and wet cleaning behaviour. A relatively recent outlet for these fibres is their manufacture into non-woven products, especially for end uses such as carpet underlay and medical applications. The non-woven structure is thermally bonded, avoiding the use of chemical adhesives and offering an absorptive product made from inert material that is also resistant to bacterial growth. In all of the abovementioned applications the low cost of production is one of the main factors promoting end use.

Finally, recent technology has developed an ultra-high strength form of polyethylene fibre (for example Dyneema SK60 and Spectra-900) for high performance applications. In the manufacture of this fibre, the PE chains are of extremely high molecular weight, and the gel-like polymer solution is 'pulled' through the spinneret (rather than pumped), after which the filaments are subjected to an extremely high draw ratio (20:1–30:1); the very flexible molecular chains of PE are suitable for such extensive drawing out and alignment. The resultant fibre has a strength and modulus equivalent to that of aramid and carbon fibres (see sections 1.4.3.3 and 1.4.4), and is used in similar fibre/resin composite high performance end uses.

1.4.3.7 Polyvinyl derivatives

1.4.3.7.1 Acrylics and modacrylics (acrylics: Acrilan, Courtelle, Creslan, Crylor, Dralon, Orlon, Sayelle) (modacrylics: Crylor, Dynel, Kanecaron, SEF Modacrylic, Teklan, Verel)

Polyacrylonitrile (X=CN, Fig. 1.4) is the most important addition polymer with respect to level of production. It is composed of repeating acrylonitrile units [—CH_2CHCN—], with a very high degree of polymerization (2000). There are two groups of acrylic fibres: one based on at least 85% by weight of acrylonitrile units, termed *acrylic* fibres, and the other based on at least 35% but not more than 85% of acrylonitrile units, termed *modacrylic* fibres.

The acrylonitrile units are largely non-polar, and the polymer system is therefore held together predominantly by van der Waals' forces. Despite the fact that the extruded polymer system is atactic (lacks steroregularity), the extremely long molecular chains tend to maximize the van der Waals' interactions and render the system highly crystalline (70–80%). In addition, it is thought that the nitrogen atoms of the nitrile side groups are slightly electronegative, thus possibly enabling hydrogen bonding to occur with methylene groups on adjacent chains.

The homopolymer of polyacrlonitrile (100%) is extremely difficult to dye, being non-polar and highly crystalline. Therefore, acrylic polymers are generally copolmerized with up to 15% of different monomers which serve to increase the dyeability of the resultant fibre. The other monomers (comonomers) are commonly drawn from the following: acrylamide, methacrylate, vinyl acetate, vinyl pyridine, acrylic acid, sodium vinylbenzene sulphonate. The purpose of these various comonomers is to open up the polymer structure and/or to incorporate anionic or cationic groups within the polymer system, allowing dyes to become attracted to and penetrate the polymer system. Basic dyes are commonly used on acrylic fibres, producing bright shades with good light fastness.

Acrylic and modacrylic fibres are solution spun (either wet or dry spun), delustred, and processed into staple fibre. Stretched or drawn acrylic fibres possess a distinctive characteristic of pronounced shrinkage when subjected to steam (more so than any other textile fibre). This longitudinal instability of acrylic fibres is thought to result from the straining of interchain forces in the stretched state, so much so that in the presence of heat and moisture they readily relax and revert to the unstretched configuration. Fibre contraction in this way is used to advantage to produce 'bulked yarns'. Yarns spun from a blend of stretched fibre and fibre that has not been stretched, are heated to induce contraction of the stretched fibre; the unstretched fibre is consequently caused to bulk considerably, giving the overall appearance of an extremely bulked yarn. This effect is more recently being achieved by manufacturing bicomponent fibres where the two components exhibit different shrinkage behaviour.

Acrylic fibres have only moderate tensile strength compared to nylon and polyester (see Table 1.3); a 10–20% loss in strength when wet indicates some penetration of water molecules into the polymer network. The weak van der Waals' interactions between polymer chains permit some slippage to occur when under strain, as evidenced by wrinkling of the textile material when subjected to distortion.

Acrylic fibres have a low moisture regain, as a result of the predominantly hydrophobic character and highly crystalline nature of the polymer system. The slight electronegativity of the nitrile nitrogens and the ionic groups introduced during copolymerization evidently attracts some moisture to the polymer structure. The hydrophobic nature of acrylic fibres also leads to a build-up of static charge.

The highly crystalline structure of acrylic fibres prevents penetration of acids and alkalis, with only hot concentrated alkalis causing surface hydrolysis. Acrylic fibres are insoluble in most organic solvents, but will dissolve in boiling dimethylformamide and dimethylsulphoxide. The resistance of acrylic fibres toward degradation by sunlight is excellent (they are the most resistant of all common textile fibres). This is thought to result from the formation of very stable ring structures within the polymer system by reaction of the nitrile groups and the main chain, initiated by sunlight radiation.

Acrylic fibres are very heat sensitive and are readily flammable on close approach to a flame; they burn with a characteristic smoky flame. This major drawback is overcome by copolymerization with vinyl chloride monomers, the most common modacrylic textile materials. The chloride-containing polymers do not support combustion and are thus considered to be flame resistant. This is because the degradation of the C–Cl bond, induced by heat energy, is an endothermic process (heat absorbing) and therefore has the effect of extinguishing the flame rather than propagating it. However, these chlorine-containing modacrylics have lower heat stability than acrylics in that they soften more readily from heat application (150°C) and shrink markedly in boiling water. They can be easily distinguished from other acrylic fibres by their chlorine content.

Acrylic fibres are most commonly blended with other fibres in knitted outerwear, furnishings, carpet, and fabrics for outdoor use (awnings, etc.). These applications make use of their high bulk characteristics, bright colours, and high light stability. Modacrylics are also usually blended with other fibres to impart flame retardancy in apparel and furnishings.

1.4.3.7.2 Chlorofibres (Fibravyl, Leavil, Rhovyl, Saran, Thermovyl, Velan)

Chlorofibres are manufactured from either polyvinyl chloride (PVC, X = Cl, Fig. 1.4) or polyvinylidene chloride (PVDC, $—CH_2CCl_2—$), and are defined as containing at least 50% of either polymer type. The most common chlorofibre now in production is 100% PVC, synthesized from the starting monomer vinyl chloride, CH_2CHCl. The polymer may be either melt spun or dry spun into fibres from a solvent mixture of acetone and carbon disulphide, and processed into either filament or staple fibre for blending with cotton and wool yarns.

The extruded fibre has a disordered polymer system as neither the isotactic nor syndiotactic stereoregular structures predominates. This factor, combined with the weak van der Waals' interpolymer forces, result in a fibre with only medium tenacity (see Table 1.3), high elongation, and a high degree of shrinkage (up to 40%) at relatively low temperature (100°C); examples of this fibre are Rhovyl (filament) and Fibravyl (staple). Some heat stability may be introduced by subjecting the oriented fibre to a thermal treatment at 100°C, where further shrinkage will then only occur above this temperature; however, the tenacity drops significantly and the extensibility increases to $>100\%$ (for example Thermovyl). The high shrinkage characteristic of PVC fibre is used to advantage in the production of high bulk yarns blended with other fibres such as wool.

PVC fibres have virtually zero moisture regain, owing to the completely non-polar nature of the polymer. For this same reason the fibres have excellent resistance to corrosive chemicals such as acids, alkalis, and oxidizing agents. Coloration is, however, a problem, and is usually achieved by mass pigmentation or disperse dyeing. A further limitation of PVC fibres is associated with their solubility in chlorinated hydrocarbons and aromatic solvents, rendering them unsuitable for dry cleaning. PVC fibres are also highly photostable and can withstand prolonged exposure to sunlight with only gradual strength loss.

The most distinctive characteristic of PVC, which determines most of its fibre applications, is its inherent flame retardancy associated with its high chlorine content. Although the fibres soften and shrink easily they do not support combustion. End use applications therefore use the stability of PVC toward corrosive chemicals (for example in filter fabrics) and its intrinsic nonflammability (for example in a variety of woven fabrics such as drapery, blankets, and underwear.)

Chlorofibres from polyvinylidene chloride are usually copolymerized with PVC to give fibres with similar properties to those mentioned above.

Another type of chlorofibre, known as *Nytril*, is defined as containing at least 85% of a long polymer chain of 1,1-dichloroethene (vinylidene dinitrile) where the vinylidene dinitrile content is no less than every other unit in the polymer chain.

1.4.3.7.3 Fluorofibres (Teflon)

Fluorofibres are manufactured via the polymerization of tetrafluoroethylene gas (CF_2CF_2). under pressure, in a special dispersion medium. The polymer, polytetrafluoroethylene [—CF_2CF_2], does not melt and is insoluble in all normal solvents; it is therefore spun as a polymer dispersion into a coagulating bath, from which the filaments are sintered at high temperature (385°C) for a few seconds and then quenched in water.

The molecular chains of PTFE form a highly organized close packed arrangement within the polymer system, as evidenced by the density of PTFE fibres (2.3 g/cm^3) which is the highest of any fibre of organic origin. The inertness of the fluorine atoms combined with the packing symmetry, offers a very effective barrier to corrosive chemicals, giving a fibre with outstanding resistance toward heat, chemicals, and solvents.

The fibre is completely stable up to 215°C, and decomposes only slowly above 300°C. It is not attacked by strong acids, alkalis, or oxidizing agents even at high temperatures. Other properties include a very low coefficient of friction and zero moisture absorbency.

One of the main applications of PTFE fibres is in the filtration of liquids and gases, having the capability to perform continuously at temperatures up to 260°C.

1.4.3.7.4 Vinylals (Cremona, Kuralon, Vinylon)

Vinylal fibres are composed of polyvinyl alcohol chains (X=OH, Fig. 1.4) having various degrees of desired acetylation. They are produced by the polymerization of the stable vinyl acetate monomer, where subsequent hydrolysis of polyvinyl acetate yields the polyvinyl alcohol polymer. An aqueous solution of PVA is then wet spun into a coagulation bath of sodium sulphate, and the resultant fibres, which are at this stage water soluble, are stretched and heat treated to improve their mechanical properties and to increase their stability in hot water. This procedure maximizes interchain hydrogen bonding and increases the orientation (90%) and crystallinity (70%) of the final fibre. PVA fibres may be rendered even more resistant to boiling water by acetylation of the OH groups, or by an after-treatment with formaldehyde which introduces both intramolecular and intermolecular chain crosslinking.

These various treatments provide fibres with various degrees of molecular orientation and a wide tenacity range of approximately 200–600 mN/tex (see Table 1.3). PVA fibres generally, however, have poor elastic recovery and wrinkle easily. The presence of polar hydroxyl groups increases the moisture regain of vinylal fibres to $>5\%$, a relatively high value for synthetic polymers, accompanied by an expected reduction in fibre strength in hot water. Vinylal fibres that have been acetylated or formaldehyde treated are quite stable in cold dilute acids and alkalis, but will dissolve in hot acids, hydrogen peroxide, and phenols.

PVA fibres are produced in the form of both water soluble and water resistant filaments. The latter fibre type possesses many of the desirable properties of cellulosic fibres such as dyeability and comfort. They are manufactured and used chiefly in Japan for many textile apparel uses (for example kimonos) and industrial applications, particularly as a biologically resistant substitute for cellulose in fishing nets, ropes, packaging materials, etc. The water soluble fibres have several other important uses which include: temporary base cloths in the manufacture of papermakers' felt; linking threads and scaffolding fibres to be conveniently dissolved in a subsequent scouring operation; and PVA bonded non-woven fabrics.

1.4.3.7.5 Other polyvinyl derivatives

Polystyrene fibre is a synthetic linear polymer of styrene. A synthetic terpolymer, known as *trivinyl* fibre, is defined as being composed of acrylonitrile, a chlorinated vinyl monomer, and a third vinyl monomer, none of which represents as much as 50% of the total mass.

Anidex fibre is defined as containing at least 50% by mass of one or more esters of a monohydric alcohol and propenoic acid (acrylic acid).

1.4.4 Man-made fibres from natural polymers

1.4.4.1 Regenerated cellulosic fibres

Man-made cellulosic fibres are regenerated from wood cellulose. Pine, spruce, and eucalyptus are used, in which cellulose constitutes approximately 50% of the weight. By comparison, flax consists of 60–85% cellulose, while cotton is almost pure (>90%) cellulose.

The production sequence usually involves dissolution of the purified wood cellulose in the form of a soluble cellulose derivative, extrusion of the viscous solution through a spinneret, and immediate solidification of the cellulose filaments by coagulation or solvent evaporation (wet of dry spun). At the final solidification stage, the filaments may be regenerated back to the form of cellulose, as in *viscose*, *modal* (or *polynosic*) and *cuprammonium rayon* (or *cupro*), or remain as derivatized cellulose, as in *acetate* and *triacetate*. The latter fibre types are termed 'cellulose esters.'

1.4.4.1.1 Viscose

Viscose rayon is the most important regenerated cellulose fibre with respect to its production level. In the preparation of viscose, cellulose pulp is initially allowed to steep in alkali in the presence of air, a process known as 'ageing'. During this first stage partial oxidation of the polymer chains takes place, reducing their average degree of polymerization by at least one third. This is essential in order to achieve the correct viscosity of the spinning dope to be prepared after derivitization. In the next stage the cellulose pulp is solubilized by forming the carbon disulphide derivative of cellulose, known as cellulose xanthate. This solution is allowed to 'ripen' to an appropriately viscous consistency, and is then extruded through a spinneret. The emerging filaments are immediately coagulated in an aqueous bath containing sulphuric acid and salts of sodium and zinc sulphate. The coagulation bath completely regenerates the xanthate derivative back into the form of cellulose. The filaments thus formed are drawn from the bath at a controlled rate with some degree of applied stretch.

During the regeneration process in the coagulation bath, a uniform deposition of cellulose forms an outer layer, creating a skin effect. As the core of the fibre becomes further regenerated the fibre shrinks, causing the skin to appear wrinkled or serrated. As a consequence of the extrusion process, the fibre has a smooth surface which reflects most incident light. This results in a harsh lustre which is overcome by the addition of a delustring agent to the spinning dope, causing the fibre to have a rather subdued lustre. Viscose is manufactured primarily as staple fibre.

Viscose is a cellulosic polymer, identical to cotton, with six hydroxyl groups per cellobiose unit capable of participating in hydrogen bonding between adjacent polymer chains. These hydrogen bonds are the most important forces of attraction between polymer chains in cellulose, contributing to the overall strength of the fibre. The much lower degree of polymerization of the molecular chains of viscose (200–400) compared to cotton (3000–4000), however, results in a number of significant differences in the physical and chemical properties between the two polymers. The polymer system of viscose has a much lower degree of crystallinity (35–40%) compared with cotton (65–70%), resulting in a weaker fibre with lower abrasion resistance and a limp handle. In particular, the wet strength of viscose is very poor as water molecules readily penetrate the largely amorphous polymer network, disrupting the existing hydrogen bonds and pushing polymer chains further apart (see Table 1.4). Further, the short molecular chains with fewer interchain hydrogen bonds can readily slide past each other when the fibre is placed under strain. When the strain is removed, polymer chains tend not to return to their original configuration, rendering the textile material distorted or wrinkled in appearance.

Table 1.4. Properties of regenerated cellulosic fibres

	Viscose	Modal	Cellulose esters
Specific gravity (g/cm^3)	1.51–1.52	--	1.30–1.32
Degree of polym. (%)	200–400	600–800	260
Degree of cryst. (%)	35–40	40–50	40
Tenacity dry:	20	40–50	10
(cN/Tex) wet:	10	35	
Extensibility dry:	15–30	5–10	25–45
(%) wet:	20–40	7–12	
Moisture regain (%) at 20°C and 65% RH	12–14	10	2.0–4.5 (TRIACETATE) 6.5 (ACETATE)
Thermal properties	Heat sensitive	Heat sensitive	Extremely heat sensitive
Softening range (°C)	130		175–190
Melting range (°C)	Does not melt	Does not melt	260
Handle	Medium–soft, limp	Crisp	Very soft limp

Viscose and other regenerated cellulosic polymers tend to have lower heat resistance and poorer heat conductivity than cotton. The thermal properties of viscose are otherwise similar to that of cotton, in that the fibre is not thermoplastic. This results from its hygroscopic nature, where the presence of water molecules within the polymer network disrupts the hydrogen bonds formed during a heat setting treatment. The lack of retention of the new heat set configuration thus results in a loss of the set on cooling.

The amorphous nature of the viscose polymer system combined with the large number of polar hydroxyl groups results in a fibre that is even more absorbent than cotton (Table 1.4). The large surface area and serrated cross-section of viscose fibres also assist the penetration of water. For this reason viscose is even more susceptible to degradation by acids and bleaches than is cotton. Dry cleaning solvents do not attack viscose. Being a cellulosic polymer, viscose is also subject to attack by mould and mildew which will discolour and weaken the fibre.

Regenerated cellulosic fibres may be readily dyed and printed with the same dyes as for cotton, although the greater reflection of incident light from the regenerated fibre, even when delustred, results in a brighter colour with a subdued lustre. The fine, flexible fibres, with a subdued lustre, give fabrics a silk-like appearance and handle. The main uses for viscose fabrics therefore include dress fabrics, underwear, and furnishings. It is the cheapest man-made fibre, and is very often blended with cotton, nylon, polyester, or wool, contributing its absorptive properties, soft texture, and static free character. 100% viscose fabrics are often resin coated to decrease wrinkling; the resin (based on cyclic ethylene urea) cannot be detected microscopically as a surface coating, but may be detected by the soda-lime test indicating the presence of nitrogen). Non-woven products from viscose fibres include bedding and personal hygiene products, where moisture absorption is important.

Crimped viscose fibres (for example Eulan, Sarille) may be produced by a slight modification of the coagulation process, causing the fibre skin to burst on one side, releasing a portion of the core at the surface. The asymmetric or crimped effect thus formed lends a bulky character to viscose staple, simulating the aesthetic appeal of wool. This fibre has been successful in carpet staple, although its use in carpets is declining.

Higher tenacity viscose fibres (for example Tenasco) are manufactured by retarding the rate of coagulation and applying a higher degree of fibre stretch. The slow regeneration produces a filament with an 'all skin' structure and greater strength, used in tyre reinforcement, conveyor belts, etc., where the reinforcement yarns are covered in rubber or plastic.

Hollow viscose fibres (for example Viloft) are another modification offering bulk and higher moisture absorption properties compared with normal viscose. The hollow effect is obtained by the addition of sodium bicarbonate to the spinning dope, which results in the generation of carbon dioxide gas within the fibre during coagulation in the acid bath. Other modified viscose fibres include flame retardant viscose (for example Darelle) and deep-dyeing viscose which is manufactured by the incorporation of appropriate modifiers in the spinning dope.

1.4.4.1.2 Modal (polynosic) rayon

The properties of viscose fibres may be modified on a more fundamental level to increase their wet strength and modulus. A very slow regeneration in the presence of dilute acid and a high concentration of zinc sulphate, accompanied by a very high degree of applied stretch (200%), results in a polymer system that is more crystalline than that of viscose. The highly oriented structure imparts a high degree of strength in both dry and wet states, lower extension at break, and lower moisture regain (10%). This high-wet-modulus viscose in termed 'modal' (or 'polynosic'). In addition, the manufacturing process of modal involves a less extensive alkali steeping stage, maintaining a higher degree of polymerization compared with viscose (see Table 1.4). The extensive fibre stretching results in an 'all core' fibrillated structure which affords a fibre with a crisper handle, closer to that of cotton.

1.4.4.1.3 Cuprammonium rayon

As early as the mid-1800s it was known that cellulose could be dissolved in an aqueous solution of ammonia containing copper salt and caustic soda, and extruded through the holes of a spinneret into filaments. It was not until the early 1900s, however, with the development of the stretch spinning process, that a rayon with satisfactory tensile strength could be manufactured, offering the first commercially useful man-made fibre, namely cuprammonium rayon.

The fibre has very similar physical and chemical properties to those of viscose; one minor difference is that the pore size of cupro fibres is larger than that of viscose, so that cupro fibres stain and dye more readily. However, the high cost of cupro production, involving the use of copper salts, resulted in cupro being superseded by viscose in the 1940s. In fact, cupro fibre is now virtually obsolete.

1.4.4.1.4 Cellulose ester fibres

Both acetate and triacetate are ester derivatives of the natural cellulose polymer, in which the hydroxyl groups have been acetylated to form the ester of acetic acid. In the manufacture of the cellulose esters, purified wood pulp is steeped in acetic acid to open up the molecular structure in preparation for chemical reaction. In the next stage, more acetic, acetic anhydride, and sulphuric acid (as catalyst) are added, and complete conversion of all the OH groups into acetate groups ($-OCOCH_3$) takes place. *Triacetate* is therefore produced first, where all six hydroxyl groups of each cellobiose unit are acetylated; this derivative is also known as 'primary cellulose acetate'. *Acetate* is formed by the partial hydrolysis of the fully acetylated polymer, leaving 74–92% of the hydroxyl groups acetylated. This polymer is thus referred to as 'secondary cellulose acetate'. In either case the polymer is solidified by dilution, washed, dried, and ground up into white flakes. Acetate and triacetate fibres are dry spun from volatile solvent systems and processed into staple fibre.

In comparison to viscose, the cellulose esters, with amorphous polymer systems (60%) and many fewer hydroxyl groups, are weaker fibres (see Table 1.4) which become even weaker when wet, as water molecules disrupt further the interchain forces within their amorphous regions. The weak forces of attraction readily give way to polymer slippage when subjected to strain, hence acetate and triacetate materials distort or wrinkle easily.

The moisture absorbencies of acetate and triacetate fibres are lower than that of viscose, as might be expected from the replacement of a significant number of hydroxyl groups with relatively non-polar acetate groups in the cellulose esters. The amorphous polymer network nevertheless allows penetration of water molecules to a significant extent, so that the cellulose esters are more absorbent than synthetic fibres. The moisture absorbency of triacetate is reduced even further on heat setting the fibre (2% regain); this change presumably results from the greater orientation of the heat set state.

Both fibres are thermoplastic, that is they may be permanently set or pleated by the application of heat. Particularly in the case of triacetate, the predominant presence of non-polar ester groups as opposed to polar hydroxyl groups allows the retention of interchain (van der Waals') forces in the heat set configuration. Triacetate yarns are often texturized.

In addition to being susceptible to acid hydrolysis, as are other cellulosic polymers, the acetyl moiety of the cellulose esters are also sensitive to alkali attack, where saponification (cleavage) of the acetyl moiety can lead to fibre yellowing. Acetate, being only partially acetylated, is soluble in relatively polar solvent media (for example 70% acetone) and insoluble in non-polar solvents (for example chloroform), while the reverse applies to the fully acetylated triacetate fibre.

Acetate and triacetate fibres are used in most fabric and garment end uses for their silk-like appearance, soft handle, and easy dyeing. Their tenacities are adequate for dress wear, linings, and furnishings fabrics, and they are often blended with cotton and synthetics for use in similar products.

1.4.4.2 Regenerated protein fibres (Merinova)

Regenerated protein fibres are manufactured from casein by extruding an alkaline solution of the protein followed by coagulation in acid. Further treatment with formaldehyde improves resistance to alkalis. The fibres may be distinguished from silk by their insolubility in cold concentrated hydrochloric acid in which silk dissolves. Regenerated protein fibres are usually incorporated into blends with other fibres such as cotton and wool (worsted, woollen, and felts).

1.4.4.3 Miscellaneous

1.4.4.3.1 Alginate fibres

Alginate fibres are prepared by spinning an aqueous solution of sodium alginate, extracted from seaweed. The presence of a calcium salt in the coagulation bath precipitates the fibre as calcium alginate. The fibre is nonflammable owing to its high metal content, and it is soluble in alkaline solution. Alginate may be used in similar end uses as water soluble PVA (see section 1.4.3.7.4). The principle application of alginate fibre, however, is in surgical end uses, where the water soluble form of sodium alginate is readily absorbed in the blood stream.

The fibre is characterized by its high ash content, and solubility in sodium carbonate solution.

1.4.5 Other man-made fibres

Carbon fibres are prepared from existing man-made fibres, most commonly polyacrylonitrile (PAN), although rayon and aramid based fibres are also produced.

Manufacture involves a controlled three-stage heating process during which the fibre is pyrolysed and carbonized (heteroatoms are removed) leaving a pure carbon fibre. The final heating temperature is in the range of 1500–3000°C, and the material is maintained in a fibrous form at all times.

Carbon fibres are black and lustrous, and are very resistant to heat, although they will tend to oxidize in air at 450°C. The polymer structure is built up of carbon atoms arranged in parallel layers, well oriented in the direction of the fibre axis. The outstanding performance of carbon fibres relates to their stiffness (high Young's modulus); when compared to extremely stiff steel fibre, on a weight-to-weight basis, the former is far superior. Carbon fibre complements Kevlar in performance, as the merit of the latter relates to its high tenacity as opposed to modulus.

The very brittle character of carbon fibre, as evidenced by the very low breaking extension, requires the fibres to be embedded in a matrix support for all end use applications. These fibre reinforced resin composites are used in the aerospace industry (for example skin structures) and high technology sporting goods (golf clubs, skis, etc.), where stiffness and light weight are essential criteria for optimum performance, easily overriding the high cost of production.

Glass fibre, composed of sand (SiO_2), soda ash (Na_2CO_3), and limestone (Ca_2CO_3), is manufactured in continuous form by mechanically drawing a molten stream of glass vertically downwards at high speed. Filament diameters are in the range of 3–16 μm, and multiple filaments (400 or 800) are brought together in a strand and processed into yarn, roving, woven roving, or chopped strand mats.

Glass fibres possess excellent tensile strength, and on a strength-to-weight basis their strength is far superior to conventional constriction material such as steel and aluminium. Their low cost compared to other high strength fibres such as aramids and carbon fibres has rendered them the best all-round fibre for the reinforcement of plastics. Applications include boat building, transport, pipes and tanks, etc. Woven glass fabrics are primarily used as filters, chosen for their temperature and chemical resistance. Several variations of continuous filament glass exist, the principal one being E-glass. Others include C-glass and S-glass which respectively have greater chemical resistance and higher strength.

Ceramic fibres are a range of inorganic fibres specifically developed for their high temperature resistance. They generally contain either silica or alumina, or proportions of both. Other inorganic fibres also include silicon carbide and boron fibres. These fibres may be spun at very high speed from molten material and obtained as fine as glass fibres. Alternatively, the carbonizing approach may be used where a precursor fibre is pyrolysed at high temperature, leaving the metallic oxide residue. The fibres are used for their extremely high operating temperatures (>1000°C) and chemical stability in filtration media. Fabric applications include hot flue gas filtration, high efficiency air filtration in hospitals, etc., and sterilizing filters in the preparation of beverages. One of the main outlets of bulk fibre is in thermal and acoustic insulation.

Other inherently flame retardant fibres include polybenzimidazole (PBI), phenolic (novoloid) fibre and polyphenylene sulphide (PPS). The first two have been developed for their excellent heat resistance combined with the advantage of high moisture regain (equivalent to or better than that of cotton) which makes them particularly suitable for protective clothing. The merit of PPS fibre lies in its chemical resistance under extremely adverse conditions.

Ultra-fine **steel** fibres (2–12 μm) may be incorporated into synthetic textiles for the purpose of dissipating static electricity (for example in carpets). The blend ratio is approximately 1 (metallic): 100 (normal), hence the metallic fibre remains virtually invisible. In the extrusion process, a bundle of fine steel fibres is sheathed in a dissimilar alloy, spun, and stretched. The sheath is then chemically removed to give the ultra-fine steel fibres.

Metallized yarn is manufactured by coating polyester film with a thin layer of vaporized aluminium. The film is subsequently slit into 0.25–1.0 mm wide ribbons and is commonly used for textile decoration (e.g. Lurex).

ADDITIONAL READING

In this relatively short contribution is has not been possible to cover all aspects of textile fibres. Interested readers may find the following texts useful.

(1) *Identification of textile materials*, 7th ed. The Textile Institute, Manchester, UK.
(2) *Textiles*. The Textile Institute, Manchester, UK.
(3) E. P. G. Gohl & L. D. Vilensky. '*Textile science*', 2nd ed. (1983). Longman Cheshire Pty Ltd.
(4) *Textile term and definitions*, 8th ed. The Textile Institute, Manchester, UK.
(5) *Handbook of fibre science and technology*, Volumes I–IV. Marcel Dekker Inc., New York.
(6) J. A. Maclaren & B. Milligan. '*Wool science:the chemical reactivity of the wool fibre*'. (1981). Science Press, Sydney.
(7) E. T. Trotman. *Dyeing and chemical technology of textile fibres*. Griffin, 1986.

ACKNOWLEDGEMENTS

The authors gratefully acknowledge the use of the teaching resources of the School of Fibre Science and Technology.

2

The forensic examination of fibres: protocols and approaches—an overview

James Robertson
Assistant Secretary and Head, Forensic Services Division, Australian Federal Police, PO Box 401, Canberra, ACT 2601, Australia

2.1 INTRODUCTION

The many different aspects of the management of a case, in which fibre examination may play a part, are each considered in detail in individual chapters of this book. In this chapter I shall attempt to coordinate, to fill gaps, and to discuss the management of fibre cases; as the title of this chapter states, to provide an overview. I also wish to discuss, as somewhat broader concepts, the vital and I would say pivotal pre-laboratory parameters of crime scene and evidence handling. In this way I hope to provide the reader with the essential skeleton on which my fellow authors provide the flesh. Whilst attempting to give an academic overview of fibre protocols it is important to discuss what are the minimum acceptable standards. In this context it is worth remembering that not all laboratories have access to the most up-to-date, sophisticated (and always expensive) equipment.

Finally, I wish to emphasize the need to maintain realistic expectations for fibre evidence. All too often the potential value of forensic evidence is lost as a result of over expectations resulting in the perception by the user of failed expectations. Whilst fibres may be highly significant in a case, more often they will be but one element of evidence put before a jury.

2.2 CRIME SCENE ASPECTS AND EVIDENCE HANDLING

2.2.1 Introduction

There is a much hackneyed phrase which puts in succinct terms the effect that the crime scene examiner can have on subsequent forensic examination of exhibits: '*Rubbish in, rubbish out*'. Yet this is the unavoidable truth. The forensic scientist is constrained by the INPUT, and as the discovery and collection of potential evidence is normally a police responsibility, then it follows that they, and to be more specific the scene examiner, have the critical role in determining the effectiveness of the subsequent laboratory investigation. It is outwith the scope of this chapter to consider crime scene examination in detail. I wish only to stress some of the principles which I consider apply to the work of a crime scene examiner and which can impact on fibre examinations.

2.2.2 Contamination

It is a rare event for the crime scene examiner to be the first person at a scene. The scene examiner will usually attend a scene discovered by someone else, a member of the public or a generalist police officer; often a relatively inexperienced officer. It is vital that all police personnel know about contamination and how it can be avoided. A contaminant can be anything added to the scene that was not present at the time of the incident. This can range from fingerprints, sweat or other body secretions, dropped cigarette butts, hairs, clothing fibres, a footprint, and so on. The simplest way to avoid contamination is not to touch anything, and, as far as is practicable, keep as far away as possible from the scene. However, in the real world this can be difficult to impossible to achieve. If a body is found, clearly the person who finds it will have to check for the presence of vital signs; first aid or resuscitation measures may be necessary. These take priority over any other consideration. Nonetheless, the body should be disturbed as little as possible, and the discoverer should move some distance from the scene as soon as possible and stop further entry to the scene by unauthorized persons, including other police officers. It may be very difficult to define the boundaries of a scene, especially outdoors, and really the only good advice is, be as cautious and thoughtful as possible; also be aware of possible routes used to enter and exit the scene.

It is good practice in a major enquiry for an incident room to be set up and the names of all persons entering the scene recorded. The thought of possibly having to go to court at a later time is remarkably effective in deterring 'professional' sightseers!

The specialist scene examiner is not expected to make elementary errors with regard to contamination, and his or her advice should be listened to and acted on by investigating officers and others. In more serious incidents, the scene examiner will have protective clothing to reduce the chance of accidental introduction of trace materials. The investigator will not. Thus, although the investigator has a natural desire and sometimes legitimate need to enter the scene, great care should be taken and the guidance of the expert (scene examiner) sought.

Of course the major crime scene is really the easy part of the job because resources will be martialled and directed to that scene. Mistakes, if they are to be made, are more likely to occur at simpler scenes or where it is not immediately recognized that it is in fact a crime scene.

In the case of a deceased person the 'evidence' is not going to walk away, and subject to adverse weather, there is time to deal with the scene in a well organized, systematic way. This in itself should reduce the chance of mistakes being made.

The potential for accidental contamination may not be obvious (Cook 1981, Moore *et al.* 1984). For example, studies by Lim (1984) have shown that, where fibres are transferred onto car seats and seat belts, a small number of fibres may remain for many days after the transfer. These fibres are a 'pool', available for secondary transfer.

Also it is worth noting that the bags used to transport bodies can themselves be a source of contamination.

With regard to fibres it is especially important to be aware of the potential for contamination in order that the necessary precautionary measures are taken. Some general rules which can assist are:

- anyone taking part in the examination of a scene should not take part in the examination of a suspect;
- protective clothing should be worn by crime scene personnel, by scientific staff, and by medical doctors taking part in any aspect of searching;
- items collected at a scene or from individuals should be immediately packaged;
- in the laboratory there should be separate areas for searching items relating to different people; complainants and suspects items must be kept separate at all times;
- ideally different examiners should examine items from a complaint and from a suspect;
- items from complainants and suspects should not be stored together.

2.2.3 The GIFT principle

In the ideal world it would be possible to secure a scene and then examine it, taking whatever time was necessary. Of course, one would also have a clear picture of the events alleged to have taken place which would help one to decide what examinations were necessary and what items to collect. Unfortunately we do not live in that ideal world. Very rarely can a scene be secured for a lengthy period. In an outdoor scene the weather may present problems. With an indoor scene the relatives may want access to the room or house. A vehicle may be impounded, but for how long? Thus, it is vital that the scene examiner works quickly and systematically to examine the scene and collect items of potential importance. The letters GIFT stand for GET IT FIRST TIME. There may not be a second chance to collect evidence. It is rarely satisfactory to go back to a scene. Once touched by human hand, it is altered for ever.

How does the examiner decide what to collect? Again, in the ideal world with full knowledge about the alleged incident, this would help the examiner decide. In the real world, since the examiner is likely to be called in close to the start of the investigation, very little background may be known. Clearly, it is important at this time that everyone concerned communicates effectively. It is a team effort. The scene examiner needs to think about the likely examinations which will follow and what samples will be necessary, paying particular attention to relevant control samples.

The police are sometimes accused of having blinkered or tunnel vision in their investigations, focusing on a particular suspect to the exclusion of all others. Whether or not this is a biased perception, it is vital that the scene examiner displays an impartial attitude, collecting samples for their possible scientific value. An impartial attitude on the part of everyone concerned is sound common sense. If the suspect turns out to be the offender (in the eyes of the jury), the scene examiner having been seen by the jury as open minded will have helped.

Realistically, it is not possible to collect everything, and the laboratory would not thank the crime scene examiner for it anyhow. There will be occasions when something is not collected which later turns out could have been important: nobody is perfect. In deciding what to collect it is important to be systematic, to think about why an item is being collected, and what information is likely to emerge from its examination. Whilst it is important that legal practitioners ensure that adequate standards are followed, they too should have realistic expectations assessed not with too large a dose of hindsight.

With regard to fibres, the importance of this principle lies particularly in:

- recognizing that a transfer of fibres may have taken place;
- taking the appropriate steps to collect necessary samples;
- ensuring that adequate and representative known and control samples have been taken.

2.2.4 Packaging and labelling

Previous mention has been made of the importance of putting items into appropriate packages. These should then be labelled as soon as possible. The appropriate package will depend on the nature of the item.

What is the purpose of a package? Primarily, the package is to preserve the integrity of the item, to avoid loss of materials which may have evidentiary value, and to stop contamination. Whether the package looks good really is not important. For clothing I personally favour paper bags, preferably wet strength. The bags should then be folded back and sealed with adhesive tape and signed across the seal.

Fibres or any small item can be placed on a sheet of clean folded paper which in turn can be placed in a plastic bag or a paper envelope. With trace evidence, if it can be seen, it is preferable to pick it off and package it as above rather than use a tape-lift.

The purpose of labelling is, firstly, to enable the item to be identified at some time in the future and, secondly, to establish a chain of custody. Whatever the design of label used, it should give the following information:

- to whom or what the item relates (a complainant, a suspect, a scene);
- a description of the item (a shirt, a tyre, etc.);
- the time and date of collection;
- the name and signature of everyone taking part in the collection and then subsequent custody of the item.

2.2.5 Conclusions

- The quality of product delivered by the forensic scientist is determined by the input.
- The police are normally responsible for the discovery and collection of items; these along with case information are the input factors.
- To maximize the effective use of the laboratory, it is important to provide that information and to constantly update it. Effective delivery requires effective communication.
- Even with good communication, cases can go wrong. A major cause of this can be contamination. All police officers need to be aware of this problem, and none more so than the scene examiner. The scene examiners' experience should be used as a resource by less experienced personnel.
- At a scene there is only one chance to collect items. Get it first time. Knowing what to collect takes experience. Think about the reasons for submitting an item to the laboratory. It may not be necessary.
- Use appropriate packaging and labelling to ensure that evidence is not lost or altered.

2.3 TRANSFER AND PERSISTENCE OF FIBRES

2.3.1 Transfer

> *Wherever he steps, whatever he touches, whatever he leaves, even unconsciously, will serve as silent witness against him. Not even his fingerprints or his footprints, but his hair, the fibres from his clothes, the glass he breaks, the tool marks he leaves, the paint he scratches, the blood or semen he deposits or collects—all of these bear mute witness against him. This is evidence that does not forget. It is not confused by the excitement of the moment. It is not absent because human witnesses are. It is factual evidence. Physical evidence cannot be wrong; it cannot perjure itself; it cannot be wholly absent—only its interpretation can err. Only human failure to find it, study and understand it, can diminish its value.*
>
> *(Harris v. United States, 331 US 145, 1947)*

Edmond Locard, Founder and Director of the Institute of Criminalistics at the University of Lyons, is credited with postulating the theory simply stated that *every contact leaves a trace*. This is often quoted as the Locard exchange principle. To the best of my knowledge no such statement actually appears in the writing of Locard, but the concept has stuck. It does not follow that when a transfer has taken place, it can be detected. In some cases the amount of material transferred may be so small that it is not detected or identifiable by current technology. Also, the rate of loss of some materials after transfer may be so great that the transfer cannot be detected at a very short time after transfer.

For fibres, proof of the theory of transfer first became available in 1975 when studies were reported from a group working at the then Home Office Central Research

Establishment (HOCRE) at Aldermaston, England. Key papers and the conclusions reached are presented below.

Pounds & Smalldon (1975a), using wool and acrylic garments, showed that the number of fibres transferred depended on a number of factors, including:

- the area of the surfaces in contact;
- the number of contacts, although here repeated contacts over the same area were found to cause transfer of some fibres back to the garment of origin;
- the force or pressure of contact with more fibres being transferred with increasing pressure;
- the nature of the recipient garment: a cotton laboratory coat with a smooth surface proved to be a poor recipient (the coarseness of the recipient seems to be important);
- as high pressure and coarse recipient garments produced a greater proportion of short fibres than low pressure and smooth recipient garments, it was suggested by the authors that fragmentation of fibres during contact may be an important mechanism in fibre transference.

Kidd & Robertson (1982) carried out similar experiments but used a wider variety of fibre types as both donor and recipient garments. Their work showed:

- the importance of the nature of both the donor and recipient garments with respect to both fibre composition and texture of the fabric;
- whilst force or pressure of contact was important, the number of fibres transferred increased with increasing pressure only up to a point beyond which increasing pressure led to no further increase;
- far fewer polyester and viscose fibres were transferred than cotton, acrylic, or wool fibres, using donor fabrics composed of these fibre types;
- the proportion of polyester and viscose fibres transferred from a mixed fabric was close to the ratio of these fibres in the donor fabric;
- most (over 80%) of the polyester or viscose fibres transferred were under 5 mm in length.

Further work by Parybyk & Lokan (1986) and by Salter *et al.* (1984) have not supported one of the above findings. Both these groups have shown that in blended fabrics the number of fibres transferred of the different types are not proportional to the stated fibre composition of the garment. For example, Salter *et al.* (1984) showed that with 65/35 polyester:cotton, 55/45 polyester:wool and 70/30 polyester:viscose donor fabrics, most fibres were shed from the minor component. Indeed, in some experiments with the polyester:wool mix only wool fibres were transferred. Parybyk & Lokan (1986) found that the shedding ratio of synthetic polymer blends approximated to their composition only when expressed by fibre number. It is important to realise that on garment labels the proportions refer to weight composition.

This issue of *differential shedding* is important as situations can arise where the number of fibres of different types recovered from a case item are not in proportion to the stated composition of an alleged donor. It is vital to understand the dynamics of both fibre transfer and subsequent persistence in attempting to interpret real life findings. Sometimes the explanation can be simple. For example, cord trousers are often a polyester : cotton mixture. Polyester fibres may not be transferred because the surface of the garment presents only cotton fibres with the polyester threads in the underlying construction of the fabric.

In attempting to interpret whether or not recovered fibres could have come from a questioned item, other factors can also come into play. For example Cordiner *et al.* (1985) showed that the diameter of wool fibres influenced their fragmentation under pressure. In a donor fabric containing wool fibres of varying diameter, more fine fibres may be recovered after a transfer, Sometimes there may be a considerable time gap between the commission of a crime and a suspect being nominated. It is important to realize that clothing collected in these circumstances may have altered because of wear, washing, or other treatment.

Studies on the mechanisms of fibre transfer by Pounds & Smalldon (1975c) led them to propose that three mechanisms may be involved:

- transfer of loose fragments already on the surface of the fabric;
- loose fibres being pulled out of the fabric by friction;
- transfer of fibre fragments produced by the contact itself.

Transfer is a dynamic process, and the fine detail of fibre transfer (that is, the number and size of fibres) will be modified as the garment ages or becomes more worn. This is borne out by the work of Kriston (1984).

All of the above studies were carried out by using artificial or simulated contact. This is necessary to control the many variables which would otherwise make the interpretation of results more difficult. However, one can question whether or not it is valid to extrapolate these 'laboratory' based findings into real life situations. In recognizing this fact Grieve *et al.* (1989) carried out a series of experiments in which greater effort was made to simulate real life conditions. This study was carried out in an attempt to help interpret real case findings where a large number of fibres were found on a bedsheet, nightdress, and wrist binding of a deceased. The authors wished to investigate whether or not these fibres could have originated by casual contact or by secondary transfer.

A great many results were obtained, and it is beyond the scope of this review to examine them in fine detail. Suffice to point out that analysing the number of fibres transferred, relating this to the nature of the recipient fabrics and size of area in contact, it was possible to say with reasonable certainty whether or not fibres had been transferred by direct, primary contact or by a secondary or later transfer.

In other words, fibres may be transferred from, say, a jumper onto a car seat—the *primary transfer*. At some time later a second person sits in that same seat, and fibres from the jumper are transferred to the clothing being worn by the second person—the *secondary transfer*.

Whilst it is possible to carry out quite sophisticated experiments to test an hypothesis in a specific case, it is naïve to expect that the results will always give an unequivocal answer. It is more likely that the number of possibilities will be narrowed. There are very real problems in 'real life'. It is often not possible or appropriate to use the case exhibit whilst pressure of casework in most laboratories would make this type of research work impossible in all but the most critical cases.

The value in the work published to date lies in its ability to assist the scientist to interpret case findings, being aware of the factors involved and the limitations in reaching a conclusion.

Most of the work described above relates to garment to garment transfers. However, other work has been published dealing with:

- transfer from automobile carpet fibres to clothing fabrics—see Scott (1985);
- transfer from clothing fabrics to car seats and seat belts—see Grieve *et al.* (1989), Lim (1984);
- transfer from carpets to footwear—see Robertson & Gamboa (1984).

In concluding this discussion on transfer my view has not changed substantially since I first published in this area in 1982, that is, whilst subtle differences should not be ignored, overemphasis is equally bad in obscuring the overall trends which emerge from studies such as those cited above. Highly sophisticated analysis is possible, but in real life the variables which will contribute to the number of fibres which may be transferred are so numerous and unknown that simulated trials can give only general guidance.

2.3.2 Persistence

Persistence is the other half of the equation which will determine whether or not fibres will be found after a transfer. Depending on the nature of the donor and recipient garments and many other factors, the number of fibres transferred at the time of contact may range from only a few to many hundreds. Whatever the nature of the transfer, it has been clearly established that there is rapid initial loss of fibres, with the early work of Pounds & Smalldon (1975b) showing an initial loss of about 80% after only 4 hours and only 5 to 10% remaining after 24 hours.

The number of fibres lost and the rate of loss depend on numerous factors, some of which were spelled out by Robertson *et al.* (1982) who found that fibre persistence was lessened by four factors:

- the continued wearing of the recipient garment—loss of fibres depends on the item being moved or worn;
- by other garments being worn over or on top of the recipient;
- if the transferred fibres are situated on an area of the recipient more prone to contact with other surfaces;
- if the pressure of the original contact was low.

These authors also found that longer fibres were lost more quickly than shorter fragments below 2.5 mm.

Pounds & Smalldon (1975c) propose that there are three states of transferred fibres, *loosely bound*, *bound*, and *strongly bound*. As time elapses after the initial transfer, then it is the loosely bound and bound fibres which are lost first, with the strongly bound fibres being those which become physically trapped in the weave of the recipient fabric. This is important in determining the way in which items are searched. If it is likely that only strongly bound fibres are likely to remain, the most efficient recovery technique would be required in these circumstances.

Once fibres have been transferred to a particular area of a garment they can also be redistributed over that garment and indeed onto other garments. (Robertson & Lloyd 1984). In real life situations one often does not know whether or not all the clothing worn has been submitted for examination. If small numbers of fibres are found on items it may be because:

- there has been a long time gap between contact/transfer and examination;
- the fibres have arrived on these garments by redistribution;
- of a secondary transfer;
- transfer is coincidental and not real.

Caution is necessary when interpreting the finding of small numbers of fibres especially to items such as underclothing. Such potentially damaging 'evidence' may have less significance than at first glance! Note that the failure to package items in separate bags can also lead to redistribution between garments.

Another factor which is worth bearing in mind is the potential for differential loss of fibres in a fabric blend. For example, smooth polyester fibres have been shown to be lost more rapidly than rougher viscose fibres (Kidd & Robertson 1982). Salter *et al.* (1984) have pointed out that if one type of fibre from a blended fabric is lost much more rapidly than the other, this could result in no fibres of one type being recovered.

The effect of garment cleaning on fibre persistence has also been studied (Grieve *et al.* 1989, Robertson & Olaniyan 1986). From both of these studies it is clear that whilst cleaning does result in both loss and redistribution of fibres, it is still worthwhile examining garments which have been cleaned.

Fibres can persist for periods of many days and even weeks when transferred to car seats and seat belts, as shown by the work of Lim (1984). The rate of loss was shown to depend on the nature of the seat covering with greater retention for fabric seat covers than for vinyl coverings.

Ashcroft *et al.* (1989) have shown that fibres transferred from ski masks onto head hair can persist for up to six days where hair was not washed, and for up to three days where hair was washed.

On the other hand the persistence of fibres on shoes is extremely poor in normal circumstances, with few fibres remaining after minutes. The pattern of the shoe sole and its composition are important. However, for fibres to stay on shoes for any length of time it would seem they have to be physically trapped or adhering to sticky deposits. (Robertson & Gamboa 1984).

2.3.3 Conclusions

In summary, the implications of all of the above are that:

- it is important to collect clothing from complainants and suspects as soon as possible after an alleged offence;
- because fibres are so readily lost and retransferred, undue significance should not be placed on the exact distribution of a small number of fibres;
- unless a suspect is apprehended fairly quickly, subsequent to an incident, failure to find fibres matching the complainant's clothing does not necessarily imply lack of contact;

- evidence of contact and hence association found through comparison of transferred fibres will generally be of recent ones;
- it is vital to the integrity of fibre evidence that good contamination prevention procedures are in place;
- as the time of wear increases, those fibres which do remain will be very persistent and difficult to remove, hence efficient methods of recovery need to be used.

2.4 RECOVERY OF FIBRES

2.4.1 Introduction

It should be clear from the preceding section that the opportunity for fibres to be transferred are great and that the fibres will often be too small to readily detect with the naked eye. The potential for the use of this type of evidence should be apparent at nearly all major crime scenes and in all sexual assault cases. In handling items and in making decisions about what should or should not be collected one should always keep in the forefront as a major consideration the GIFT (get it first time) principle.

Scenes and items should be treated as though the case will eventually depend on physical evidence, its effective recovery and subsequent examination. If, in the fullness of time, it becomes clear that the facts do not warrant the examination of the physical evidence, then nothing has been lost. If, however, the scene or items have been incorrectly dealt with and altered or contaminated, then meaningful subsequent examinations will not be possible.

A final word of introduction regarding the nature of physical evidence: while much of it is too small to be readily detected, there will be cases where there are more or less obvious visible tufts of fibre, for example on a broken window, attached to a wire fence, or on a knife or other weapon.

Less obvious, but well worth looking for, are 'fused' fibres on the plastic trim inside a vehicle. Fibres can become fused as a result of the high temperatures caused by violent impact in a vehicle accident (Masakowski *et al.* 1986). Fabric impressions may also be created.

2.4.2 Methods of retrieval

Where fibres are visible a decision needs to be made whether or not to recover where they are found. Where possible the evidence should be photographed *in situ*, removed, and protected before being sent to the laboratory. If there is any chance that the evidence may be lost during transport it should be removed at the earliest opportunity. Fibres can be collected by using fine forceps aided by the use of low magnification with a hand lens. The fibres can then be placed into a folded sheet of clean paper and put into a paper or plastic envelope. With a dead body, consideration should be given to collecting fibres at the scene before removal to the mortuary. Apart from picking off visible materials, the main methods which can be used to recover fibres are discussed below.

2.4.2.1 Tape lifts

In this method, pieces of clear adhesive tape are pressed onto the surface of the item being examined, and the whole item is systematically searched in a grid fashion. Tapes with different adhesive qualities can be obtained. The 'stickier' the tape the more effective it is in recovering fibres. However, the cost is that the fibres comprising the item being searched are also recovered more efficiently, creating an often dense background of fibres which can hinder subsequent searching of tape lifts for 'target' fibres.

2.4.2.2 Vacuuming

By using special purpose vacuum equipment, large areas can be quickly searched. The technique is useful for the recovery of particulate materials (glass, paint, soil) from difficult areas such as car boots and interiors. For fibres the technique can cover large areas quickly. Its efficiency depends on the strength of the vacuum equipment, and it suffers from a potentially serious drawback in that it can recover a lot of ancient history. The material collected is extremely difficult and time consuming to search.

Other techniques such as brushing or shaking have, I believe, such serious drawbacks that they should **not** be used to recover extraneous fibres.

The advantages and drawbacks of the above techniques have been discussed in detail by Pounds (1975) and more recently by Lowrie & Jackson (1991). It is quite clear that the method of choice for fibre recovery is the use of the *tape lifting* technique.

McKenna & Sherwin (1975) have described the use of seeded combs for the recovery of fibres from hairs. Fibres are transferred during the wearing of hats, balaclavas, or masks often worn during robberies.

2.4.3. 'To tape lift or not to tape lift'.

In my view all serious, and many less serious, cases should be considered for their potential to yield fibre evidence, and scenes/items must be treated accordingly. Thereafter the items worth searching can be decided only when information becomes available. To use an analogy, it is rather like working initially surrounded by mist which gradually clears. It has been suggested that the forensic scientist should work in the dark with no preknowledge of the alleged circumstances surrounding a case. Presumably this is in case such knowledge will lead to a biased approach by the scientist. I do not subscribe to this view. Impartiality is not gained through ignorance! It is essential that the scientist has the fullest possible knowledge. This will result in resources being channelled where they are needed with the effective use of available expertise.

Thus, it is very important to evaluate the case history before starting laboratory examinations. A case conference involving the scientist, crime scene, and investigating officers can be invaluable. There is no point in conducting a lengthy search to show that fibres are present in a location where they may reasonably be expected to be as a result of a legitimate transfer. Equally futile is a lengthy search which fails to reveal the presence of fibres, only for it to turn out that there is good reason to believe that the suspect's clothing was not involved or not worn during an incident under investigation. The type of information which should be sought should include:

- what is alleged to have taken place, who is involved, and how?
- where is the incident said to have taken place? If it was in a house or in a car, who was the occupier or owner?
- with a sexual assault, did it occur on a bed, on the floor? Is it possible to reconstruct the sequence of events? Were bed covers present and were they moved?
- when did the incident take place, and was there any delay before the scene was examined?
- did any person involved have legitimate access to the scene or legitimate contact with the other person or persons before the incident?
- are reliable descriptions available of what was being worn by the offender?
- were items of clothing removed during the incident?

This type of information will enable the scientist to concentrate on what is likely to be productive.

Once a decision has been made to proceed on the basis of the case information available, it then becomes necessary to evaluate which possible transfers are worth following up.

2.5 LABORATORY EXAMINATIONS

2.5.1 Case scenario

To illustrate the approach and techniques used, the following case scenario will be used.

Joan Smith, a single mother, lives in a ground floor unit. She is in the habit of sleeping in a rear bedroom and does not shut the window. In the early hours of the morning entry has been gained to her bedroom by cutting open the fly screen on the window. It is alleged that the offender has then threatened the complainant and had vaginal intercourse before leaving the scene. At the time of the alleged assault she was wearing a white cotton nightdress which the offender roughly pulled up around her breasts. The bedsheets are a very pale yellow colour, and the label indicates the fibre composition to be polyester/cotton. A blue coloured blanket was also present on the bed. Although the complainant cannot be certain, she thinks the offender was wearing blue denim jeans and a red coloured windcheater. A suspect has been apprehended within twelve hours of the alleged incident.

2.5.2 Examination of fibres

2.5.2.1 Protocol

Fig. 2.1 shows a flow sheet for fibre examination. The search for extraneous fibres is based on the assumption that contact has occurred between the offender and the complainant. The purpose of the examination is to see if there are any fibres which would associate the suspect with the complainant and/or her environment and vice versa.

The first step is to consider the clothing, bedding, and carpets, etc. relating to the complainant and the suspect or scene from which fibres may have been transferred. These are donor items from *known* sources.

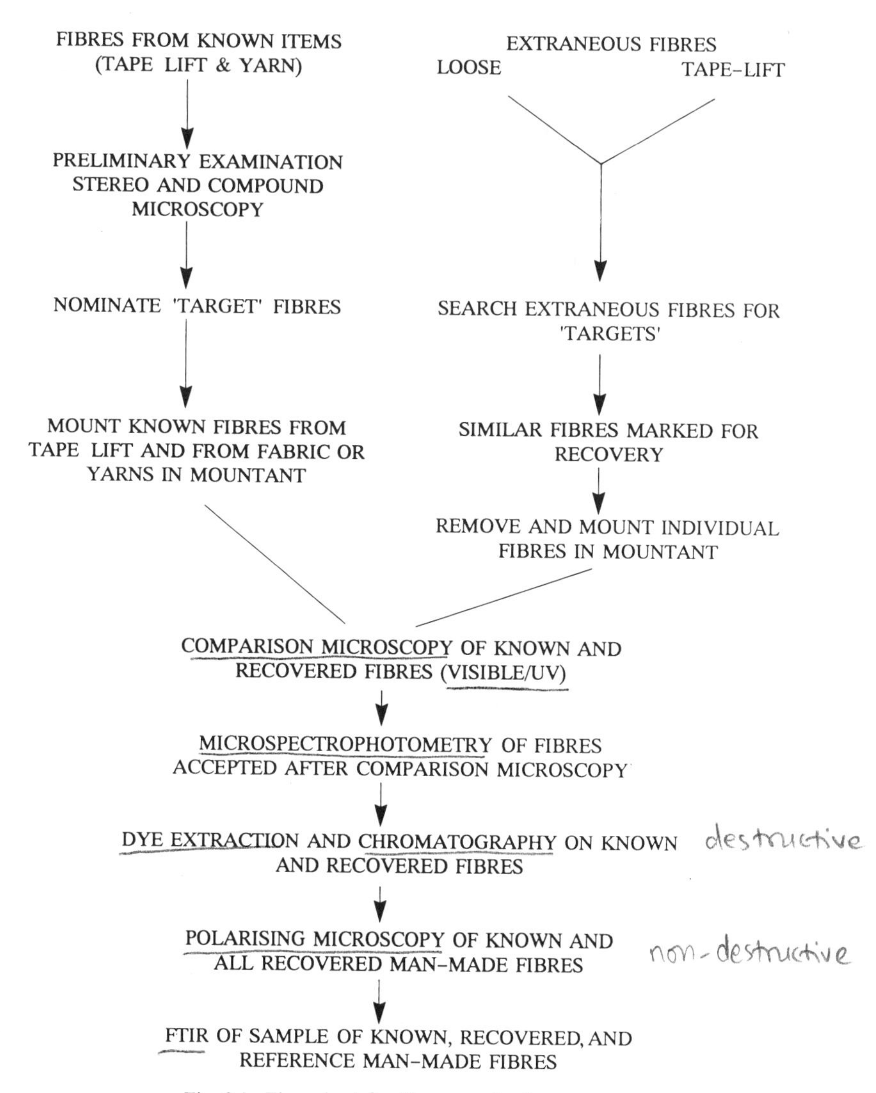

Fig. 2.1. Flow sheet for fibre examination.

All items are visually examined. An illuminated low power magnifier can also be used. Visible sized extraneous material should be collected before items are tape lifted, and fully examined for the presence of hairs, stains, damage, etc. Items from the complainant should be searched in a separate location from those of the suspect. A series of tape lifts is thus created for each item. It is important to also collect known fabric/fibre samples from these known items. A small sample of the fabric to include both warp and weft yarns should be taken. A tape lift of the surface should also be taken; this serves two purposes, giving an indication of the shedability of the fabric (that is how likely it is that fibres will be shed or transferred) and giving more accurate information about the fibres available on the surface for transfer. With carpets and items which cannot be readily or easily moved, yarns can be cut from the item. As for any known item, the samples must fulfil two criteria. They must be large or adequate enough and they must be representative, that is, include all variation in the item.

The next stage in the examination is to carry out a preliminary study of the known fibre samples, using both a low magnification stereo microscope (Fig. 2.2) and where necessary a compound light microscope (Fig. 2.3) with transmitted light and higher magnifications. I prefer, at this point, to use tape lifts for the stereo microscopic examination in the same way that the recovered tape lifts are later examined.

In this case the items submitted are:

COMPLAINANT	SUSPECT
white cotton nightdress	blue denim jeans
pale yellow sheets	red windcheater
blue blanket	white underpants

For the sake of this discussion this preliminary examination reveals the following information:

ITEM	COLOUR	FIBRE DESCRIPTION
cotton nightdress	microscopically colourless	twisted or convoluted
sheets	almost colourless	mixture of convoluted cotton fibres and delustred man-made fibres
blanket	mid blue	lightly delustred man-made fibres
jeans	pale to dark blue	short convoluted cotton fibres
windcheater	pale and mid red	pale convoluted cotton fibres and darker delustred man-made fibres
underpants	microscopically colourless	convoluted cotton fibres

With this information so-called *target fibres* are selected. There are some rules which assist in making this selection. Some types of fibre are not well suited as target fibres. These can be divided into four groups.

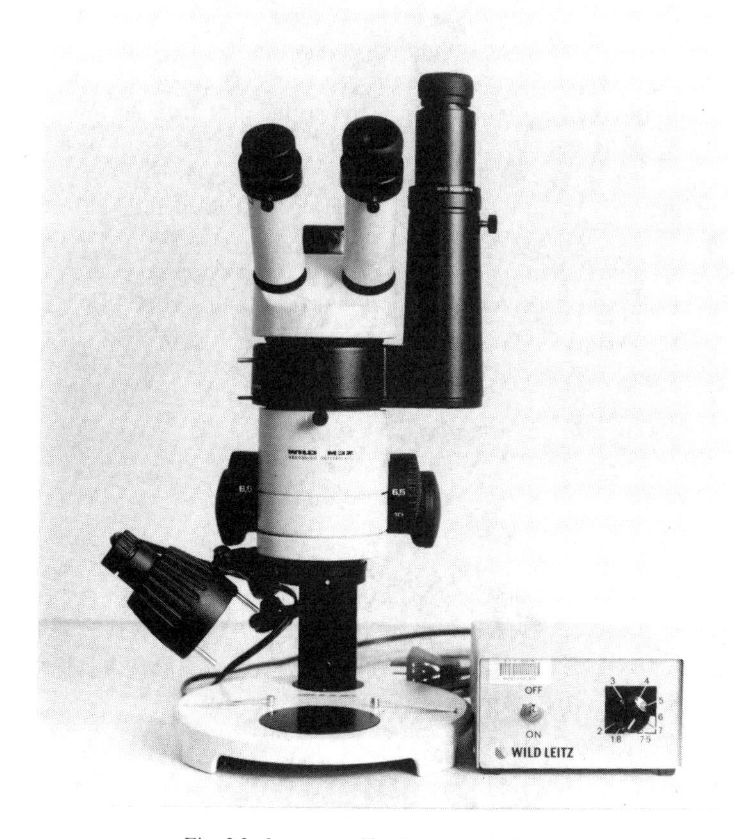

Fig. 2.2. Low magnification stereo microscope.

- Fibres which are extremely common or ubiquitous. Examples are, colourless cotton and, in most cases, blue cotton fibres from denim fabric. In some situations work related clothes will yield fibres which, in normal circumstances, would be of value but are ubiquitous in the specific case.
- Garments which are constructed from very smooth, shiny fabrics are unlikely to shed or to retain fibres in any quantities. Shedability can be tested by using controlled fabric to fabric experiments (Parybyk & Lokan 1986).
- Fibres which are undyed or those where the dye concentration is so low that the fibres appear to be microscopically colourless.
- Fibres from items constructed of shoddy (the terms refers to fabric made from reprocessed fibres which are so variable that it is not possible to establish what is a proper known sample).

It cannot be overemphasized that each case must be assessed on its merits.

There will be circumstances where colourless or almost colourless fibres can provide useful information. Some discrimination is also possible with blue cotton from denim

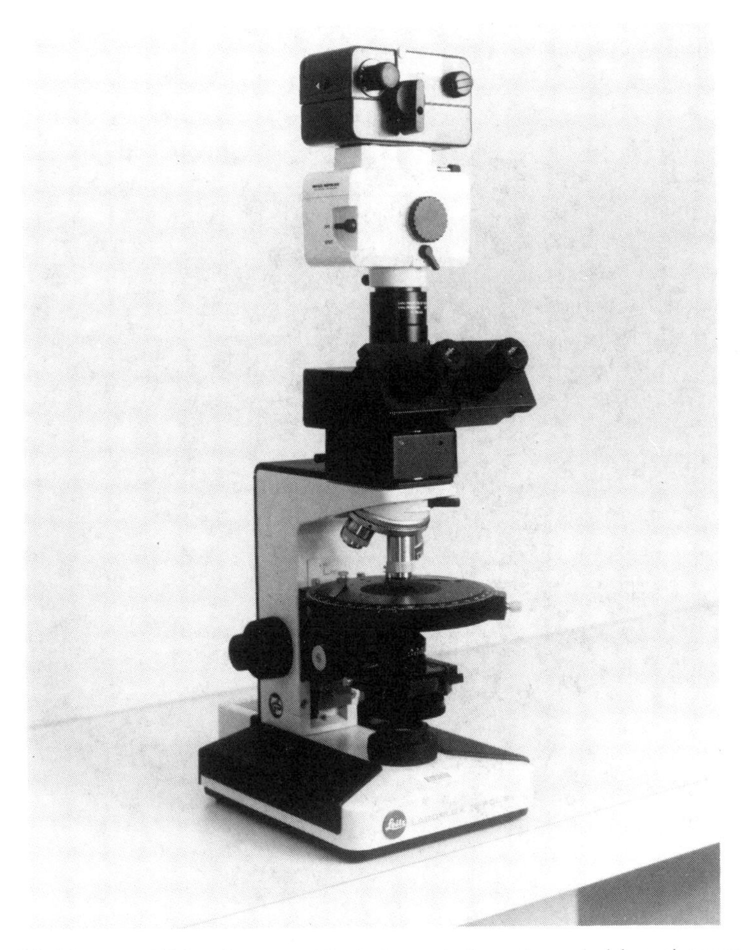

Fig. 2.3. Compound light microscope (the instrument shown is a polarizing microscope).

fabric. In one murder case large numbers of blue denim cotton fibres were found on the pubic hair of the female deceased. She had not been wearing jeans, and, although on its own it was not powerful information, at the least it would have been reasonable to expect the offender could have been wearing blue jeans.

Gaudette (1988) has also suggested a series of criteria which should be considered at this point in the examination:

- highly shedable garments are difficult and time consuming to tape, but are likely sources of questioned fibres;
- garments of low shedability are good for taping but are not likely sources of questioned fibres;
- the more colour contrast there is between the questioned fibres and the garment taped, the better;

- fibres that fluoresce under ultraviolet radiation make good questioned fibres, particularly if the garment to be taped does not fluoresce, or fluoresces to a lesser extent;
- the less common the fibre type, the better it is as a questioned fibre;
- the darker the fibre, the better it is as a questioned fibre, since the chances of a successful microspectrophotometric or dye analysis will be greater;
- the coarser the fibre, the more identification and comparison tests can be preformed on it, hence the better it is as a questioned fibre;
- if a garment is damaged, it is a likely source of questioned fibres;
- special attention should be paid to important areas of the garment, as indicated by the circumstances of the case (for example the seat of a pair of slacks where it is believed that the person was sitting on a seat cover);
- if possible, it is desirable to establish two-way contact or association.

When all of these factors are taken into consideration it may be that even if the overall circumstances of the case warrant fibre examination, this may not be possible owing to a lack of meaningful target fibres.

In our case scenario, taking these criteria into account, the likely target fibres would be as follows:

- blue man-made fibres—blanket relating to complainant;
- red cotton and man-made fibres—windcheater relating to suspect.

The next step would be to:

- search the tape lifts taken from the items relating to the suspect for blue fibres similar to those comprising the blue blanket;
- search the tape lifts taken from the items relating to the complainant for red fibres similar to those comprising the red windcheater.

Tape lifts from clothing or other items will have background fibres from the items themselves along with a mixture of other fibres picked up over time from all the contacts the wearer has made. Searching for target, fibres is like looking for the proverbial *needle in a haystack*. Where fibres are present there will usually be a small number found in a background of up to hundreds of extraneous fibres on a single tape lift. The analyst should note the presence of any large number of outstanding fibre types and colours. Other than in exceptional circumstances it is quite ludicrous to expect that it would be practical, or serve any useful purpose, to attempt to determine the origin of all the extraneous fibres found to be present. This is sometimes suggested in Court as a practical and sensible option.

Fibres similar to the targets are marked on the slide with a permanent marker. These are the *questioned* fibres. They may be removed by cutting a small window on top of the tape lift, placing a drop of xylene on the fibre, and picking the fibre of interest off with fine forceps (Fig. 2.4). All of this is done by using low magnification microscopy. Other methods of searching and fibre recovery can be used, and some of these are described by Grieve (1990). The *recovered* fibre is placed in a suitable mountant and a cover slip added. Choice of mountant is important as the mountant

(a)

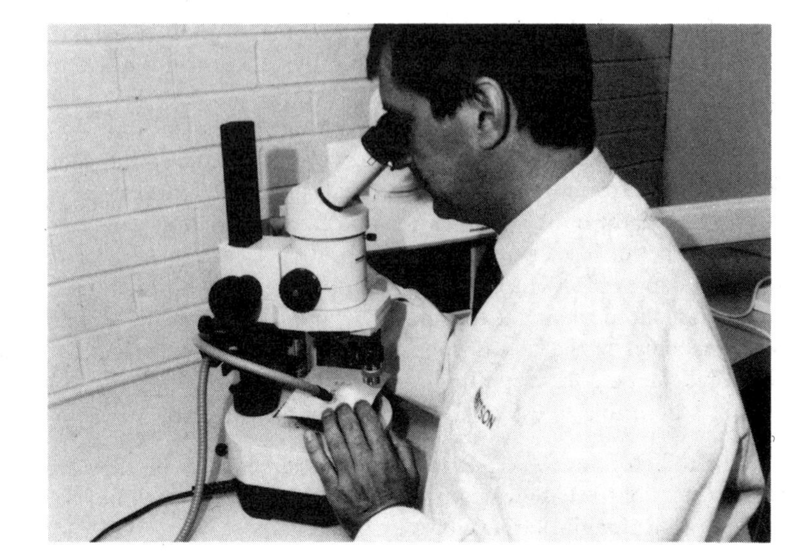

(b)

Fig. 2.4. Examination of tape lift. (a) using a Kombiscope, (b) cutting a window in the tape, (c) lifting the flap of tape, (d) placing a drop of xylene on the tape, (e) placing a drop of XAM on the slide, (f), (g) transferring the 'recovered' or questioned fibre in the XAM, (h) finally, placing a cover slip over the now individually mounted fibre.

(c)

(d)

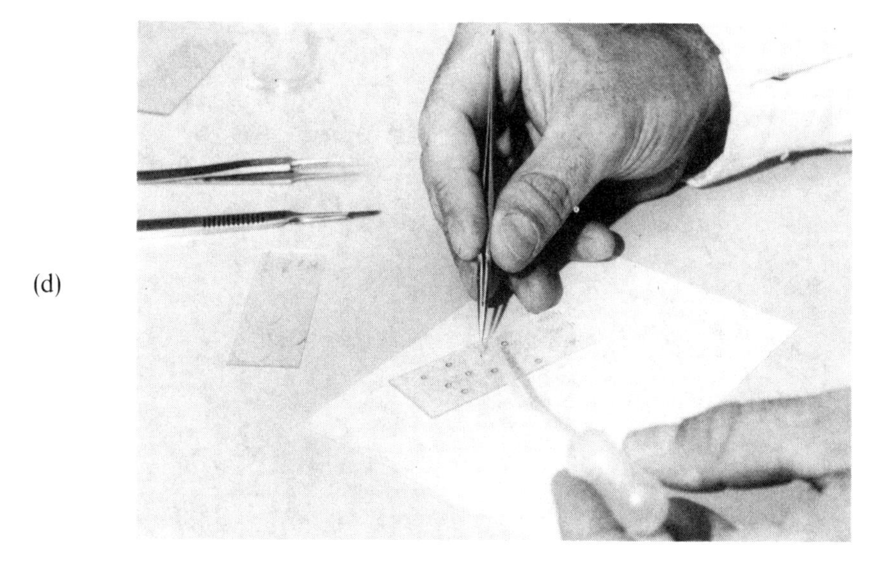

Fig. 2.4.—*continued*

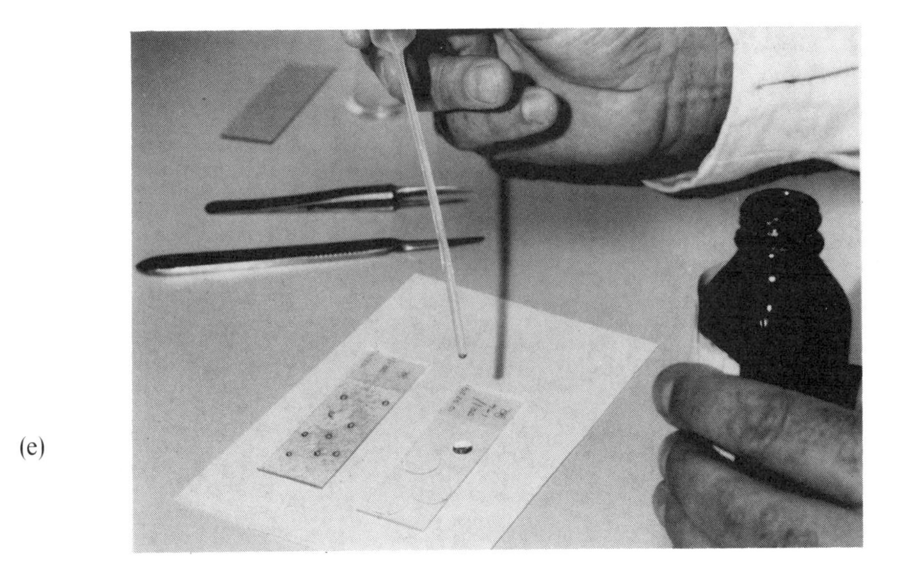

(e)

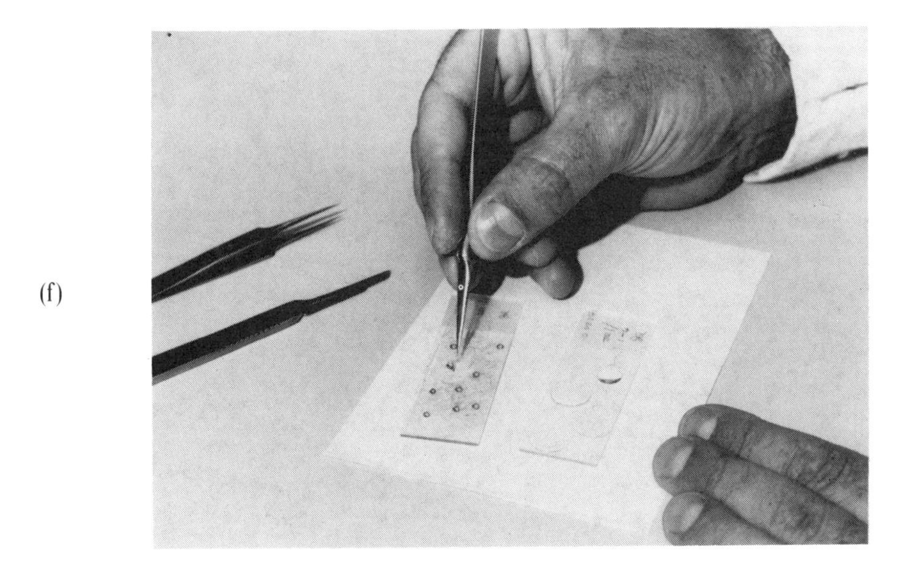

(f)

Fig. 2.4.—*continued*

(g)

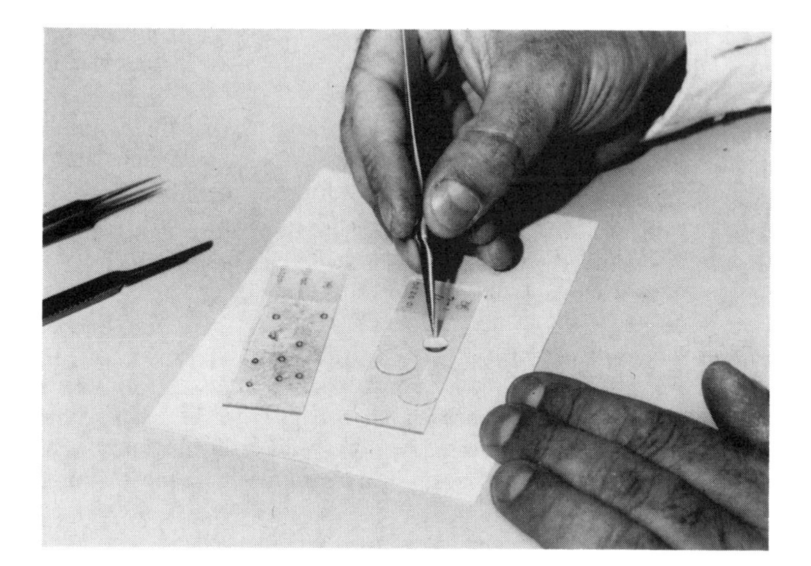

(h)

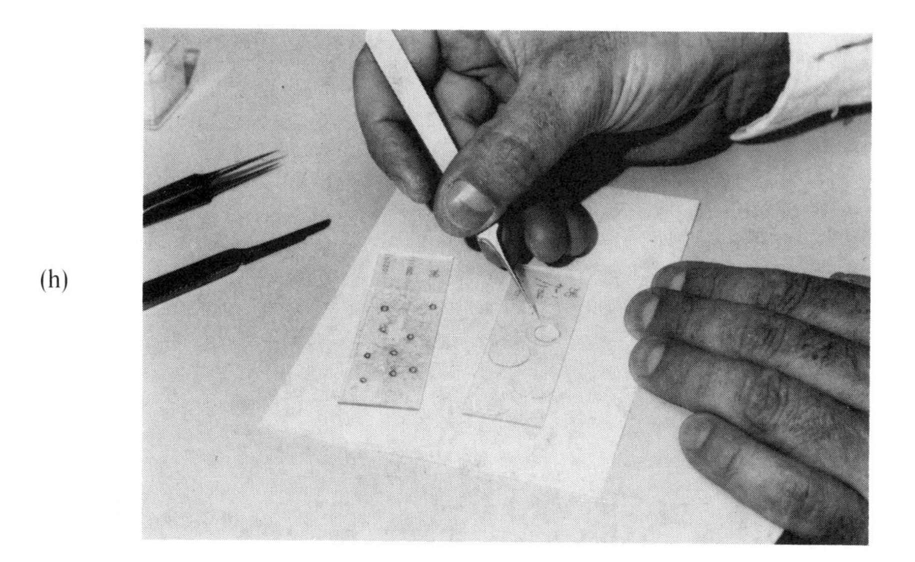

Fig. 2.4.—*continued*

should meet some essential criteria. These have been discussed by Cook & Norton (1982). *There is no single 'correct' mountant, but XAM Neutral Improved Medium White* is widely used.

Fibres from the known sources will also be prepared in permanent mountant. These should include fibres recovered from tape lifts of the known item and also fibres teased out from yarns or threads (warp and weft) from the fabric.

Subsequent examination consists of three aspects: *microscopic comparison, colour analysis* and *fibre identification.* In this Chapter I shall not deal with these in depth, as they are considered in the remainder of this book.

2.5.2.2 Microscopic comparison

The side by side comparison of recovered/questioned fibres with fibres from known sources is conducted using a *comparison microscope* (Fig. 2.5). This consists of two compound light microscopes connected by an optical bridge. A number of manufacturers market comparison microscopes; Leica instruments are most commonly found in forensic laboratories.

Using the comparison microscope, fibres can be compared side by side. It is critical to look for both points of identity (the same features present), but equally points of difference. A lot is often made of this aspect of the comparison, that the examiner

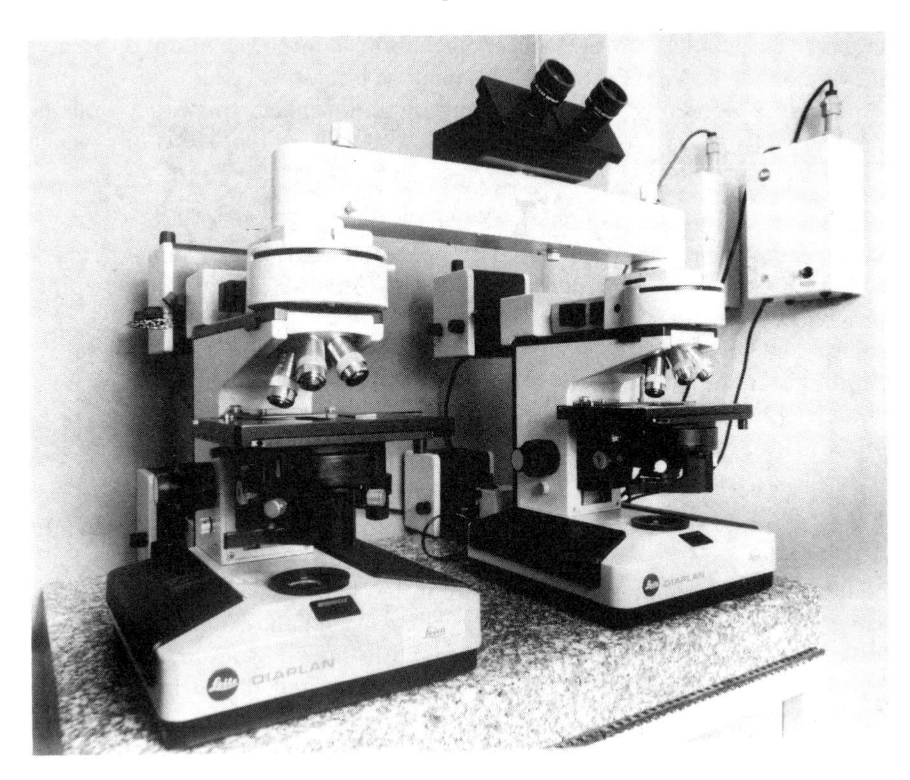

Fig. 2.5. Comparison microscope.

may concentrate too much on points of identity and ignore or give insufficient weight to differences. There is no simple answer to this criticism, but the following discussion may help.

With natural fibres it is probably fair to say that no two fibres are identical. It is generally accepted that man-made synthetic fibres are highly uniform; at most any variation is within narrow limits. However, the fibres and items examined by the forensic worker are not straight off the production line. The donor item may be old, it may have been exposed to weathering, or it will have been washed or dry cleaned. All of these factors, and possibly others, mean that there is a much greater potential for variation in the population of fibres which are available on any item for transfer. After transfer, changes may take place in transferred fibres to further complicate the comparison process. Thus it is facile to believe that, even with man-made fibres, it will always be possible to find identical fibres in the known sample. Clearly, the way in which the known sample is taken is important. *Every effort needs to be made to take a sample which reflects what would transfer during contact.*

Thus, examiners talk of *no significant differences.* There is an element of interpretation as to what is a significant difference, and this may be expected to vary, depending on many factors, especially experience and inherent discriminating ability of examiners. Experience does not always equate, however, with ability.

Returning to the case scenario, at this comparison stage the examiner will decide whether or not each of the recovered/questioned fibres show no significant differences from one or more of the known fibres. Use is made of bright-field microscopy (normal compound light microscopic conditions), and, if the microscope is fitted with the appropriate accessories, incident fluorescence microscopy. The types of features which will be seen include:

- fibre diameter and variation along fibre length;
- fibre shape—to accurately study this, cross-sections may have to be made;
- surface features may be seen, depending on the refractive index (RI) of the fibre surface relative to that of the mountant;
- internal detail including the presence, amount, size, shape, and distribution of delustrant (titanium dioxide is used as a delustrant, the effect being to reduce the brightness of man-made fibres);
- colour;
- fluorescence.

Assuming that all other characteristics are common, colour is the critical feature. In many garments a range of shade will be present in the known fibre sample. Each individual recovered fibre must match a fibre in the known sample for colour (hue and shade). Where a number of fibres have been recovered, the examiner would also wish to be satisfied that the range of shade found in the known sample and in the recovered fibres is comparable. In my view that cannot and should not be viewed from a mathematical or statistical viewpoint. However, where a large number of recovered fibres are found it may be possible.

Fluorescence, mentioned above, is also valuable. Fluorescence is detected when the absorbing fibre is excited with energy of a given wavelength and it emits light of equal or longer wavelength (equal or lower energy). Fluorescence can arise from a number of different sources, from the fibre itself, from dyes, or from optical brighteners and other finishing agents added to the fibre. Fluorescent properties of fibres may also be altered, for example by the use of optical brighteners from detergents or wear and tear. These factors need to be considered when assessing what is or what is not a significant difference.

Gaudette (1988) lists four properties of fluorescence which can be compared:

- presence or absence of fluorescence;
- excitation wavelength producing maximum fluorescence;
- colour of fluorescence under various excitation conditions;
- intensity of fluorescence.

Fluorescence is not studied over a continuous wavelength spectrum. Use is made of a combination of excitation filters with defined wavelength characteristics. Fluorescence properties can increase the chances of discrimination, particularly for colourless fibres.

Because there is a subjective element in the comparison process, *quality assurance* procedures are a must. As a minimum requirement a percentage of recovered/questioned fibres should be independently checked by a second examiner.

2.5.2.3 Colour analysis

Three approaches are used in the analysis and comparison of colour:

- microscopic comparison;
- visible spectroscopy;
- chromatography.

2.5.2.3.1 Microscopy

Microscopy was considered in the previous section. It is the first method used to assess colour, and other methods should not be employed until the examiner is satisfied that recovered/questioned fibres show no significant differences from fibres from the known source. I have seen instances where examiners have proceeded to chromatographic analysis straight from an examination with a low magnification stereo microscope. This is not acceptable as you could easily be dealing with quite different fibres which coincidentally have been coloured by the same dye.

At this point it is worth mentioning a feature called *metamerism*. This is where two coloured objects (here fibres) appear to be the same under one set of lighting conditions but not under a different set of conditions. For example, two garments may be very similar in colour to the observer under daylight conditions but are different when seen under fluorescent lighting. Comparison microscopes usually have one type of bright-field light source. Hence, metamerism will go undetected unless the colour is analysed spectroscopically.

2.5.2.3.2 Visible spectroscopy

The assessment of colour by microscopy is limited by the colour sensitivity of the human eye, and of course the result cannot be recorded in an objective way. That

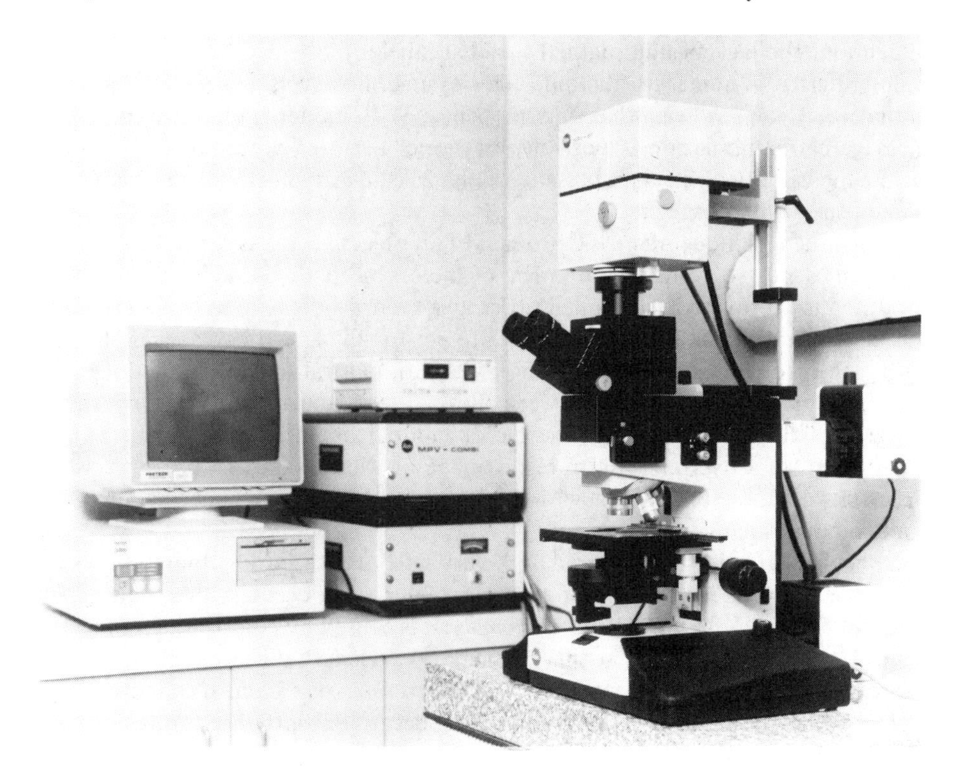

Fig. 2.6. Leica microspectrophotometer.

two fibres match (or cannot be distinguished) is the opinion of the observer or observers. It is now common practice in many laboratories for some or all of the recovered/questioned fibres and known fibres to be examined by using a microscope mounted visible spectrophotometer or *microspectrophotometer*.

Fig. 2.6 shows one instrument available on the market. At least four makes of instrument are being used in forensic laboratories, marketed as:

Micro-Colorite by Rofin, Australia;
Nanospec 10S by Nanometrics;
MPM 400/800 series by Zeiss;
MPV series by Leica.

The fibre to be examined is aligned with a window through which the light passing through the fibre enters the spectrophotometer. The essential components of a microspectrophotometer system are a good quality microscope, an interface with the microscope and the spectrophotometer itself.

Data from the spectrophotometer may be displayed on a chart, and/or it may be converted from analogue to digital form and displayed on a computer screen. Increasing instrument sophistication since the introduction of the Nanopsec in the late 1970s has seen enhanced computer control and presentation of the data. A word of caution: the lack of automatically added labelling of early Nanopsec charts has been criticized in some jurisdictions. This system, however, had a major advantage in that what you saw was what had happened. With modern computer control, data massage or manipulation is possible, and strict audit of procedures is essential to ensure the validity of the data presented. Fig. 2.7 shows typical spectra obtained from a microspectrophotometer.

Whilst the technique offers the promise of an objective assessment like most things in life, it is not so simple. A number of factors need to be addressed if the data produced for recovered/questioned fibres and known fibres are to be compared. For the known fibres it is common practice to record the spectra of at least five fibres with man-made fibres and at least ten fibres with natural fibres. This is done to take account of possible variation between fibres. There are a number of reasons why variation will occur, including possible differential or non-homogeneous uptake of components of a dye mix into fibres, damage or wear characteristics such as twists or cracks in the fibre, fibres with highly irregular shapes, and the presence of contaminants such as blood or oil.

Difficulties are also encountered with very pale or very dark coloured fibres. With very pale coloured fibres a specific problem which can occur is an enhanced signal at about 500 μm. Known as *Wood's anomaly*, it is caused by polarization of the energy from the diffraction grating at this particular wavelength (Grieve 1990).

Spectra can be presented as *absorption* or *transmission* data. It is more common to find spectra presented as absorption data, because absorbance is directly proportional to the concentration of dye. This is critical if the spectral data are to be subjected to mathematical and statistical treatment. This brings us to the subject of the interpretation of spectral data, or, when is a match not a match!

Visual comparison. The primary data are in the form of absorption or transmission spectra (see Fig. 2.7). Transmission spectra may show quite large differences in the size of the spectral area because of (although not in a quantitative manner) differences in dye concentration—visibly *shade.* Some scientists argue that this is the best type of data for visual comparison of spectra. With transmission spectra comparison has to be between fibres of very similar shade because spectral shape is modified by the area under the spectral curve (see Fig. 2.7). Certainly this approach has the benefit that fibres of similar shade are being compared, thus eliminating any small effect that this may have on spectral shape. This is perhaps more critical with natural fibres, especially cotton. In the transmission mode it may be necessary to produce more than the recommended five or ten spectra of known fibres. Grieve (1990) suggests that more accurate comparisons can be made if the spectra from the known sample are run on the same fibres used in visual matching. These can be precisely located on the case microscope slide by using a device known as an England Finder (a microscope slide marked with a reference grid).

Whilst there are arguments for the use of transmission spectra for visual matching of spectra, there is in my view an internal scientific flaw in the argument. The purpose in using spectroscopy to measure colour is that it is numerical and hence more objective than microscopic comparison. Furthermore, the spectra produced arise

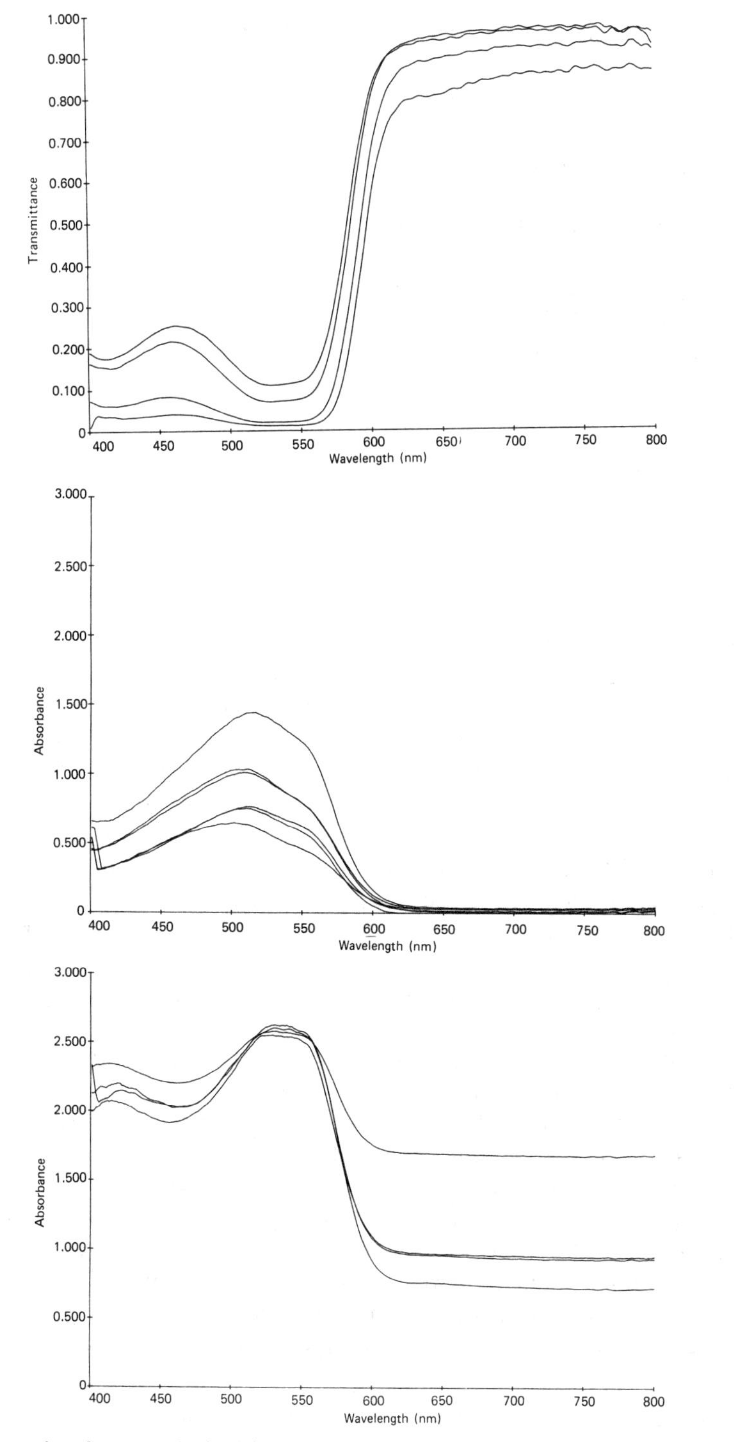

Fig. 2.7. Examples of spectra obtained from a Leica microspectrophotometer: (a) absorbance spectra for six different red wool fibres; (b) transmittance spectra for different shades of red wool fibres showing effect of shade on spectral shape; (c) normalized absorbance spectra for different shades of red wool fibres.

primarily because of the dyestuffs used, irrespective of the fibre type or the nature of the fibre. It is then surely more logical to normalize the spectral area producing spectra which can then be compared directly by, for example, overlaying spectra on a light box. The detail to be considered would then be true spectral information such as the overall shape of the spectra, maxima, minima, and the presence of shoulders or plateaus. This approach is used in the United Kingdom Home Office Laboratories and in the Metropolitan Police Laboratory.

Whatever form of data is used, a match between a questioned fibre and a known fibre can be called only when there is very close agreement between the two spectra being compared. Note that I have not said identical, as some variation may be expected, especially for natural fibres, Grieve (1990) points out:

> *There is no certainty that two samples are from the same textile even if the spectral curves are very similar. We can only be sure they do not have a common origin if the spectra are different.*

Mathematical/statistical comparison. Absorption has a linear relationship with dye concentration, and absorption data can be used for statistical treatment. Two different approaches have been used. Firstly, it is possible to use an analysis of the variation in data where the questioned fibre spectra are compared against date obtained from the collective known fibre data. Interpretation of what is and what is not a significant difference caused this approach to be dropped by the Home Office Forensic Science Service (HOFSS) who were the only people, to my knowledge, to have actually attempted this approach.

The second method of analysing spectral data is based on the use of the CIE (Commission Internationale de l'Eclairagè) system. This is an internationally recognized system used to describe and define colour, and it finds application wherever colour is measured. It is based on a mathematical model of the eye where daylight, as perceived by the human eye, is considered to comprise a mixture of equal amounts of red, green and blue light. These are called the reference, matching, or primary stimuli, and, as there are three, the amounts of each primary stimuli which make up any colour are then defined as the *tristimulus values*.

Any colour can be described by its tristimulus values, nominally referred to as X, Y, and Z. This topic is dealt with in detail in Chapter 4. The values obtained depend on the viewing conditions, including the illuminating source. In commercial colour applications the viewing conditions are specified and defined to allow acceptable reproducibilty between manufacturers; that is, so that two manufacturers can produce garments of the same colour. However, for forensic applications it is not possible to compare tristimulus values between laboratories unless the same microscope and illumination source are used. What is more important is to ensure internal reproducibility with the laboratory protocol.

Tristimulus values can be normalized as follows:

$$x = \frac{X}{X+Y+Z},\ y = \frac{Y}{X+Y+Z},\ z = \frac{Z}{X+Y+Z} \text{ where } x + y + z \text{ must equal 1.}$$

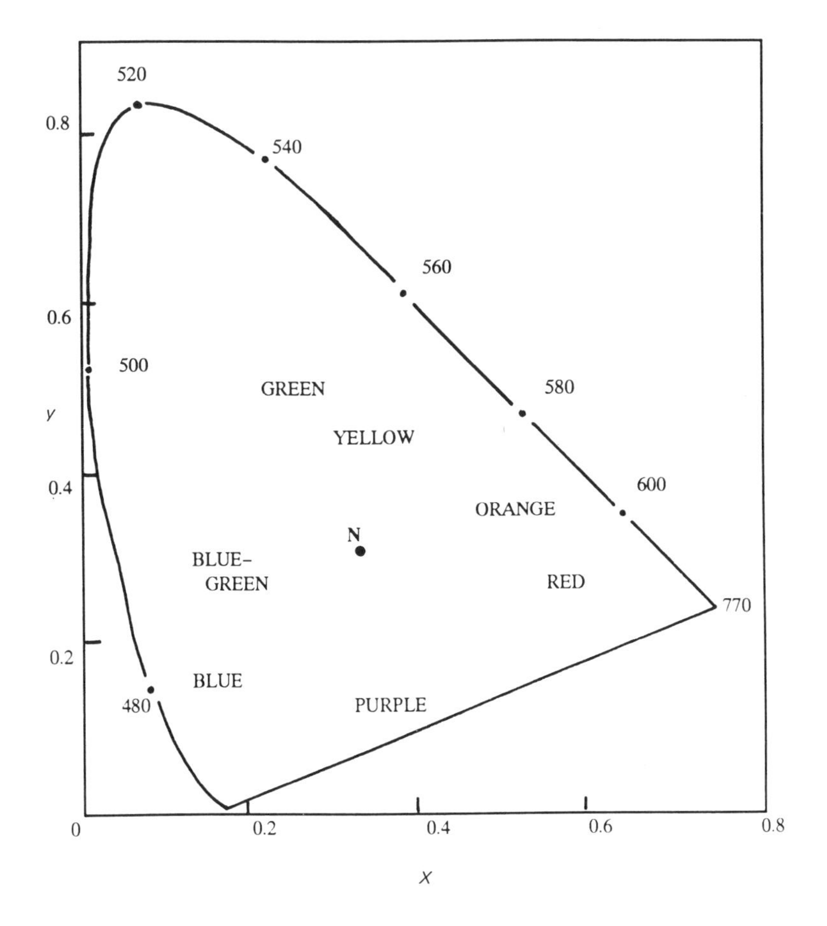

Fig. 2.8. CIE chromaticity diagram.

Therefore only two of these parameters need to be known to specify colour, usually x and y. These are called *chromaticity coordinates*. A plot of x against y is called a *chromaticity diagram* (see Fig. 2.8). This diagram is two-dimensional, and whilst defining colour it does not take into account shade. Shade or lightness, as defined by white to black, can be shown in a so-called colour solid.

It has been shown that the relationship between tristimulus values and colour concentration is nonlinear, resulting from the fact that tristimulus values are based on transmittance. Because textile fibres may exhibit variation in dye uptake and fibre thickness there will be differences in shade. Thus, two fibres with the same dye components could give different tristimulus values. To overcome this, absorbance measurements are used instead of transmittance. The wavelengths of light transmitted are equivalent but not identical to the colour absorbed. Thus, the colour seen is said to be complementary to the absorbed colour, and instead of having chromaticity coordinates we have *complementary chromaticity coordinates* (CCC) (see Fig. 2.9). This

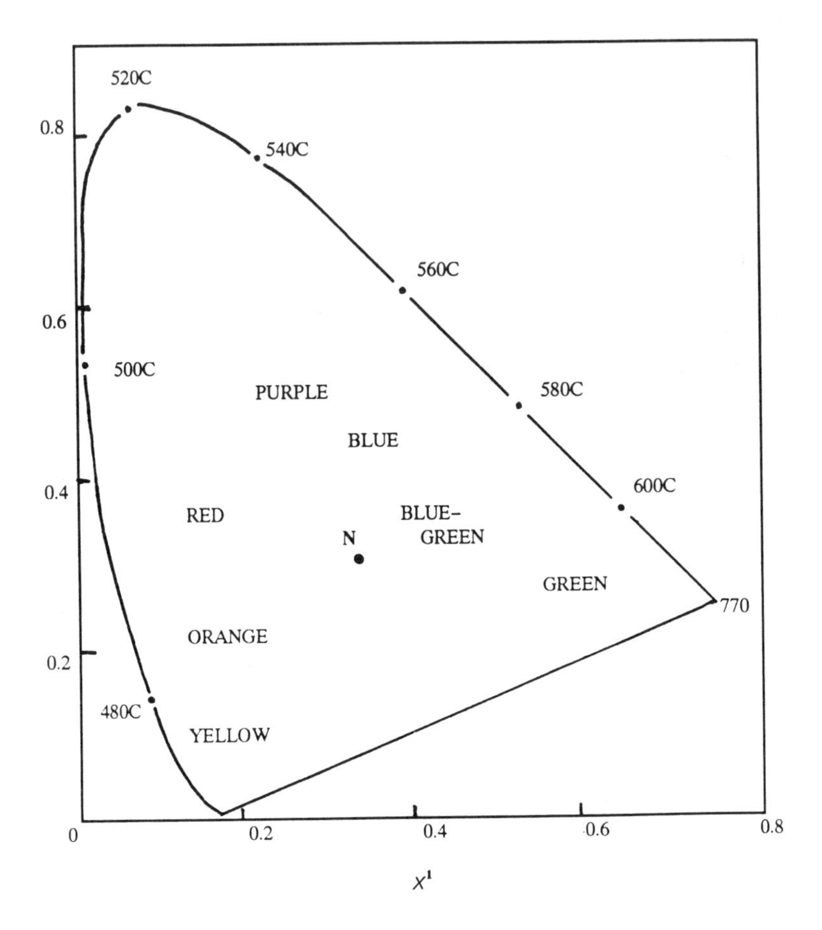

Fig. 2.9. Complementary chromaticity diagram showing approximate positions of colours. The colours are 180° away from their positions in the CIE diagram (Fig. 2.8).

approach was developed by the HOFSS for handling spectral data. By using their approach, both tristimulus values and CCC can be recorded for fibres.

To the best of my knowledge, tristimulus values and CCCs are not in use in any operational laboratory for the comparison of spectra. They have been used in the accumulation or production of data from a reference collection to assist in the interpretation of the significance which should be attached to a recovered fibre match. Again, the HOFSS, in conjunction with the MPFSL, have produced such a databank. Interrogation of the databank will yield information on how often a particular fibre/colour combination appears. Actual figures are not presented in Court reports, and there still seems to be some debate about how best to make use of the databank. At this time it would appear the HOFSS may develop a series of statements such as 'very strong evidence', 'strong evidence', and so on down the scale, the use of these being based in part on the interrogation of their databank. Incidentally, this databank contains much more information than merely the spectral characteristics.

Finally, a word of caution; when the application of spectral analysis was being developed and introduced I think it is fair to say that there were over expectations that it was the solution to all problems. Properly applied, and with conservative interpretation of the data produced, the technique *is* a useful additional tool in the fibre examiners armoury. It is *not* a substitute for comparison microscopy, nor is it the complete answer to colour analysis.

2.5.2.3.3 Chromatography

Whilst spectral analysis gives information about the colour of a fibre it does not tell us anything about the dye components used to produce that colour. It is quite possible to produce the same colour (as defined by CIE nomenclature) with a different dye or combination of dyes. Chromatographic analysis is complementary to spectral analysis. Two methods can be applied to the chromatographic analysis of dyes, *high pressure liquid chromatography (HPLC)* and *thin layer chromatography (TLC)*. These are considered in detail in Chapter 5.

High-pressure liquid chromatography (HPLC). Figs. 2.10 and 2.11 show a typical set-up for HPLC. The basic components are a metal column with a sample injection port, a pump, electronic components to run the system, a detection system and a recording system. The dye must first be extracted (more of which later) and applied to the column. It is then moved through the column under pressure. The time taken

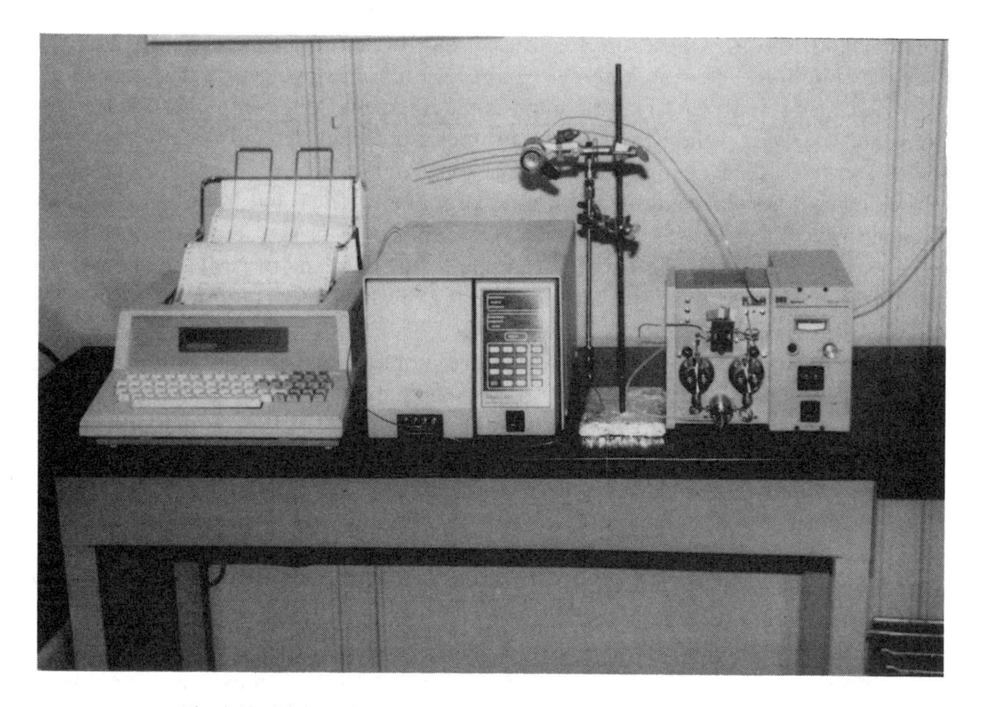

Fig. 2.10. High performance liquid chromatograph (HPLC) apparátus.

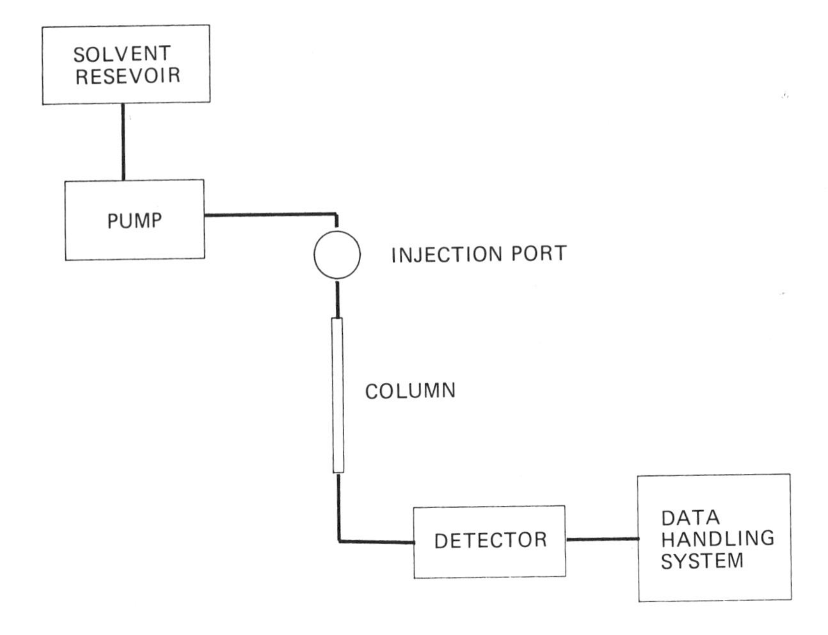

Fig. 2.11. Typical HPLC set-up.

for a dye to move through the column and to separate components (where there is a mix) depends on the solvent or eluent used and the nature of the packing material inside the column. The eluent emerges from the column carrying with it components of the dye being analysed. These components emerge at different times, hence the term *retention times*. Whether or not anything is detected depends on the detector used. Older HPLC detectors were simple spectrophotometers usually capable of monitoring at a set wavelength. Thus, any component which did not absorb at that wavelength was not detected. While it is probably true to say that detectors have been a limiting factor in the use of HPLC, multiwavelength and other types of detector are now becoming available which enhance the level of information which can be achieved with HPLC.

Most HPLC systems use eluents which are aqueous (water) compatible. Herein lies a problem in using HPLC for fibres dyes, as the dyes can often be extracted only with organic solvents which are not easily compatible with HPLC columns. Difficulties are encountered in solubilizing the dye into a solvent suitable for or compatible with the HPLC column material.

Limited detector capability and solvent compatibility problems have, thus far, limited the practical application of HPLC in forensic laboratories. Several papers have been published on the use of HPLC (see Chapter 6), but it could be by no means described as being a routinely used technique for fibre dye analysis at this time.

It is probable, however, that the problems associated with the use of HPLC for dye analysis will be resolved in the next few years. HPLC has many potential advantages over TLC. Some of these are:

- the results can be measured and are objective;
- increased sensitivity allows the analysis of smaller fibres or fibres with paler shades;
- some colours which are poorly visible on TLC, such as yellows, will be detected by HPLC;
- a permanent record of the results is obtained;
- data should be reproducible and may be stored in a databank in the same way as spectral data.

Thin layer chromatography (TLC)

TLC *is* routinely used in most forensic laboratories. It is a simple technique in which the extracted dye is applied as a small spot to a thin layer plate (a glass or aluminium sheet on which is a coating of silica gel powder). The plate is placed in a glass tank with solvent which moves up the plate by capillary action and in so doing causes movement of the dye and separation of components if present. The solvent is allowed to move to a set distance, and the plate is then removed from the tank and the residual solvent is allowed to evaporate. Visible coloured spots can then be assessed. Some components can be seen only by using ultraviolet radiation. The distance moved by a component relative to the distance moved by the solvent is the *Rf* value. Separation of components depends on mainly the choice of solvent or solvents used as an eluent.

The technique is extremely useful, but there are limitations. TLC in general is not highly reproducible by comparison to instrumental techniques. Thus, inter-laboratory comparison of results, which would enable data collection, is not feasible. It is not always possible to extract dye components from fibres. If one thinks about it logically, this is not surprising as one of the most important aims of a dyer is to produce a fast dye which will not be washed out or fade.

The fibres found in casework are usually very short, often less than 0.5 cm. Insufficient dye may be extracted from short fibres to give detectable results. This depends on the nature of the fibre, the efficiency of dye extraction, and the shade of colour.

Thus, in summary, whilst TLC is an extremely useful technique, there are practical reasons why it will not always be possible to obtain results.

Dye extraction, classification and chromatography will now be dealt with in a little more detail.

Dye extraction and classification. Two approaches are commonly used for dye extraction. The HOFSS have developed a series of sequential schemes for the extraction of dyes which have the advantage that they also give information about the class of dye used to colour the fibre.

Some degree on interpretation is needed in deciding what dye class is present, and my personal view is that the classification should be viewed as a screening test in which the dye classification is merely indicative. It's main value lies in that if one has two fibres and they react differently to the extraction scheme, then they are clearly different and do not require further analysis.

Most laboratories do not use the HOFSS system, or at most use shortened versions. Many laboratories merely maintain a range of extractants which are used for the known fibres to find a successful extractant before analysing the recovered fibres.

Chromatography. In casework it is important to use similar sizes of known and questioned fibres and to analyse these on the same plate at the same time. This may seem obvious, but it does not always happen. The use of a standard reference dye or dyes is also to be encouraged, as this validates that the system is working in the way it should on a day to day basis. The choice of eluent system is again often based on the trial and error approach where extracts from known fibres are analysed, using several empirically derived solvent mixes.

The alternative approach is that developed by the HOFSS. A series of eluent mixes have been developed, the choice of which is dependent on the class of dye identified from their sequential extraction schemes. Even here it is usually better to try a number of alternatives with the known fibre extract before any attempt is made to work with the recovered fibre.

In conclusion, TLC is a complementary technique to spectroscopy. Conclusions are normally reached by visual examination of the plates, comparing the pattern of spots or a single spot for both known and questioned fibres. Because it is a comparison process and is to a degree subjective, it is vital to emphasize again the importance of quality assurance to ensure meaningful results.

Looking to the future, the technique of capillary electrophoresis may find use in the analysis of fibres dyes. Its application to ink dyes has been reported in the literature (Fanali & Schudel 1991), and it has been applied to the analysis of food dyes (Wells personal communication).

Case scenario. When we left our case scenario we had reached the point where fibres had been nominated and their presence marked on the tape lifts. These fibres would be removed from the tape lifts and mounted individually onto glass slides, using XAM mountant. The known fibres would be prepared in the same way and all fibres analysed by comparison microscopy. Depending on the number of fibres recovered they would then all be examined by spectroscopy, or a percentage of them would be examined. Standard scientific practice would be to analyse 10% of the recovered fibres where more than ten fibres are recovered. This is after the comparison microscopic examination. With ten or fewer fibres all fibres would be examined.

The same rules would apply to TLC, although, because this is destructive, in the sense that the colour is removed from the fibre, not all fibres would necessarily be examined even for small numbers of recovered fibres.

Information would have been gained about the possible identity of the known and recovered fibres at the comparison microscopy stage. In the case of man-made fibres the fibres would require further testing for identification.

2.5.2.4 Fibre identification

The procedures used to identify fibres depend on:

- the nature of the fibre, natural or man-made;
- the equipment available.

2.5.2.4.1 Natural fibres

For natural fibres such as cotton, wool, other animal fibres, mineral fibres, and vegetable fibres, microscopic examination is the most important method. These fibres should be examined as follows:

- fibres should be mounted in a suitable mountant and viewed with bright field illumination. For vegetable fibres pretreatment of the fibres is almost invariably necessary;
- scale casts can be made by a number of methods to reveal the surface pattern of animal fibres;
- cross-sections may be made by using a microtome or by improvised means: cross-sections have some value in the identification of animal hairs and vegetable fibres; for man-made fibres they are a useful comparison feature and can give some indication of the method of manufacture of the fibre, but do not really help too much in identification;
- determination of optical properties; the optical properties of a fibre arise from two sources:

  the chemical or molecular composition of the fibre;
  changes in orientation and spacing of the constituent molecules or other structural units that occur during spinning, stretching, and other treatments.

Optical properties are studied by using a polarized light microscope.

Vegetable fibres. It is beyond the scope of this chapter to deal with the detailed examination and identification of vegetable fibres. A scheme of analysis for vegetable fibres is given in Fig. 2.12 and a description of features in Table 2.1. The reader is referred to Catling & Grayon (1982) for a detailed treatment of this topic. With the exception of cotton, which is almost ubiquitous, vegetable fibres are now seen only rarely in forensic case work. Their most common end product use is in cordage, ropes, and mats.

It is important to understand the potential for confusion in the use of the term *fibre* for vegetable fibres. For example, one of the more common fibres is jute. The commercial fibre varies greatly in the coarseness or thickness of the 'fibre'. In fact what is called a fibre is a bundle of true botanical fibres; the term fibre in strict botanical use refers to an elongate cell with pointed ends. These individual cells are often referred to as *ultimates*. In the identification of vegetable fibres it is important to study features associated with both the groups of ultimates and the individual ultimates themselves. The fibre bundles require pretreatment to separate the individual cells. Polarising microscopy is then a very useful tool in observing features of these separated cells. The identification of vegetable fibres is quite specialized, and few, if any, forensic scientists see these fibres often enough to be confident of identifying the precise fibre. It may only be possible to identify a group such as leaf or bast fibre.

Table 2.1. Summary of features of common bast and leaf fibres

Botanical Origin	Species	Common Name	Length of Ultimates or Fibre Cells (mm)	Extraneous Material	Cross Markings	Lumen & Cell Wall	Pits	Crystals & Silica Ash
Bast fibres	Linum usitatissimum	flax	1.6–24.0	Epidermis with paracytic stomata. Parenchyma. Xylem elements.	Rare. When seen, often very regular along whole length of fibre cell.	Cell wall occasionally striated. Cell wall thick, lumen narrow, regular.	Very fine, not obvious. Can be seen by using polarised light.	No crystals reported or seen.
	Corchorus capsularis & olitorius	jute	0.6–5.3	Few. Mostly parenchyma, sometimes with cubic or cluster crystals. Very occasional vessels.	Few, faint. Occasonal marks from chambered cells. Scalloped edges to fibres cells.	Lumen of varying width, often varying regularly along whole length of cell.	Bordered, funnel shaped in side view.	Cubic crystals in chains sometimes mixed with occasional cluster crystals. Single cluster crystals.
	Cannabis sativa	hemp	1.0–34.0	Laticiferous elements in unmacerated fibre. Parenchyma of various types. Cluster crystals free or in cells. Hairs. Epidermis. More rarely, blocks of tissue and xylem elements.	Variable. Some cells with fine regularly spaced marks in every specimen. Marks from chambered cells. Several series on one fibre cell. More frequent than in flax. Occasional remains of cells attached to marks.	Cell wall striated. Lumen most commonly three to five times the width of the cell wall.	Slit-like, parallel to the long axis of cell, sometimes coalesing.	Cluster crystal in short chains, often three or four together. Single cluster crystals. *Very* occasional cubic or rhombic crystal in some specimens.
	Boehmeria nivea	ramie	13.0–82.7	Parenchyma. Cluster crystals free or in parenchyma. Cluster crystals in chambered cells. Few hairs and vessel elements.	Common, fine, nearly always with attached remains of parenchyma cells. Several series on one fibre cell. Occasional marks of chambered cells.	Lumen difficult to see because (a) it varies, (b) cell wall is striated, (c) fibre cells tangle. Lumen commonly two to three times the width of the cell wall.	Elongated, slit-like, parallel to long axis of cell, sometimes coalesing.	Cluster crystals in chains. Single cluster crystals.

Table 2.1. – *continued*

Botanical Origin	Species	Common Name	Length of Ultimates or Fibre Cells (mm)	Extraneous Material	Cross Markings	Lumen & Cell Wall	Pits	Crystals & Silica Ash
Leaf fibres	Agave sisalana	sisal	3.0–6.7	No epidermal cells or stomata seen. Mesophyll. Blocks of tissue. Tracheary elements with annular thickening or opposite to scalariform pitting. Spirals, 'Butt' fibre cells.	No cross marks seen. Scalloped edged to some cells.	Lumen has maximum width five to six times the thickness of the cell wall, most commonly four to five.	Obvious, slit-like, at approximately 40° to long axis of cell.	Large acicular crystals, sometimes with organic matter annealed to the surface.
	Musa textilis	abaca	2.0–9.3	Parenchyma. Stegmata. Eqidermis with stomata and subepidermal sclerotic layer. Occasional blocks of tissue. Tracheary elements with spiral or annular thickening.	Variable, never very common, occasionally regularly spaced.	Lumen has width two to seven times the thickness of the cell wall, most commonly three to five. Wide variation within and between specimens.	Slit-like, at slight angles to long axis of cell.	Silica bodies, *Very* occasional small acicular crystals in some specimens.

(Data adapted from Catling D and Grayon J (1982).)

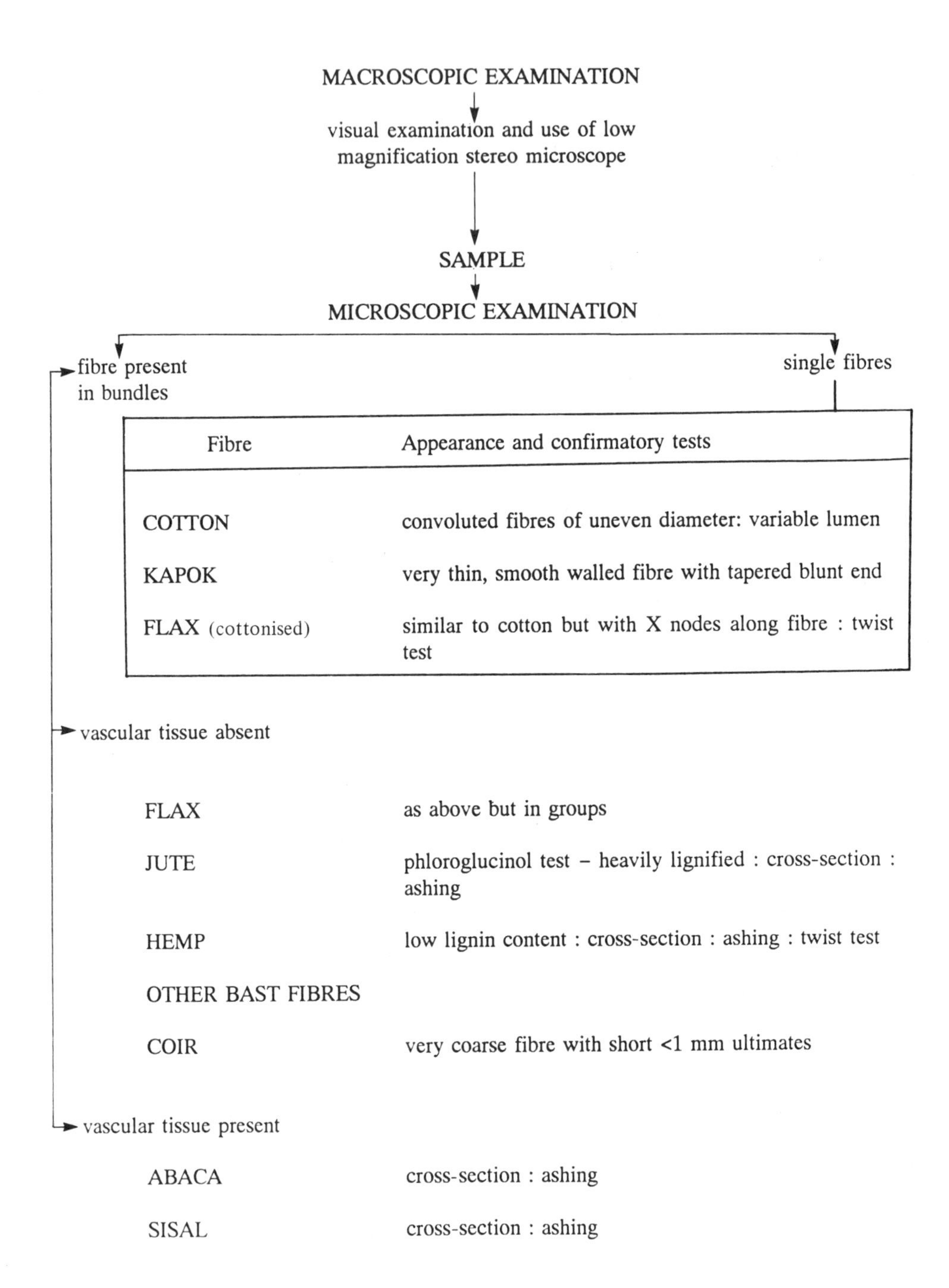

Fig. 2.12. Scheme of analysis for plant fibres.

Animal fibres. Animal fibres are composed of either *keratin* or *fibroin* protein. Only silk is made of fibroin, and in the true sense is not an animal fibre. Of the keratin based fibres or hairs the examination of those from humans is by far the most important in forensic applications. The examination of human hairs is nevertheless only a specialized example of the general examination of hairs. For most hairs the main purpose is to identify (as far as is possible) the species of animal from which the hairs originated. Little individualization is possible. Clearly hairs from a poodle are different from the hairs of a German Shepherd, but hairs from all German Shepherds are likely to be indistinguishable. Recent research has shown that some individualization is possible within a single breed of dog (Suzanski 1988). The approach to the examination of all animals hairs has common elements and requires a knowledge of the structure of hairs.

The basis of all hair work is microscopic examination, and there is no substitute for a detailed and systematic examination of individual hair shafts. With respect to methods, the procedure to be taken is visual examination, followed by examination by a stereomicroscope and then with a compound light microscope. Hairs may be mounted on microscope slides in temporary mountant, usually glycerine, water (9 : 1 v/v) or in a suitable permanent mountant. Roe *et al.* (1991) have evaluated mountants suitable for use in forensic hair examination.

If permanent mountant is to be used, them impressions of the hairs surface—*scale casts*—should be made before mounting the hair.

The selection of mountant is important, and there are points in favour of and against temporary and permanent mountants. A permanent mountant should have a refractive index (RI) in the order of 1.52–1.54, similar to that of hair at about 1.55. Glycerine water has a much lower RI which tends to make the outside of the hair more visible but makes the study of the cortex more difficult.

The first question which arises in any hair examination is whether the hair is of human or non-human origin. The differentiation of human hair from other hairs is usually quite straightforward. Table 2.2 lists a comparison of features found in the two groups.

Once it has been established that the hair is human or non-human, different protocols are followed.

Non-human hair. Hairs are composed of three anatomical regions, the *cuticle*, *cortex*, and, when present, the *medulla* (see Fig. 2.13). The attempted identification of non-human hairs requires that the fine detail associated with these regions is studied in a systematic way.

With non-human hairs, different types of hair can be present in the fur or peltage. These are usually visually clearly different, based on their degree of coarseness. Whilst different classification systems exist the most commonly found types are *guard* and *underhairs*. Guards hairs are longer and coarser than the underhairs, which are usually fine. Guard hairs also display the widest range of microscopic features, which makes them the most useful for identification.

Both guard hairs and under hairs should be examined separately if present. The following features should be assessed.

1 PROFILE (general shape)
Shield
Straight
Symmetrically thickened
Wavy

2 CUTICLE OR SCALE FEATURES

2.1 **Scale margin**
Smooth
Crenate .... Sharp pointed teeth
Rippled .... Indentations deeper than above and rounded.
Scalloped .... Margins with broad rounded teeth

2.2 **Distance between scales**
Close
Near
Distant

2.3 **Scale pattern**
Mosaic .... { Regular mosaic / Irregular mosaic }
Wave .... { Simple regular wave / Interrupted regular wave / Streaked wave / Irregular waved, mosaic / Regular waved, mosaic }
Chevron .... { Single / Double }
Pectinate .... { Coarse / Lanceolate }
Petal .... { Irregular / Diamond }

3 MEDULLA

Note whether present or absent. Where present it may be **continuous, interrupted**, or **fragmented**.

In non-human hairs it is often continuous with a defined structure. The structure can be of two main classes, **ladder** or **lattice**.

A ladder medulla is so called because it looks like the rungs of a ladder. Where there is a single row of 'rungs' this is a *uniseriate ladder*, and with several rows, a *multiseriate ladder*.

A lattice medulla is so called because it has the appearance of a lattice made up of 'struts' of keratin which outline polyhedral shaped spaces each of which is continuous with its neighbours. A special type of lattice medulla, called an *aeriform lattice*, differs in that the shapes giving the appearance of the lattice have arisen because of cell collapse leaving air filled gaps roughly polyhedral in shape.

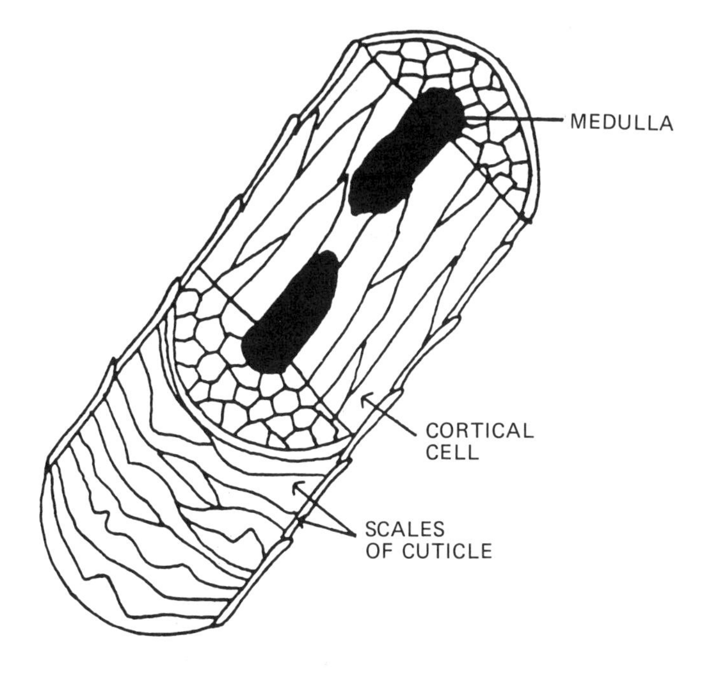

Fig. 2.13. Schematic diagram of animal hair.

Table 2.2. Human vs non-human animal hair

FEATURE	HUMAN	OTHER
Colour	Relatively consistent along shaft	Often showing profound colour changes and banding
Cortex	Occupying most of width of shaft - greater than medulla	Usually less than width of medulla
Distribution of Pigment	Even, slightly more towards cuticle	Central or denser towards medulla
Medulla	Less than $\frac{1}{3}$ width of shaft	Greater than $\frac{1}{3}$ width of shaft
	Amorphous, mostly not continuous when present	Continuous, often varying in appearance along shaft, defined structure
Scales	Imbricate, similar along shaft from root to tip	Often showing variation in pattern along shaft from root to tip

COLOUR. The colour of hairs results from pigment particles deposited in the cortex. Overall visual and macroscopic colour can be important in non-human hairs, but the pigmentation in the cortex is less important than in human hair. Pigment should be assessed with respect to:

a. **amount** .... sparse or dense
b. **distribution** .... along the shaft .... even
across the shaft .... denser near centre
denser near cuticle

CROSS-SECTION. This is not always carried out, because it is destructive. Information from cross-sections is threefold:

a. a good appreciation of pigment distribution across the shaft is gained;
b. the position of the medulla which can be in the middle (*centric*) or off to one side (*eccentric*);

(Both of these features can be assessed by optical sectioning when fibres are viewed in longitudinal plane).

c. the shape of the hair.

Non-human animal hairs should be examined from their root end to their tip end, and variation in the above features recorded. In particular, changes in medulla and scale pattern appear to be consistent enough to have value in identification.

It is helpful to use a check list to record features. It also encourages systematic observation. An example of such a check list is presented in Fig. 2.14.

Where possible, questioned animal fibres should always be compared with authenticated standards.

There are many scientific papers containing descriptions of the microscopy of animal hairs, but few are of real assistance. Usually they deal with only a very small group of species and take into consideration only those features which have most value in discriminating the members of that group. Often descriptions also assume that the scientist has a bulk sample whereas in forensic work each hair has to be treated separately. Thus, the identification of all but the more common hairs is often not possible, and analysis becomes comparative.

Some specific examples of common animal hairs

1. Wool type fibres. Wool fibres come from different breeds of sheep. Whilst an experienced wool grader can distinguish different breeds by examining unprocessed bulk samples, the identification of breed from single fibres or even very small samples is beyond the expertise of the forensic examiner. Once wool has been processed there is no accurate way of identifying the breed. Raw wool fibres are classed into four groups based on the degree of coarseness, *kemps*, *outercoat*, *coarse*, *fine*. In textile end products usually only coarse and fine fibres can be recognized, and the only features readily assessed are the presence or absence of medulla and scale features. Other fibres are used in similar end product uses and, in particular, goat hairs. The scale features of fine wool and goat hairs are compared in Table 2.3.

FORENSIC SERVICES DIVISION				CASE NUMBER:	
ANIMAL HAIR EXAMINATION 1				ITEM NUMBER:	
Date: Time:				Page of	
MACROSCOPIC	HAIR NUMBER				
FEATURE	1	2	3	4	5
LENGTH CM					
SHAFT PROFILE SHIELD					
STRAIGHT					
SYMM THICK					
WAVY					
COLOUR					
GENERAL DESCRIPTION:					
EXAMINER:	NOTES CHECKED BY:				

Fig. 2.14. (a) Checklist for examination of animal hairs.

FORENSIC SERVICES DIVISION ANIMAL HAIR EXAMINATION 2 Date: Time:			CASE NUMBER: ITEM NUMBER:		
MICROSCOPIC	HAIR NUMBER				
FEATURE	1	2	3	4	5
PIGMENT DENSITY: NONE / LIGHT / MEDIUM / HEAVY					
PIGMENT DISTRIBUTION: TOWARDS CENTRE / EVEN / TOWARDS CUTICLE					
MEDULLA DISTRIBUTION: NONE / FRAGMENTED / INTERRUPTED / CONTINUOUS					
MEDULLA TYPE: AMORPHOUS / LADDER / LATTICE / AERIFORM LATTICE					
SCALE EDGE: SMOOTH / CRENATE / RIPPLED / SCALLOPED					
DISTANCE BETWEEN SCALES: CLOSE / NEAR / DISTANT					
SCALE PATTERN: MOSAIC / SIMPLE WAVE / INTER. WAVE / WAVED MOSAIC / SINGLE CHEVRON / DOUBLE CHEVRON / COARSE PECTINATE / LANCEOLATE PECTINATE / IRREGULAR PETAL / DIAMOND PETAL					
OVOID BODIES					
EXAMINER:			NOTES CHECKED BY:		

Fig. 2.14. (b) Checklist for examination of animal hairs.

Table 2.3. Comparison of wool and goat hairs

SCALE FEATURE	SHEEP WOOL	ANGORA GOAT MOHAIR	CASHMERE GOAT
Margin	near prominent	near to distant shallow	distant shallow
Pattern	irregular mosaic	irregular waved mosaic	regular waved mosaic

The most distinguishing feature of wool is the prominent scale margin. It is not easy to absolutely identify fibres as wool, especially when the fibres transferred during contact have indistinct or damaged scales. For this reason it is safer to refer to fibres as **wool-like**.

2. Cat v dog. There is wide variation in colour, length, and profile in hairs from cat and dog breeds. General class characteristics are given in Table 2.4.

It must be stressed that these are general characteristics, and there is a great deal of variation within and between cat and dog hairs. As previously stated, the progression of features along the hair shaft provides useful information, and according to Peabody *et al.* (1983) the following scale progression, found in cats, does not occur in dog hairs.

	Irregular mosaic	smooth scale margins
	Petal	smooth scale margins
ROOT	Regular mosaic	smooth scale margins
TIP	Regular mosaic	rough scale margins

Dogs showed a wider variation in scale pattern progression.

Two further features can be of value in discriminating cat and dog hairs.

(a) ROOT SHAPE

CAT	DOG
elongated, not distinct shape, fibrils often frayed at base of root	spade or arrow shaped

(b) MEDULLARY INDEX

A graph of medulla width against hair width shows that the medullary index for cat hairs is in general higher than for dog hairs. Not *all* cat and dog hairs can be correctly identified, but the error is small. (see Fig. 2.15).

CONCLUSION. This is a very general guide to the approach used in the examination of animal hairs. As I have stated previously, it is often possible to carry out only a comparative examination.

Table 2.4. Features of cat and dog hairs

Animal	Whole Mount			Cross-section				Scale Pattern		
	Profile	Medulla	Pigment distribution	Contour	Medulla	Cuticle	Pigment distribution	Base	Mid-length	Tip
CAT *Felis ocreata catus*	*Fine* Regular diameter, scale margins fairly prominent	Ladder	Sparse to dense, even	Circular to oval, some angular	Concentric	Thin	Some sparse and near to medulla, some dense	Regular waved, smooth; near margins	Coarse pectinate →regular mosaic, smooth; near to distant margins	Regular waved mosaic, smooth; near margins
	Coarse Regular diameter, fairly smooth	Continuous, fine lattice	Sparse to dense, even	Circular to oval	Concentric	Thin	Some sparse and near to medulla, some dense	Regular mosaic, smooth; near to distant margins	Irregular mosaic, smooth; near margins	Irregular waved, crenate; close margins
DOG *Canis familiaris*	*Fine* Regular diameter, scale margins fairly prominent	Some none, some fragmental, some ladder	Varies from none to dense, even or streaky	Almost circular	Some none, some circular and narrow	Fairly thin	Some even, some in large aggregates	Regular mosaic, smooth; near to distant margins	Alternating diamond petal and waved, rippled; near margins	
	Coarse Regular diameter, scale margins prominent	Continuous, sometimes ladder	Varies from none to dense, even or streaky	Almost circular	Concentric, some narrow	Fairly thin	Some even, some in, large aggregates	Regular mosaic, smooth; distant margins	Diamond petal →irregular mosaic	Irregular waved, rippled; near to close margins

Adapted from Appleyard (1978)

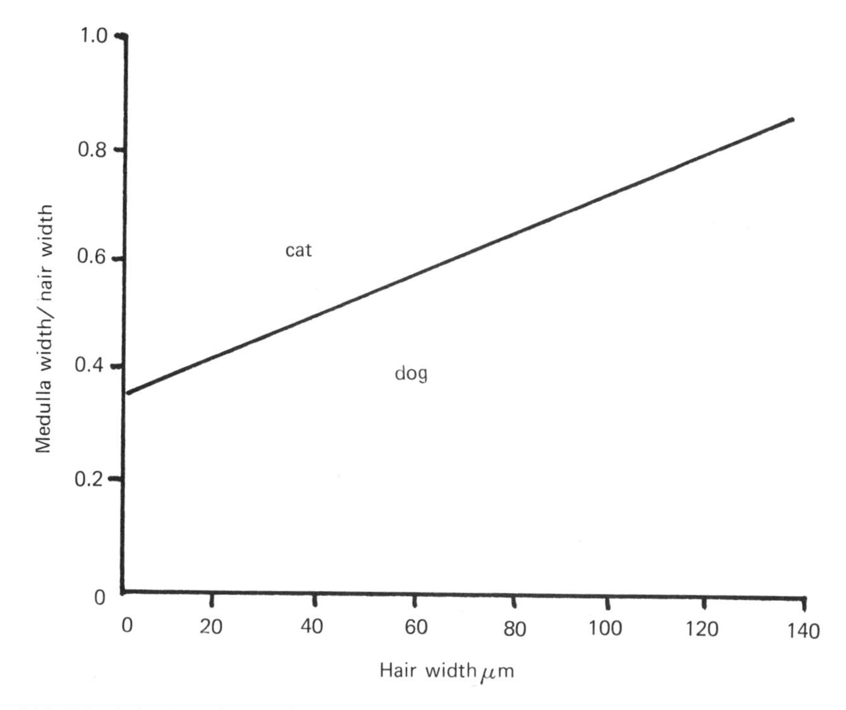

Fig. 2.15. Discrimination of cat and dog hairs by medullary fraction analysis (Peabody *et al.* 1983).

A comprehensive reference collection of hairs is invaluable, but there really is no substitute for experience. Animal hairs are ubiquitous, but this is not reflected in their use as evidence of contact. Paradoxically, this may be because they are ubiquitous, and forensic workers lack the ability to discriminate sufficiently to give then evidential significance. It is my belief that there are many more cases in which animal hairs could provide more useful information than is currently realized. However, it has to be said that few forensic laboratories have the people with the necessary expertise.

The examination of human hair is a specialised topic, and is not considered in this book.

2.5.2.4.2 Man-made fibres

Microscopy—bright field. Man-made fibres should always be examined firstly by microscopic means. Many of them will have features which will be readily seen when using bright field microscopy. Cross-sections will give more information about the shape of the fibre. However, it is often possible to interpret shape from surface examination alone, and few laboratories routinely make cross-sections. Details are not presented here of microscopic features for man-made fibres because although there are features commonly associated with generic groups (acrylics, nylons, polyesters, etc.), these do not uniquely identify the fibre type. The overall surface features (for example, a smooth surface *versus* striations *versus* grooves), arise mainly from the method or extrusion of manufacture of the fibre. Whilst it is usual for a particular fibre type to be produced by one method of spinning, it is not a unique feature. The shape of the fibre is also determined by the shape of the hole through which the fibre forming substance is extruded. Certain shapes are produced for different end product uses. The process is not hit and miss. For example, a very thick trilobal fibre would indicate a fibre produced for carpet manufacture. The bean or peanut found in acrylic fibres is aimed at producing a fibre which can be used in knitwear to give a bulky appearance or texture. Fundamentally, the maker of man-made fibres is attempting to produce a copy of a natural fibre.

Microscopy—polarized light. Man-made fibres can be identified from their optical properties. These arise because of the way in which the fibre-forming polymer chains are highly organized within the final hardened fibre. Man-made fibres have amorphous and crystalline areas, but overall may be considered crystalline. The optical properties that are most useful for fibre identification are *sign of elongation* and quantitative *birefringence*. The measurement of these is dealt with in detail in Chapter 3. Optical properties can also be used in a qualitative way incorporating the use of a first order red place and a compensator (quartz wedge or Berek type).

By using a first order red plate it is possible to separate fibres with low birefringence from those with higher birefringence. Low birefringence fibres appear blue or orange, depending on their orientation relative to the first order plate and polarized light. Fibres with low birefringence include acrylic, cellulose acetate, and viscose fibres. In fact acrylics have a negative birefringence which, used with caution, is a valuable generic identification feature.

Fibres with high birefringence include nylon and polyester fibres. Viewed with a first order plate these fibres have a range of colours across their breadth. More accurate determination of birefringence can then be calculated if necessary by using a compensator.

In some laboratories, polarized microscopy is used to identify man-made fibres. It is most often used as a screening test to give broad generic identifications.

Infra-red spectroscopy (IR). IR spectroscopy is a well established technique for the identification of organic materials. Its limitation for the examination of fibres has always been the size of sample required for analysis. With older dispersive instruments the advent of anvil type *diamond cells* meant that it was possible to work with individual fibres of about 0.5 cm in length. With the advent of *Fourier Transform Infrared Spectroscopy (FTIR)* (see Fig. 2.16) with associated infrared microscope attachments sample preparation has become simpler, although use may still be made of a micro-diamond cell. With current technology, sample size is determined by what is large enough to be recovered and handled successfully. Thus, it is possible to remove one or two millimetres from a larger fragment of several millimetres, leaving the rest of the fibre for further testing.

It is important to understand that whilst IR spectroscopy is considered to be a technique which can be used to identify a fibre, there are limitations. Primarily IR

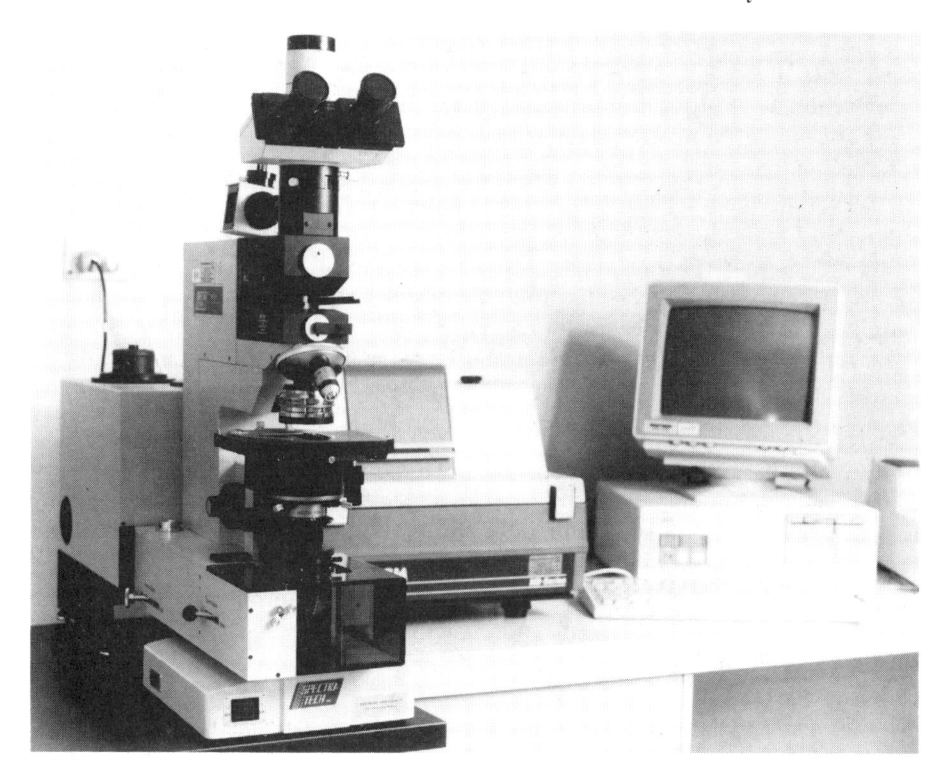

Fig. 2.16. Fourier transform infrared spectrophotometer (FTIR) showing Spectratech microscope.

spectroscopy gives information about molecular structure, particularly the recognition of functional groups (such as OH, CH_3 NH_2) and their molecular environment. The way in which a fibre is prepared does influence the fine detail of the spectra obtained. The main aspect here lies in the extent to which the fibre is squashed. This squashing is essential in order to get sufficient energy transmitted through the fibre substance. To get reproducible results the technique of squashing needs to be reproducible. Thus, in any one laboratory it will be necessary to build up a library of spectra from authenticated samples against which spectra from case materials can be empirically compared.

With FTIR spectroscopy it is possible to go further than identifying the generic class of fibre (acrylic, nylon, polyester). It is now possible to separate or differentiate fibres based on differences in copolymer composition.

In summary, for any laboratory which has access to FTIR with a microscope attachment this is the technique of choice for identification of man-made fibres.

The topic of infrared spectroscopy is dealt with in detail in Chapter 6.

Pyrolysis gas chromatography (Py–Gc)

The application of this technique to fibre analysis goes back to the late 1960s. It is undoubtedly a very useful technique for a range materials encountered in forensic work. It has not found universal favour for fibres, for a number of reasons.

- The technique requires more sample than FTIR—estimates vary, but it certainly cannot deal with mm size fragments.
- The technique is destructive—in order to analyse a fibre it is rapidly burned or pyrolysed at high temperature to produce volatile components which can then be subjected to gas chromatographic analysis.
- Although supporters of Py–Gc will debate this point, there are some technical difficulties in getting highly reproducible results which can lead to problems in producing a reliable collection of reference data.
- Unless the components of the pyrogram (the name given to the Gc chromatogram) are identified by mass spectrometry it is not possible to absolutely identify the components within the pyrogram which identify the original fibre.

On the positive side, proponents of Py–Gc argue that more discrimination within generic groups is possible with this technique. The one limitation which cannot at this time be overcome is the need for a much larger sample than is needed for FTIR. Thus, whilst laboratories with expertise and a history of applying Py–Gc to fibres will continue to use this technique. FTIR will probably be used in most laboratories for fibre identification. Challinor discusses the use of Py–Gc in Chapter 7 of this book.

Other techniques. The literature is full of techniques which have been used for fibre analysis. Whilst some of these have produced interesting results in specific instances, few, if any, have found broad application by forensic practitioners. However, it is worth pointing out that before the introduction of modern instrumental techniques, *solubility tests* were commonly used to assist in identifying fibres. Although now used to a much more limited degree, solubility tests can provide useful back-up data to instrumental techniques and can be a very simple method to differentiate fibres having similar microscopic features. Solubility tests are still worth considering in any protocol for fibre identification. The reader is referred to the *Identification of textile materials* published by the Textile Institute for solubility tests and schemes. Thermal data such as melting point, can also be useful.

2.6 INTERPRETATION OF FIBRE EVIDENCE

In our case scenario a number of fibres were found which could have come from items relating to the complainant on items relating to the suspect, and vice versa. The test procedures available revealed no differences which were considered significant.

Assuming that the technical/scientific aspects of the analysis are satisfactory, the question remains of what significance ought fairly to be placed on the *could have* statement. Whilst this is a question for the jury, assisted by direction from the judge, clearly it is the role of the lawyers, prosecution and defence, to ask questions which will bring out in a fair and unambiguous way the weight which can properly be given to the evidence.

Interpreting the value of fibre evidence is the most difficult task facing the scientist. Grieve addresses these issues in Chapter 8. Grieve gives a list of factors which need to be considered:

- **factors that are known** (or that can usually be determined);
    - the circumstances of the case;
    - the time that has elapsed before collection of the evidence;
    - the suitability of the fibre type for recovery and comparison;
    - the extent of comparative information derived from the samples;
    - the number of types of matching fibres;
    - whether or not there has been as apparent cross transfer of fibres;
    - the quantity or number of matching fibres recovered;
    - the location of the recovered fibres;
    - the methods used to conduct the examination;

- **factors that are unknown**;
    - the degree of contact that has occurred and the pressure;
    - the degree of certainty that specific items were definitely in contact;
    - shed potential;
    - frequency of occurrence of the matching fibre types.

Whilst the latter factors are difficult to measure there are a number of ways in which they can be assessed:

- transfer studies;
- 'target fibre' studies;
- studies on differential shedding;
- experiments on the alteration of fibre characteristics in a manner consistent with localized conditions specific to a case;
- manufacturers enquiries;
- use of data collections;
- use of statistical treatment.

It needs to be stressed that the answers to many of the above questions are quite case specific, and would require a research project to derive the relevant information. Clearly, this is only practical in the largest and most critical cases. Whilst general assistance can be gained from the transfer and persistence studies discussed earlier, the limitations of this type of work to a specific set of circumstances must be borne in mind. When it comes to manufacturers enquiries it is critical that the correct questions are asked.

As I have mentioned earlier, it is the view of some lawyers that it is preferable that analysts should be in a total ignorance of the circumstances of a case to ensure complete impartiality. I consider this to be dangerous nonsense. I would simply put it to any lawyer reading this chapter, 'Could you prepare a case if you were given no information, and does the possession of knowledge, however imperfect, automatically bias your thought processes?' I would go so far as to consider the suggestion that the scientist lacks the intellect to recognize the danger and to deal with it accordingly, as professionally insulting. The scientist does not religiously believe every detail received, but it does provide a framework, a basis on which to work whilst keeping a completely open mind about any investigation and what may later transpire.

In general, the following factors will tend to lower the evidential value of apparent fibre transfer:

- any doubt, no matter how slight, that contamination may have occurred;
- indications of previous legitimate contact between suspect and victim; that is, that they were drinking, dining, or dancing together, or were together socially in a particular environment—time may play an important role here;
- offences which occur in a family environment, for example incest. The number of transferred fibres recovered may assume extreme important in such instances.

Conversely, certain circumstances may tend to enhance the evidential value:

- when apparent transfer can be demonstrated between individuals totally unknown to one another, particularly a cross transfer involving several colours/types of fibre;
- when the delay between the offence and collection of the evidence has been minimal—this reduces the likelihood of matching fibres, present in quantity, having originated from other putative sources;
- when transfer of fibres has occurred onto, or better still, between items of underwear in cases of a sexual nature;
- when the results of transfer analysis follow in detail a pattern that is totally consistent with the reported circumstances.

When there are only a small number of fibres it is often suggested that these could be present by coincidence, and that if anyone had been nominated as a suspect and their clothing searched such fibres would have been found. The results of several *target fibre studies* would not support this assertion. For example, when Cook & Wilson (1986) searched 335 items of clothing for the presence of fibres frequently encountered in the UK, only 11 'matching' fibres were found on ten garments. Nine were of one type, blue wool, and no more than two fibres were found on any one item. Jackson & Cook (1986) also reported the results of further studies where the front seats of 108 vehicles were searched for the presence of two fibres again commonly found in the UK. From nearly 8500 recovered fibres, 37 of one type of fibre (red wool) and 8 of the second fibre type (brown polyester) were found. The maximum number of matching fibres (red wool) found on any seat was 13 and in any one vehicle 20. Fibres matching both types of target were found in only one vehicle. Sources were identified for the examples where larger numbers of fibres were found. These studies demonstrated that the chance of finding large numbers of fibres of one type or colour as a matter of coincidence is small. Thus from this type of study, examples of which are regrettably few, it would appear that the finding of more than a few matching fibres is difficult if not impossible to explain by coincidence. Clearly, this **does** depend on the nature of the fibre. Everyone has colourless cotton and, more likely than not, denim blue cotton fibres on their clothing or environment. The other side of the coin is, however, that there are also some very rare fibres (Deadman 1984 a,b).

A further approach to assist in the interpretation of fibre evidence which has been developed by the HOFSL system is the establishment of a *data collection.* I do not wish to discuss this in detail. Again, the reader is referred to the discussion by Grieve in Chapter 6. Suffice to say that although interrogation of databases can provide some useful information, they have in-built weaknesses. They are largely historical in context as they are impossible to keep up to date, and they provide only one part of the overall picture. Data collections should be used only with caution.

Attempts have been made to assess the value of fibre evidence by *statistical analysis.* One approach, favoured by the HOFSL system, is the use of *Bayesian* statistics. Whilst there is a considerable literature on this approach, to my knowledge few if any forensic workers would use this to present their evidence. The whole issue of statistical treatment is covered in *The use of statistics in forensic science* (Ed. Aitken & Stoney 1991).

One of the leading figures in the use of the Bayesian approach for forensic problems is Ian Evett of the HOFSL. He points out that a numerical statement for the so-called *likelihood ratio*, generated by Bayesian analysis, is unlikely to convey much to the Court (Evett, in Aitken & Stoney 1991). He suggests that it would be possible to assign a verbal convention to this ratio, such as 'the evidence weakly supports' at one end of the positive scale to 'the evidence very strongly supports' at the other end.

It should be self-apparent to the reader that there is no simple method to evaluate what significance should be attached to the statement 'could have'. The weight will vary from almost none for extremely common fibres through the spectrum to a close certainty that contact has taken place. At this time it is not possible to put a mathematical number on that degree of certainty. (Note, however, that even if one is certain contact took place, this is *all* it proves. There may be perfectly innocent explanations for contact.)

BIBLIOGRAPHY

(including references cited in the text)

General

1. Books

Aitken, C. G. G. & Stoney, D. (1991) *The uses of statistics in forensic science.* Ellis Horwood, Chichester.

Catling, D. L. & Grayon, J. (1982) *Identification of vegetable fibres.* Chapman & Hall.

Gaudette, B. D. (1988) The forensic aspects of textile fibre examination. In: Saferstein, R (ed), *Forensic science handbook*, Vol 2. Prentice Hall, Englewood Cliffs.

Grieve, M. C. (1990) Fibres and their examination in forensic science. In: Maehly, A. & Williams, R. L (eds), *Forensic science progress* Volume 4. Springer-Verlag, Berlin.

2. Papers

Cook, R. & Norton, D. (1982) An evaluation of mounting media for use in forensic textile fibre examination. *J For Sci Soc* **22**, 57–63.

Cook, R. & Wilson, C. (1986) The significance of finding extraneous fibres in contact situations. *For Sci Int* **32**, 267–273.

Deadman, H. A. (1984a) Fibre evidence and the Wayne Williams trial (Part 1). *FBI Law Enf Bull* **53**, 13–20.

Deadman, H. A. (1984b) Fibre evidence and the Wayne Williams trial (conclusion). *FBI Law Enf Bull* **53**, 10–19.

Grieve, M. C., Dunlop, J. & Haddock, P. S. (1989) Transfer experiments with acrylic fibres. *For Sci Int* **40**, 267–277.

Masakowski, S., Bruce Enz, M. F. S., Cothern, J. B. & Rowe, W. F. (1986) Fibre-plastic fusions in traffic accident reconstruction. *J For Sci* **31**, 903–912.

Fibre transfer, persistence and recovery

Ashcroft, C. M., Evans, S. & Tebbet, I. R. (1989) The persistence of fibres in head hair. *J For Sci Soc* **28**, 289–293.

Cook, R. (1981) The problem of contamination. *Police Surgeon* **1**, 65–66.

Cordiner, S. J., Stringer, P. & Wilson, P. D. (1985) Fibre diameter and the transfer of wool fibres. *J For Sci Soc* 425–426.

Kidd, C. & Robertson, J. (1982) The transfer of textile fibres during simulated contacts. *J For Sci Soc* **22**, 301–308.

Kriston, L. (1984) On the evidentiary value of microtraces of textiles. *Arch. Kriminge.* **173**, 109–115.

Lim, M. S. (1984) The transfer and persistence of fibres from clothing to car seats and seat belts, MSc Thesis, University of Strathclyde.

Lowrie, C. M. & Jackson, G. (1991) Recovery of transferred fibres. *For Sci Int* **50**, 111–119.

McKenna, F. J. & Sherwin, J. C. (1975) A simple and effective method for collecting contact evidence. *J For Sci Soc* **15**, 277–280.

Moore, J., Jackson, G. and Firth, M. (1984) Movement of fibres between working areas as a result of routine examination of garments. *J For Sci Soc* **24**, 394.

Parybyk, A. & Lokan, R. J. (1986) A study of the numerical distribution of fibres transferred from blended fabrics. *J For Sci Soc* **26**, 61–68.

Pounds, C. A. (1975) The recovery of fibres from the surface of clothing for forensic examinations. *J For Sci Soc* **15**, 127–132.

Pounds, C. A. & Smalldon, K. W. (1975a) The transfer of fibres between clothing materials and their persistence during wear: Part 1. Fibre transference. *J For Sci Soc* **15**, 17–27.

Pounds, C. A. & Smalldon, K. W. (1975b) The transfer of fibres between clothing materials and their persistence during wear. Part II. Fibre persistence. *J For Sci Soc* **15**, 29–37.

Pounds, C. A. & Smalldon, K. W. (1975c) The transfer of fibres between clothing materials and their persistence during wear: Part III. A preliminary investigation of the mechanisms involved. *J For Sci Soc* **15**, 197–207.

Robertson, J. & Gamboa, X. M. de (1984) The transfer of carpet fibres to footwear. *Proc. 10th Int Assoc For Sci.* Oxford. A C Associates, Solihull.

Robertson, J. & Lloyd, A. (1984) Redistribution of textile fibres following transfer during simulated contacts. *J For Sci Soc* **24**, 3–7.

Robertson, J. & Olaniyan, D. (1986) Effect of garment cleaning on the recovery and redistribution of transferred fibres. *J For Sci* **31**, 73–78.

Robertson, J., Kidd C. B. M. & Parkinson, H. M. P. (1982). The persistence of textile fibres transferred during simulated contacts. *J For Sci Soc* **22**, 353–360.

Salter, M. J., Cook, R. & Jackson, A. R. (1984) Differential shedding from blended fabrics. *For Sci Int* **33**, 155–164.

Scott, H. (1985) The persistence of fibres transferred during contact of automobile carpets and clothing fabrics. *Can Soc For Sci J* **18**, 185–199.

Colour Assessment

1. Colour—general

Abrahart, E. N. (1977) *Dyes and their intermediates.* 2nd ed. Edward Arnold, London.

Billmeyer, F. W. & Saltzman, M. (eds). (1981) *Principles of colour technology*. 2nd ed. John Wiley and Sons, New York.

Chamberlin, G. J. & Chamberlin, D. J. (1980) *Colour: its measurement, computation, and application.* Heyden and Son Ltd., London.

Judd. D. B. & Wysyecki, G. (1975) *Colour, business, science and industry*. 3rd ed. John Wiley and Sons, London.

Patterson, D. (1976) The development of colour science. *Rev Prog Coloration* **7**, 46–54.

Venkataraman, K. (ed) (1977) *The analytical chemistry of synthetic dyes.* John Wiley & Sons, New York.

Zollinger, H. (1987) *Color chemistry: synthesis, properties and applications of organic dyes and pigments.* Weinheim, New York.

2. Chromatography

Arsov, A. M., Mesro, N. K. & Gatera, A. B. (1973) Thin-layer chromatography of basic dyes. *J Chromat* **81**, 181–186.

Beattie, J. B., Dudley, R. J. & Smalldon, K. W. (1979) The extraction and classification of dyes of single nylon, polyacrylonitrile and polyester fibres. *J Soc Dyers Col* **95**, 295–301.

Beattie, J. B., Roberts, H. R. & Dudley, R. J. (1981a) The extraction and classification of dyes from cellulose acetate fibres. *J For Sci Soc* **21**, 233–237.

Beattie, J. B., Roberts, H. R. & Dudley, R. J. (1981b) Thin-layer chromatography of dyes extracted from polyester, nylon and polyacrylonitrile fibres. *For Sci Int* **17**, 57–69.

Chang, J., Wanogho, S. O., Watson, N. D. & Caddy, B. (1991) The extraction and classification of dyes from cotton fibres using different solvent systems. *J For Sci Soc* **31**, 31–40.

Donshera, M., Arsov, A., Kostova, V. & Mesrob, B. (1976) Thin-layer chromatography of disperse dyes. *J Chromat* **121**, 131–137.

Fanali, S. & Schudel, M. (1991) Some separations of black and red water soluble fibre-tip pen inks by capillary zone electrophoresis and thin layer chromatography. *J For Sci* **36**, 1192–1197.

Hartshorne, A. & Laing, D. K. (1984) The dye classification and discrimination of coloured polypropylene fibres. *For Sci Int* **25**, 133.

Home, J. M. & Dudley, R. J. (1981a) Thin-layer chromatography of dyes extracted from cellulosic fibres. *For Sci Int* **71**, 71–78.

Home, J. M. & Dudley, R. J. (1981b) Revision of the scheme of extraction and classification of dyes from polyacrylonitrile fibres. *J Soc Dyers Col* **97**, 17.

Laing, D. K., Gill, R., Candida, C. & Bickley, H. M. (1988) Characterisation of acid dyes in forensic fibre analysis by HPLC using narrow-bore columns and diode array detection. *J Chromat* **422**, 187.

Laing, D. K., Dudley, R. J., Hartshorne, A. W., Home, J. M., Rickard, R. A. & Bennett, D. C. (1991) The extraction and classification of dyes from cotton and viscose fibres. *For Sci Int* **50**, 23–35.

Lloyd, J. B. F. (1977) Forensic significance of fluorescent brighteners: their qualitative TLC characterisation in small quantities of fibres and detergents. *J For Sci Soc* **17**, 145–152.

Macrae, R. & Smalldon, K. W. (1979) The extraction of dyestuffs from single wool fibres. *J For Sci* **24**, 109–116.

Macrae, R., Dudley, R. J. & Smalldon, K. W. (1979) The characterisation of dyestuffs on wool fibres with special reference to microspectrophotometry. *J For Sci* **24**, 117– 129.

West, J. C. (1987) Extraction and analysis of disperse dyes on polyester textiles. *J Chromat* **208**, 47–54.

Wheals, B. B., White, P. C. & Paterson, M. D (1985) High-performance liquid chromatographic method utilising single or multi-wavelength detection for the comparison of disperse dyes extracted from polyester fibres. *J Chromat* **350**, 205–215.

White, P. C. (1988) Comparison of absorbance ratios and peak purity parameters for the discrimination of solutes using high performance liquid chromatography with multiwavelength detection. *Analyst* **133**, 1625–1629.

White, P. C. & Harbin, A. M. (1989) High performance liquid chromatography of acidic dyes on a dynamically modified polystyrene divinylbenzene packing material with multiwavelength detection and absorbance ratio characterisation. *Analyst* **114**, 877.

3. Microspectrophotometry (see also 2)

Halonbrenner, R. (1978) Microspectrophotometric analysis of textile fibres. *Int Crim Pol Rev* **33**, 7–16.

Hartshorne, A. W. & Laing, D. K. (1987) The definition of colour for single textile fibres by microspectrophotometry. *For Sci Int* **34**, 107–129.

Laing, D. K., Hartshorne, A. W. & Harwood, R. J. (1986) Colour measurement on single textile fibres. *For Sci Int* **3**, 65–77.

Fibre identification

Bortniak, J. P., Brown, S. E. & Sild, E. W. (1971) Differentiation of microgram quantities of acrylic and modacrylic fibres using pyrolysis gas liquid chromatography. *J For Sci* **16**, 380–392.

Carlsson, D. J., Suprunchuk, J. & Wiles, D. M. (1977) Fibre identification at the microgram level by infrared spectroscopy, *Tex Res J* **47**, 456–458.

Challinor, J. M. (1983) Forensic applications of pyrolysis capillary gas chromatography. *For Sci Int* **21**, 269–285.

Grieve, M. C. & Cabinness, L. R. (1985) The recognition and identification of modified acrylic fibres. *For Sci Int* **29**, 129–146.

Grieve, M. C. & Katowski, T. M. (1977) The identification of polyester fibres in forensic science. *J For Sci* **22**, 390–401.

Hartshorne, A. W. & Laing, D. K. (1984) The identification of polyolefin fibres by infrared spectroscopy and melting point determination. *For Sci Int* **2**, 45–52.

Hughes, J. C., Wheals, B. B. & Whitehouse, M. J. (1978). Pyrolysis-mass spectrometry of textile fibres. *Analyst* **103**, 482–491.

Janiak, R. A. & Damerau, K. A. (1968) The application of pyrolysis and programmed temperature gas chromatography to the identification of textile fibres. *J Crim Law, and Pol Sci* **59**, 434–439.

Kirret, O., Pank, M. & Lake, L. (1980) Characterisation and identification of polyester fibres and their modification by infrared spectrometric method. *Eesti NSV Tead Akad Toim Keem Geol* **29**, 92–96.

Kirret, O., Koch, P. & Lake, L. (1981) Characterisation and identification of polyamide fibres by infrared spectrophotometric method. *Eesti NSV Tead Akad Toim Keem Geol* **30**, 287–288.

Kirret, O., Koch, P. & Lake, L. (1982) Characterisation and identification of polyacrylonitrile fibres and their modifications and modacrylic fibres by infrared spectroscopy. *Eesti NSV Tead Akad Toim Keem Geol* **31**, 197–203.

Morris, N. M., Pittman, R. A. & Berni, R. J. (1984) Fourier transform infrared analysis of textiles, *Tex Chem and Col* **16**, 30–35.

Midkiff, C. R., Washington, W.D. & Kopec, R. J. (1979) Diamond and sapphire cell infrared spectroscopy – a powerful new tool in the forensic laboratory. *J Pol Sci and Admin* **7**, 426–437.

Perlstein, P. (1983) Identification of fibres and fibre blends by pyrolysis gas chromatography. *Anal Chim Acta* **155**, 173–181.

Read, L. K. & Kopec, R. J. (1978) Analysis of synthetic fibres by diamond cell and sapphire cell infrared spectrophotometry. *J Assoc Off Anal Chem* **61**, 526–532.

Wampler, T. P. & Levy, E. J. (1985) Analytical pyrolysis in the forensic science laboratory. *Crime Lab Digest* **12**, 25–28.

Wilkinson, J. M., Richard, R. A., Locke J. & Laing, D. K. (1987) The examination of paint films and fibres as thin sections. *The Microscope* **35**, 233–248.

Animal fibres

Appleyard, H. M. (1978) *Guide to the identification of animal fibres.* 2nd ed. WIRA, Leeds.

Peabody, A. J., Oxborough, R. I., Cage, P. E. & Evett, J. W. (1983) The discrimination of cat and dogs hairs. *J For Sci Soc* **23**, 121–129.

Roe, G. M., Cook, R. & North, C. (1991) An evaluation of mountants for use in forensic hair examination. *J For Sci Soc* **31**, 59–66.

Suzanski, T. W. (1988) Dog hair comparison: a preliminary study. *Can Soc For Sci J* **21**, 19–28.

3

Forensic fibre microscopy

Glenn R. Carroll, BSc,
Acting Chief Scientist—Hair & Fibre, Royal Canadian Mounted Police,
Central Forensic Laboratory, P.O. Box 8885,
Ottawa, Ontario, Canada, K1G 3M8

Sherlock Holmes had been bending for a long time over a low-power microscope. Now he straightened himself up and looked round at me in triumph. . . . I stooped to the eyepiece and focused for my vision. 'Those hairs are threads from a tweed coat.'

Sir Arthur Conan Doyle in
'The Adventure of
Shoscombe Old Place'.

3.1 INTRODUCTION

Little did Sir Arthur Conan Doyle realize at the turn of the 20th century, I am sure, that forensic fibre examiners, almost a hundred years later, would still rely on the same analytical methods as his renowned detective. From a forensic standpoint, fibre examinations, both identifications and comparisons, start with a microscopical examination. As we shall see, the choices of microscopical methodologies are numerous and the information content is great. For example, with large diameter man-made fibres, such as monofilaments, or the so-called 'technical' fibres or fibre bundles associated with the natural cordage fibres, as well as with certain yarns, threads, and fabrics, the starting point may well be a low power stereomicroscopical examination. Regardless of the fibre type this will almost certainly be followed by transmitted high power bright field microscopy to assess morphology, measure physical features, and observe colour. Polarized light microscopy may follow to further characterize and quantitate optical features such as birefringence, fluorescence microscopy to examine fibre colour at various points in the visible spectrum, and probably comparison microscopy if questioned fibres are compared to a known sample. On occasion, scanning electron microscopy may be used to advantage, particularly to characterize surface contaminants and physical damage.

In addition to equipment requirements, two additional elements are essential for a successful microscopical examination; these being examiner training and an adequate standard reference collection of authentic samples.

3.2 FIBRE IDENTIFICATION

Fibre identification may be a stand-alone examination, as often happens in the early stages of a major investigation—no suspect (or from a physical evidence viewpoint no suspect textile material) has yet come under question. Any information which can be generated concerning trace material at the crime scene may be of value and may direct the course of the investigation. In the later stages of an investigation, a preliminary fibre identification is of most value as a screening tool to rapidly sift through numerous exhibits. Ultimately, a comparison with known and questioned materials may be warranted.

The information generated from a fibre examination was summarized by Gaudette (1988) as follows:

- whether the fibre is natural or man-made;
- generic and subgeneric type;
- colour and shade of a fibre;
- the expected usage or source application of the fibre;
- the manufacturer;

and later by Grieve (1990) to include additional microscopical features such as:

- diameter;
- the amount of delustrant, the particle size and distribution;
- cross sectional shape;
- birefringence;
- refractive indices;
- melting point;
- fluorescence;
- absorption spectrum and colour coordinates (a numerical nomenclature precisely defining the colour with the use of a microspectrophotometer).

Now, with the implementation of databases (Carroll *et al*, 1989) based on authentic reference samples and those based on actual casework samples (Laing *et al*, 1987), additional information may be generated:

- in some cases a time period of manufacture;
- the expected frequency of occurrence or commonness.

Case examples are numerous in the literature where such identifications have been of assistance (Mitchell & Holland 1979, Grieve 1983a, Petraco 1985); Deadman (1984a) gave an excellent account of one high profile serial homicide case. The reader is also directed to the work of Scott (1985) with automotive fabrics which pointed to the potential of sourcing and ageing textile materials given good reference samples.

3.2.1 Natural fibres—overview of morphological features

The key to successful identification of natural fibres is their morphology—those subtleties which nature has so ingeniously introduced into them. Three broad subdivisions of natural fibres are generally accepted—vegetable, animal, and mineral.

3.2.1.1 Vegetable fibres

The literature is replete with identification keys, narrative descriptions, and photomicrographs. The collections which I have found most useful are Catling & Grayon (1982), Strelis & Kennedy (1967), and perhaps the ultimate *vade mecum*, *Matthews' Textile Fibres* (1954) (unfortunately no longer in print).

Vegetable fibres can be further grouped by their function within the plant into the following groups (with major examples):

- bast or stem fibres—flax, jute, hemp, ramie;
- leaf fibres—sisal, abaca;
- seed and fruit fibres—cotton, coir, kapok.

Almost any plant material has at some time been used for the production of textile materials, thus major commercial fibres are a function of economy, availability, familiarity, and acceptance. Their identification generally proceeds with:

- in the case of bast and leaf fibres, macroscopical and stereomicroscopical examination of the 'technical fibres'; that is, the fibre bundles;
- cross-sectioning the technical fibres by the plate method, fibre microtome, or other methods (Palenik & Fitzsimons 1990, Grieve 1990);
- macerating the technical fibres. A non-aggressive method which has been found to work well is a one hour reflux in 6% hydrogen peroxide—glacial acetic acid (2 : 1) (Strelis & Kennedy 1967);
- mounting the ultimate fibres; that is, after maceration of bast and leaf fibres or directly in the cases of seed and fruit fibres, in a temporary aqueous mountant or a permanent medium. A quick staining with Chlorazol Black E (1% aq.) will improve contrast and facilitate further identification (Isenburg 1967);
- ashing, differential staining, or chemical tests such as Billinghame's differentiation, may prove useful, on occasion, in distinguishing fibres with similar morphologies. For further information see: *Identification of Textile Materials* (1985), Test Methods of the Technical Association of the Pulp and Paper Industry (TAPPI), particularly methods T 401 and T 259 (because many of these fibre sources have at one time been used in the paper industry as well), AATCC test methods. These tests, however,

may be limited owing to their destructiveness, or if the sample is already deeply coloured.

- Polarized light twist test, or Herzog test (Heyn 1954, Hartshorne & Stuart 1970), which can distinguish 'S' twist fibres such as flax and ramie from 'Z' twist fibres such as hemp, jute, and sisal. The fibre is placed between crossed polars (that is, polarizing filters) with its longitudinal axis parallel to the vibration direction of the analyzer (East–West). With a first order red plate in place, Z twist fibres will show a small subtractive effect, that is first order orange, while S twist fibres show a small additive effect, that is second order blue.

3.2.1.2 Animal fibres

Wool, the underfur from sheep, represents by far the most important commercial animal fibre. However, other hairs, termed in the industry 'speciality fibres', can, on occasion, become forensically important. These other hairs tend to be underfur from other mammals, and are termed commercially by names which do not always refer to their biological origin. Examples include camel, cashmere (cashmere goat), mohair (angora goat), alpaca, llama, and vicuna, and are all difficult to distinguish from one another because they are underfur and hence microscopically do not possess many unique outstanding morphological features. Some progress has been made in this area by scanning electron microscopical examination of scale margins (Wortmann & Ams 1987) and more recently by DNA characterization. However, both of these approaches can be regarded as applicable only to very large samples rather than single fibres.

Other animal fibres, particularly if they include the coarser guard hairs, can be more easily distinguished microscopically by their morphology, examples being angora (angora rabbit) and other rabbits, and commercial fur-bearing mammals such as mink and muskrat.

Silk is the remaining major fibre type in this category. However, in practical terms for most forensic fibre examiners it is not routinely encountered owing to its frequent use in the more expensive segment of the garment industry. Silks, from several species of caterpillar, are characterized, in the raw state, by twin filaments of the protein, fibroin, cemented together by a gum, sericin. The yarns and fabrics are often manufactured with the sericin *in situ*, acting as a natural finish, and subsequently removed or 'degummed'. The morphologies of the fibroin fibres vary somewhat with species, cultivated silk fibres being somewhat triangular in cross-section with wild or Tussah silk fibres being elliptical.

3.2.1.3 Mineral fibres

The usage of asbestos, the most important mineral fibre, has certainly diminished with the awareness of its environmental and health hazards. The forensic specialist most often encounters this fibre originating from building materials, particularly in various forms of insulation, and most frequently as a result of break and enter or theft crimes. These fibres are characterized by their extremely fine diameter, the larger fibre bundles bifurcating into smaller and smaller units until reaching below the resolving power of the light microscope. McCrone (1974, 1978, 1979, 1985) has written prolifically on the microscopical characterization of asbestos, particularly with the polarized light microscope.

3.2.2 Man-made fibres—overview of morphological features

Man-made fibres include those fibres made from regenerated polymeric materials, such as cellulose from which rayon is derived, and the true synthetic materials with which we have become so familiar in our daily lives, such as nylon, polyester, and acrylic. The morphologies of these fibres are often as diverse and complex as those of the natural fibres, having either been intentionally designed and engineered into the fibre or are inadvertent artefacts of its manufacture.

3.2.2.1 Designed features

Most man-made fibres are produced by extrusion whereby the molten or dissolved polymeric material is forced through an orifice, called a spinneret, in much the same manner as water is forced through a showerhead. The size and shape of the spinneret controls the shape and diameter of the resulting fibres. When multiple orifices are contained in the same spinneret, the number of individual filaments in the resulting fibre bundle is determined. Fibre diameter is critically controlled, and in general, finer diameter fibres are encountered with finer or lighter fabrics.

The fibre shape in cross-section is extremely important, not only from the manufacturer's standpoint but also forensically. Some cross-section morphologies are shown in Fig. 3.1.

Some cross-section morphologies seem to be associated with generic class, for example peanut shape with acrylics, and multilobal with the rayons. Other features are intentionally built-in for performance and may indicate the end use of the fibre, the classical example being the trilobal shape often associated with carpet fibres and often recovered forensically from footwear or from bodies which have been transported in automobiles, or the variety of hollow fibres used for insulation or moisture absorptivity, often encountered on suspected murder weapons such as knives. Any uncommon feature with respect to cross-section can become extremely significant in a criminal investigation, and case examples have been well documented (Deadman 1984a, b).

The presence and amounts of dulling agents, termed delustrants, are valuable comparative features. Titanium dioxide is finely dispersed in 'dull' or 'semi-dull' fibres (terms used in the textile industry to indicate the relative amounts of delustrant). It usually appears as black particles owing to its opacity. In 'bright' or shiny fibres, no or little delustrant is present. On occasion spurious colours, as observed by transmitted light microscopy, can be produced by finely dispersed delustrant particles. White or lightly coloured fibres may appear brown because of such diffraction effects, or if very heavily delustred, a white fibre may actually appear black. If possible, the fibre should be examined by reflected light to verify its actual colour.

3.2.2.2 Inadvertent features

Some features which are microscopically observable seem to be benign in terms of a fibre's performance. Their presence is unintentional, but they can be distinctive. For

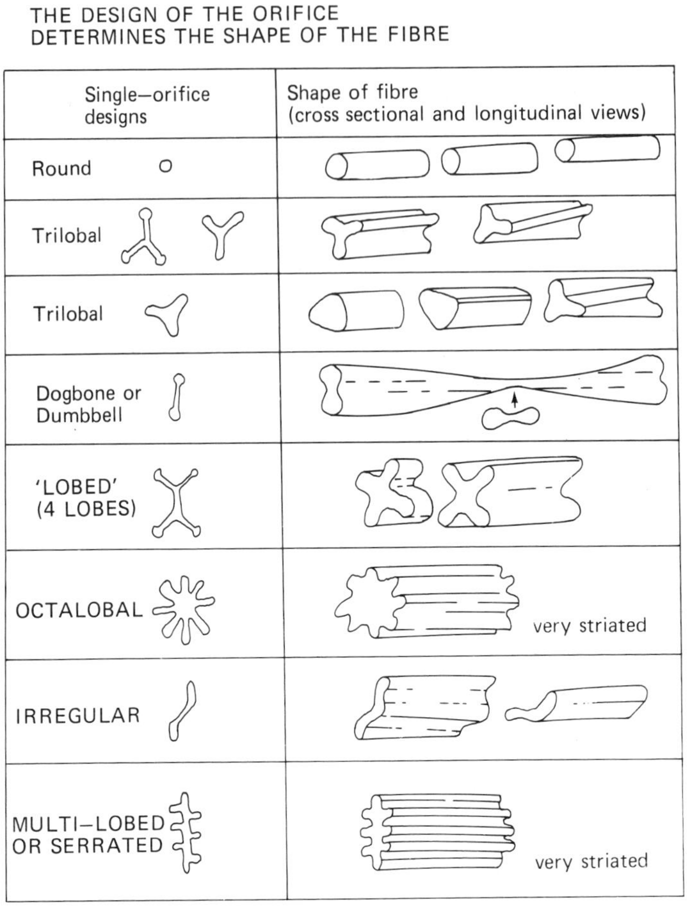

THE DESIGN OF THE ORIFICE
DETERMINES THE SHAPE OF THE FIBRE

Single–orifice designs	Shape of fibre (cross sectional and longitudinal views)
Round	
Trilobal	
Trilobal	
Dogbone or Dumbbell	
'LOBED' (4 LOBES)	
OCTALOBAL	very striated
IRREGULAR	
MULTI–LOBED OR SERRATED	very striated

Fig. 3.1. Man-made fibre cross-section morphologies (courtesy of FBI).

example, the cross-section shape of modacrylic fibres may at first appear to be an assortment of random irregular patterns, but these fibres do tend to group themselves on the basis of different manufacture (Grieve & Cabiness 1985). The fine, faint striations in KEVLAR aramid fibres are another example of a characteristic feature found only in that generic class.

Texturizing can impart characteristics in fibres indicative of the method used. A saw-tooth shape in fibres can indicate a crimped process, while smooth regular undulations may indicate that the fibre or yarn originated from a knitted fabric.

Dyeing can produce artefacts which are often characteristic. If, for example, a fabric is piece dyed, that is dyed after weaving, the colouring of individual fibres may be uneven because of uneven exposure of the individual fibres to the dye solution. In printed fabrics, the colorant is often seen as surface blotches. Some processes result in fibres which are 'skin' dyed, that is the dyestuff may uniformly coat the outside of the fibre, and in cross-section appear as a contrasting skin on the outside of the fibre. The colour of a sample in bulk form, that is fabric, yarn, or thread, may not be truly indicative of the individual fibres. Thus it is not uncommon for the individual fibres from a black fabric to be, for example, dark green or purple microscopically. Also, if a fabric is a blend of two or more components, each component may appear, microscopically, quite distinct because the fibres adsorb the dyestuff components differently.

3.3 FIBRE ANALYSIS METHODS

A fibre analysis should be an ordered stepwise examination based on an established protocol and using the examiner's knowledge, skill, and experience. In the Royal Canadian Mounted Police Forensic Laboratories, for example, the fibre examination protocol has been arrived at through group discussions by senior examiners, tested by all personnel, and codified into a working protocol. A sample of the data work sheet currently in use is illustrated in Fig. 3.2.

Even with codified protocols, the fibre examiner must still be allowed the flexibility to follow individual hunches because of the uniqueness of forensic casework and the simple fact that each case truly is different.

3.3.1 Stereomicroscopy

A preliminary stereomicroscopical examination, particularly on 'large' samples that is fabrics, threads, or yarns, can confirm the reflected light colour of the subject material, since when we look at a garment or textile material in our everyday environment, it is most often by reflected light. The direction of twist and the number of twists per unit length can be determined, as well as the type of weave or knit, and the count of cloth (the number of yarns per unit measured in both directions of the fabric). Texturizing may also be observable stereomicroscopically, even with single fibres, and characterized by evenly repetitive zig zags or undulations.

For exhibit searching and textile damage assessment, particularly with large unwieldy items, we have found a floor-mounted surgical style stereomicroscope particularly useful (Fig. 3.3). It is now possible to attach almost any stereomicroscope to such a movable floor stand, and with the variety of microscope objectives currently available for stereomicroscopes, searching and examination are greatly facilitated. Some surgical stereomicroscopes equipped with vertical or epi-illumination are disadvantageous since with shiny exhibits the reflections can be disturbing.

3.3.2 Transmitted light microscopy

Transmitted light microscopy for forensic fibre examination invariably implies the use of a comparison microscope (Fig. 3.4). This instrument should be configured to be capable of polarized light and fluorescence examinations in addition to bright field.

RCMP GRC	FIBRE DATA WORK SHEET	FEUILLE DE DONNÉES SUR LES FIBRES	Lab File No. - Dossier du lab. n°	
	KNOWN - CONNUE(S)			QUESTIONED - INCONNUE(S)
ITEM					
DESCRIPTION					
SLIDE, FIBRE LAME, FIBRE					
Class - Genre					
Colour - Macro/stereomic. Couleur - macro/stéréomic.					
Colour - Micro. Couleur - micro.					
Diam. (μm) max.					
min.					
Long. Morph. Morph. long.					
Delustr. - Débrillant					
X S Morph. Morph. - coupe transv.					
Fluor. UV					
V - Vi					
B					
G - Ve					
Pleo. - Pléochr.					
n_{iso} (mountant:) (préparation:)					
$n_{\parallel}$					
$n_{\perp}$					
Biref. - Biréfr.					
Melt. Pt. - Point de fusion					
Vis. Spec. - Spectr. vis.					
I R - Infr.					
Solubil.					
TLC - C.C.M.					
# fibres					
CONCLUSION					

2628 (87-07)

Fig. 3.2. Fibre data worksheet.

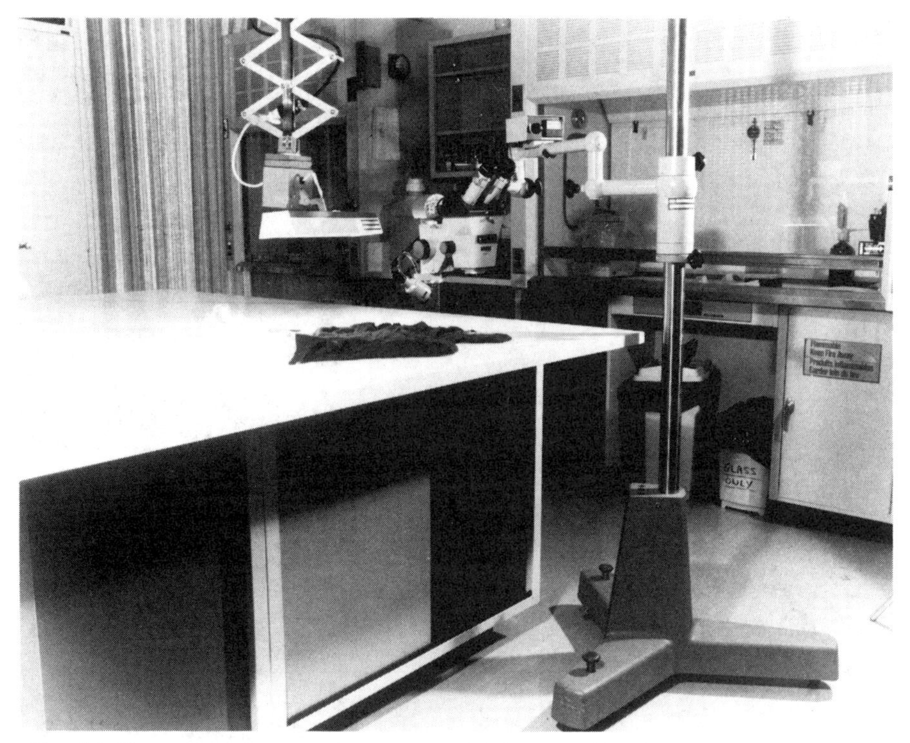

Fig. 3.3. Floor-mounted stereomicroscope, useful for searching large exhibits such as hit-and-run automobiles, or simply for routine examinations.

The examination begins with the preparation for longitudinal examination by mounting the samples on glass microscope slides. With known samples or large questioned fibre samples, such as yarns or threads, the fibres are teased apart with dissecting needles or the points of fine forceps to produce a thin layer of dispersed fibres on the slide. Single questioned fibres, usually having been recovered by taping or stereomicroscopical examination, should be mounted individually, each under its own small cover glass. These cover glasses, ranging in size from about 2 to 5 mm square, can with a little practice be made from larger commercially available cover glasses by scoring with a diamond tipped pencil and then breaking apart. Although it is possible to use a temporary mounting medium, such as glycerol–water, many examiners prefer to use a permanent mountant from the outset. Cook & Norton (1982) surveyed a variety of mountants and concluded that the proprietary product, XAM†, offers the best combination of desirable properties: refractive index of approximately 1.491, low inherent fluorescence, inertness, removability, long-term freedom from crystallization. Permanent mountants such as this can be used effectively for both day-to-day casework slide preparations and long-term standard reference samples.

---

†manufactured and distributed by BDH Ltd

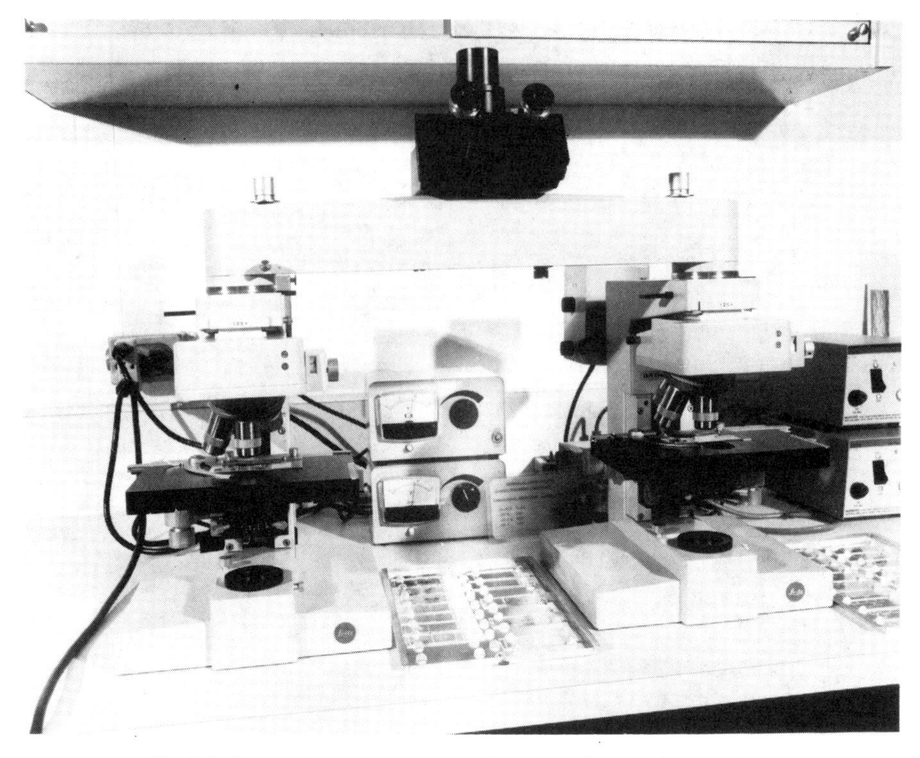

Fig. 3.4. Comparison microscope configured for forensic fibre analysis.

From Fig. 3.2 one can see that the fibre features observed include colour, diameter (both maximum and minimum), longitudinal morphology, delustrant (presence or absence, and if present a subjective quantitative assessment as to bright, semi-dull, dull), and optical properties.

Optical properties can be determined quantitatively, as discussed in the following section. A preliminary qualitative assessment can be determined on a comparison microscope fitted with simple polars (polarizer and analyser) and may actually preclude the need for later absolute determinations under the polarizing microscope.

Cross-sections primarily provide information on the fibre morphology, and can assist in its identification and comparison. In the case of a trilobal or multilobed fibre, its modification ratio can also be determined. Modification ratio is defined as the ratio of the diameter of a circumscribed circle to the diameter of an inscribed circle. A calculator or computer program, called MODRAT. BAS and written in BASIC, was described by Crocker (1986), wherein measurements can be made directly through the microscope of the distances between a trilobal fibre's apexes and the distance between troughs, and, by means of basic geometry, the modification ratio determined directly rather than by the more arduous task of measurements from photomicrographs (Appendix 1).

Added information from cross-sections can include visualization of non-uniform dyeing phenomenon such 'skin dyeing' whereby dyestuff penetration is not uniform throughout the fibre and the outer surface is dyed more intensely than its interior. Phase contrast microscopy may enhance these skin/core differences.

3.3.3 Polarized light microscopy

DeForest (1982), Gaudette (1988), and Grieve (1990) have extensively outlined polarized light microscopy as it applies to forensic fibre examinations. To add to their comments and clarify certain points, some personal observations follow. The major advantage with polarized light microscopy lies in that a fibre can be examined directly with no requirement to de-mount it. However, the major drawback, as with any of the techniques for determining optical properties, occurs if the fibre is very heavily dyed or delustred.

Equipment for quantitative polarized light microscopy can be incorporated into a comparison microscope or a stand-alone instrument, but should include strain-free optics, rotating stage, graduated rotatable polarizer and analyser, and provision for an optical compensator and retardation plates.

3.3.3.1 Refractive index

Many, but not all, fibres possess two refractive indices—one parallel to the length of the fibre (termed $n_{\parallel}$), the second perpendicular (termed $n_{\perp}$). Birefringence is the numerical difference between these two. If the greater refractive index is parallel to the fibre length, birefringence, or more precisely the sign of elongation, is said to be positive.

One may examine refractive indices separately and calculate $n_{\parallel}$ and $n_{\perp}$ independently, or examine the isotropic refractive index, n_{iso}, relative to a particular mounting medium. This latter property is a numerical combination of $n_{\parallel}$ and $n_{\perp}$ expressed as:

$$n_{iso} = \frac{n_{\parallel} + 2n_{\perp}}{3} \tag{3.1}$$

It is estimated *without* polarizer or analyser in place, involves observing the Becke line, a bright band along the fibre margin, and noting whether it is much lower, lower, equal to, higher, or much higher than the mountant. The Becke line can be observed when a difference exists between the refractive index of the fibre and the mountant. As the focusing mechanism is adjusted and the objective is moved away from the specimen, the Becke line will be seen to more toward the medium with the higher refractive index, and conversely.

The following scheme uses *Permount*†, a mountant of refractive index 1.52, which distinguishes most generic and some sub-generic classes of man-made fibres.

†manufactured and distributed by Fisher Scientific Co.

Table 3.1. Identification and grouping of man-made fibres by optical properties)
(where $\gg$: much higher
$>$: higher
$\approx$: approximately equal
$<$: lower
$\ll$: much lower
* : optical properties usually sufficient for identification)

Birefringence	n_{iso} vs mounting medium of $n = 1.52$	$n_{\perp}$ vs mounting medium of $n = 1.52$
Negative		
Saran* (−0.008 to −0.010)	$\gg$	
Acrylic (−0.000 to −0.005)	$\approx$	
Some modacrylics (−0.000 to −0.005)	$>$	
Novoloid* (−0.000 to −0.002)	$\gg$	
Isotropic or very low		
Glass* (0)	$\gg$	
Triacetate (0)	$\ll$	
Nytril (0)	$\ll$	
Low positive (<0.010)		
Vinyon (0 to 0.005)	$>$	
Triacetate (0 to 0.001)	$\ll$	
Acetate* (0.001 to 0.005)	$\ll$	
Some modacrylics (0.002 to 0.005)	$>$	
Azlon (0.002 to 0.006)	$>$	
Viscose (0.002 to 0.014)	$\approx$ to $>$	
Medium positive (0.010 to 0.050)		
Vinal (0.025 to 0.030)	$\approx$ to $>$	$<$ to $\approx$
Viscose (0.010 to 0.046)	$\approx$ to $>$	$\approx$
Cupro (0.021 to 0.037)	$>$	$\approx$ to $>$
Modal (high-tenacity viscose) (0.021 to 0.050)	$\approx$ to $>$	$<$
Polypropylene* (0.028 to 0.034)	$<$	$<$
Qiana polyamide (0.036 to 0.050)	$\approx$ to $>$	$<$
High positive (0.050 to 0.070)		
Qiana polyamide (0.036 to 0.050)	$\approx$ to $>$	$<$
Nylon 11 (0.046)	$\approx$	
Fluorofibre* (0.040 to 0.050)	$\ll$	
Nylon 6 and 66 (0.049 to 0.063)	$>$	
Polyethlene (0.050 to 0.052)	$>$	
Polyester—partially-oriented PET (0.044 to 0.056)	$>$	
Polyester—unspecified (0.048 to 0.076)[a]	$>$	

Very high positive (>0.070)		
Polyester—PCDT (0.096 to 0.104)[a]	≫	>
—PET (0.128 to 0.191)[a]	≫	>
—modified PET (0.128 to 0.160)[a]	≫	>
—PBT (0.148 to 0.150)[a]	≫	>
Aramid* (0.120 to 0.400)	≫	≫

[a]Values from Grieve (1990).

To determine $n_{\parallel}$ and $n_{\perp}$, the fibre is aligned in an extinction position under crossed polars, that is, a position where the fibre appears darkest and usually parallel or perpendicular to the polars. The analyser (or the polarizer) is then removed from the optical path, the position of the fibre noted, that is North–South or East–West, and the Becke line observed. By using a range of calibrated liquids of known refractive index a precise determination of the fibres refractive index either parallel or perpendicular to the polarizer may be made. The procedure is then repeated for the second refractive index with the fibre rotated 90° from the first. Unfortunately, this method is rather tedious and not well-suited to working with the small fibre fragments so often encountered in forensic casework since one could easily lose these small fragments in successive re-mountings in various liquids.

3.3.3.2 Optical compensators and retardation plates

Berek and Ehringhaus compensators are accessories which consist of a calcite or quartz crystal plate mounted in a calibrated tilting holder. They provide a rapid and non-destructive method of calculating a fibre's retardation or comparing relative retardation between a known and questioned sample. They are inserted into the optical path at 45° to and between the polarizer and analyser. From this determination of retardation, birefringence can easily be determined from the following relationship:

$$\text{Birefringence} = \frac{\text{retardation (nm)}}{1000 \times \text{thickness } (\mu\text{m})} \tag{3.2}$$

The fibre to be examined is centred on the rotating stage of a polarized light microscope and oriented in the position of maximum brightness (45° position). As the compensator crystal is tilted a series of coloured bands will be seen to move along the fibre. If this tilting decreases the interference colours within the fibre, the sign of elongation of the fibre is *negative* when it is aligned parallel to the slow wave of the compensator (usually engraved on the compensator as γ or Z'), or *positive* when aligned perpendicularly. The fibre must be in the 45° position which causes the interference colours to decrease, termed the *subtractive* position. The crystal is tilted until the compensation point is reached whereby the centre or the thickest part of the fibre will appear black or dark grey. From this tilt angle, measured by means of a calibrated scale on the device, the fibre's retardation is easily computed. Then, by dividing this retardation by the fibre's diameter or thickness at the point of measurement, the *birefringence* is determined as shown above in equation (3.2). In

Table 3.1 fibres are listed in ascending order by their birefringence, in addition to other optical properties.

With fibres of high birefringence, as occurs with generic classes which are inherently high, such as polyesters, or large diameter fibres of moderate birefringence such as nylon carpet fibres, false compensation bands may be encountered which are due, apparently, to the difference in refractive index of the fibre and the crystal used in the construction of these tilting compensators. Upon close observation it will be seen that these bands are not black, as occurs with true compensation bands, but darkly coloured. In these instances, while the absolute birefringence values may not be obtainable, a relative comparison should yield similar results between samples of common origin.

The calculation of birefringence from optical compensator data can be facilitated with the assistance of a calculator or computer program. One such, called BIREF.BAS and written in BASIC, automatically computes the birefringence and assigns a generic class or classes which match the data (Appendix 2).

Quarter wavelength ($\lambda/4$) and first order red (λ) retardation plates can be used for assessing a fibre's refractive index relative to the plate. The former, also known as a 'sensitive tint plate' because very small differences in retardation cause a large difference in the colours produced, can best be used with fibres of low birefringence, for example, acrylics, acetates, and some modacrylics. A first order red plate, as previously discussed, can be used to distinguish S twist and Z twist natural fibres.

A quartz wedge can be used to estimate retardation. However, more precise measurements can be obtained by using an optical compensator. When white light is used as the illuminating source, a series of repeating interference colours, caused by the increasing thickness of the wedge, will be observed when the wedge is inserted into the optical path between polarizer and analyser. When an anisotropic (or birefringent) fibre is inserted into the optical path, oriented at 45° to the polarizer or analyser, and if the combined retardations of the fibre and wedge cancel one another (the fibre is in the subtractive position), as the wedge is progressively inserted a compensation point is reached where the two cancel each other completely and the thickest part of the fibre appears black. The background colour produced by the wedge at this point may then be visually compared with an appropriate chart, and a close estimate can be made of the fibre's retardation.

With any of these methods several determinations should be conducted on different parts of a fibre or on different fibres in a sample. With fibres having cross-sectional shapes other than round, determination of retardation with an optical compensator or quartz wedge can be somewhat problematical. One must bear this in mind and measure retardation at a point where the fibre thickness can also be determined or closely estimated.

3.3.3.3 Pleochroism

When observed with only one polar in place, that is either polarizer *or* analyser, and if a fibre is rotated by means of a rotating stage, changes in colour can often be observed. This colour change is termed pleochroism. It can be a valuable comparative feature since not only can the depth of colour change, but also the shade of colour.

It is sensitive, rapid, and non-destructive.

This selective absorption can be a function of the fibre itself or the dyestuff on or in the fibre, but it is most probably due to a combined effect of both the fibre matrix and the dyestuff. If the sample is seen to change in colour in two directions, then this special form of pleochroism is termed dichroism.

3.3.4 Fluorescence microscopy

With the development of the incident fluorescence attachment, known as the Ploemopak, comparison of fluorescence properties is far more practical and convenient. Since these attachments have provision for multiple filter combinations it is possible to rapidly compare fluorescence properties in various regions of the spectrum, for example ultraviolet, violet, blue, and green. The value of adding fluorescence examination in the ultraviolet alone was reported by Macrae *et al.* (1979) as increasing the discriminating power of fibre comparisons from about 50% for white light bright field comparison microscopy to as much as 80% with the addition of UV fluorescence comparison. With these fluorescence attachments mounted on a comparison microscope, after normal white light comparison microscopy the fibres can be compared directly under a variety of fluorescence conditions. Differences may arise unpredictably in any part of the spectrum, and will be found to be consistent between fibres of the same manufacture.

For recovery of fibres from exhibit tapings or from the exhibits themselves a stereomicroscope or macroscope fitted with an incident fluorescence attachment has been found useful when the target fibres fluoresce (Fig. 3.5). If the same filter combinations are installed on this screening instrument, the same fluorescence properties will be observed in subsequent comparison microscopy.

3.3.5 Microspectrophotometry

Colour analysis is treated in-depth elsewhere in this book. However, it is sufficient to state at this point that many forensic laboratories worldwide have successfully used microscope-mounted spectrophotometers (Fig. 3.6) to compare fibres and other materials such as inks, paints, and plastics. Much of the computer software has been developed in-house in forensic laboratories particularly for textile fibre colour comparisons.

3.3.6 Thermal microscopy

Melting phenomena are well documented for polymeric materials including textile fibres. Those fibres which are thermoplastic and thus do, in fact, melt are conveniently examined via microscopical methods. Hot stages of various designs are commercially available. The ones of choice in forensic fibre examination have been the Mettler FP series (FP 5 and the more recent FP 800) which are conveniently placed on the stage of a compound microscope (Fig. 3.7). These afford reproducible, rapid results on small samples, for example single fibres less than a centimetre in length.

Sample preparation for melting determination is minimal in that fibres may be 'dry mounted' by simply placing the sample on a microscope slide and covering it with a cover glass, or preferably by using a high temperature stable silicone oil as a

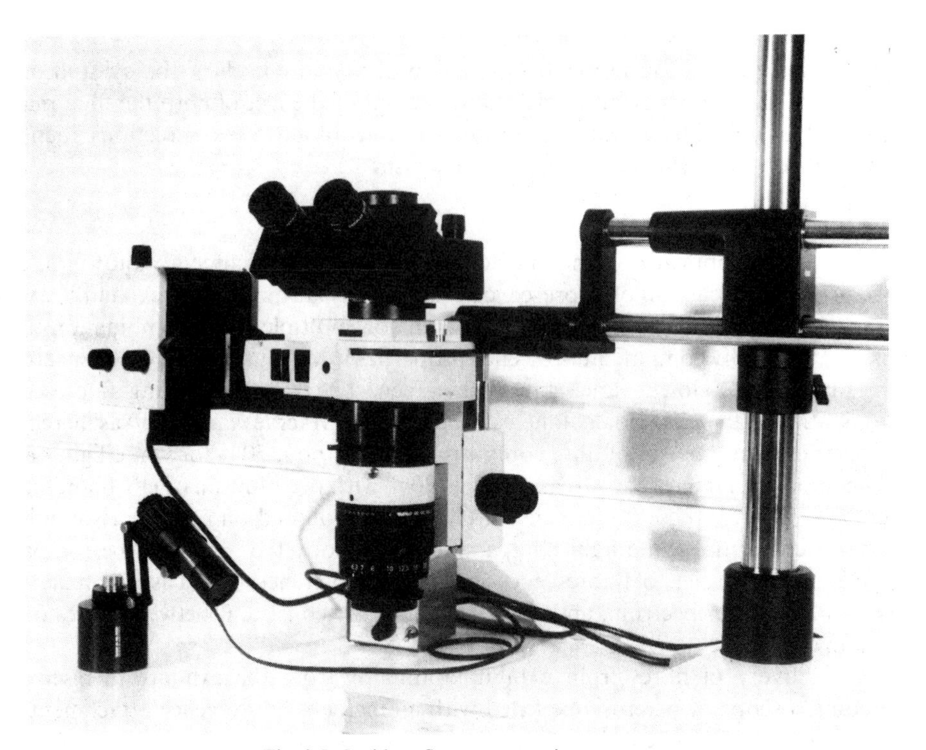

Fig. 3.5. Incident fluorescence microscope.

mountant to aid in thermal transfer from the hot stage oven to the sample. Examination of thermal behaviour can often be optimized by using a polarized light microscope whereby the fibre sample is examined under slightly crossed polars. As an anisotropic fibre (one which is birefringent) is heated, its birefringence decreases until it melts, thereby becoming a liquid, and it becomes isotropic or black under crossed polars or colourless under slightly uncrossed polars. Under bright field conditions the start and finish of melt are often easily missed, the latter because the melted polymer is very viscous and the fibre appears to retain its previous shape. It is common for workers to refer to a fibre's melting *point* when in reality the phase change usually covers a temperature *range*. This is not surprising, for two reasons:

- Textile fibres are not a single pure chemical compound but in reality a mixture of a base polymer and additives such as delustrants, plasticizers, dyestuffs, and any combination of finishes which have been added for such properties as anti-static, anti-stain, waterproofing, etc. In addition, with forensic samples, there is a host of consumer added contaminants such as detergent residues, dirt, grease and oils, and body fluids.

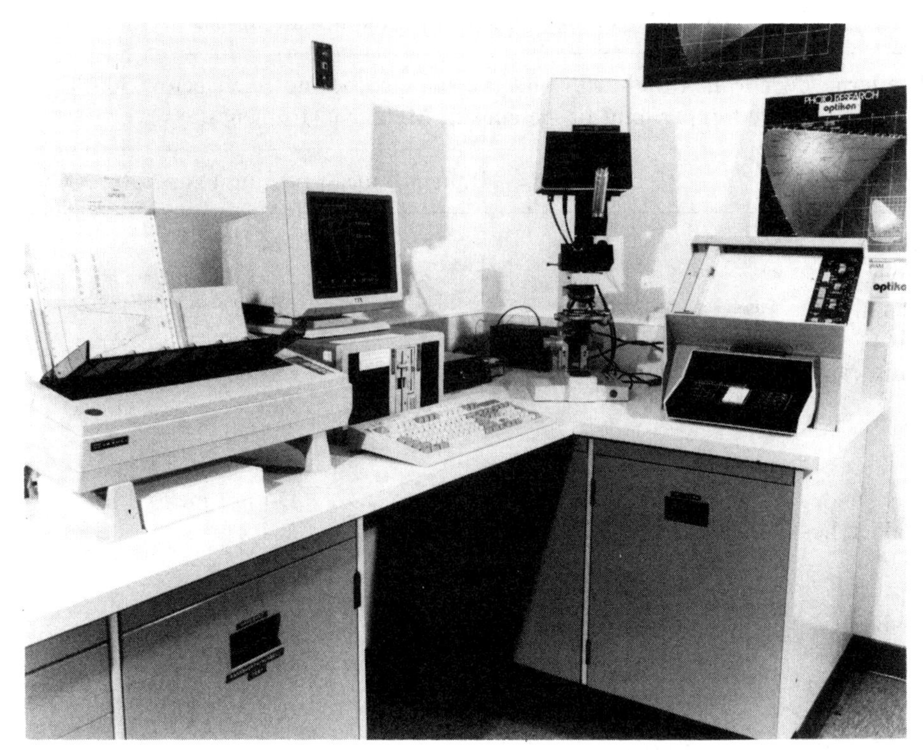

Fig. 3.6. A NANOSPEC 10S microspectrophotometer interfaced to a personal computer for spectral comparison of fibres.

- Textile fibres are not totally crystalline, but a mixture of crystalline and amorphous regions.

One should observe, then, the range of temperatures over which a sample melts, that is when melting begins and then when no further changes occur, and either refer to this range or take the mean of these two temperatures. This mean melting temperature, then, may more accurately describe the phenomenon.

Not all fibres, however, are thermoplastic. These include the natural fibres, rayons, acrylics, and some modacrylics. In addition to melting, thermoplastic fibres often exhibit other thermal behaviour, the most striking of which is contraction, which can be documented along with melting range. Melting ranges which we have calculated in our laboratory from standard reference samples supplied in the *Collaborative Testing Services Reference Collection of Synthetic Fibers*† and from our own collection are listed in Table 3.2.

---

† Collaborative Testing Services, Inc. Box 1049, Herndon, VA 22070, USA.

Table 3.2 Melting behaviour of man-made fibres.

(The term 'do not melt' indicates that the sample does not melt below 300 °C, the useful upper limit of the current Mettler instrumentation.)

Fibre type	Range of mean melting temperatures (°C)
Acetate	224–280
Acrylic	Do not melt
Aramid	Do not melt
Azlon	Do not melt
Fluorocarbon—PTFE	Do not melt
2*	300
3*	277
4*	245
Glass	Do not melt
Modacrylic—A**	204
B**	225
C**	Do not melt
Novoloid	Do not melt
Nylon—6	213
612	217–227
66	254–267
Qiana (diamine & dicarboxylic acid)	276–277
Olefin—Polyethylene	122–135
Polypropylene	152–173
Polycarbonate	161
Polyester—PET	256–268
PCDT	277–298
3***	222
4***	222
5***	230–231
6***	231–245
Rayon	Do not melt
Rubber	215
Saran	167–184
Spandex	231
Sulfar	283
Triacetate	260
Vinal	200–260
Vinyon—PVC/PVA	135

* Where:

2—Polyvinylidene fluoride

3—Ethylene/tetrafluoroethylene
4—Ethylene/chlorotrifluoroethylene
** As designated by Grieve & Cabiness (1985) where:
A—Acrylonitrile/vinylidene chloride/methacrylamide
B—Acrylonitrile/vinylidene chloride
C—Acrylonitrile/vinyl chloride
*** As designated by Grieve (1983b) where:
3—Poly(hyroxyethoxybenzoate)
4—Polybutylene terephthalate
5—Terephthalic acid/p-hydrobenzoic acid/ethylene glycol co-polymer
6—Other co-polymers

3.3.7 Microscopical accessories for infrared spectroscopy

In recent years, with the advent of Fourier transform infrared spectroscopy, microscope accessories have been shown to be a valuable analytical adjunct. These offer the ability of examining not only single fibres of several millimetres length but even discrete, isolated areas within a single fibre. Contrary to manufacturer's claims,

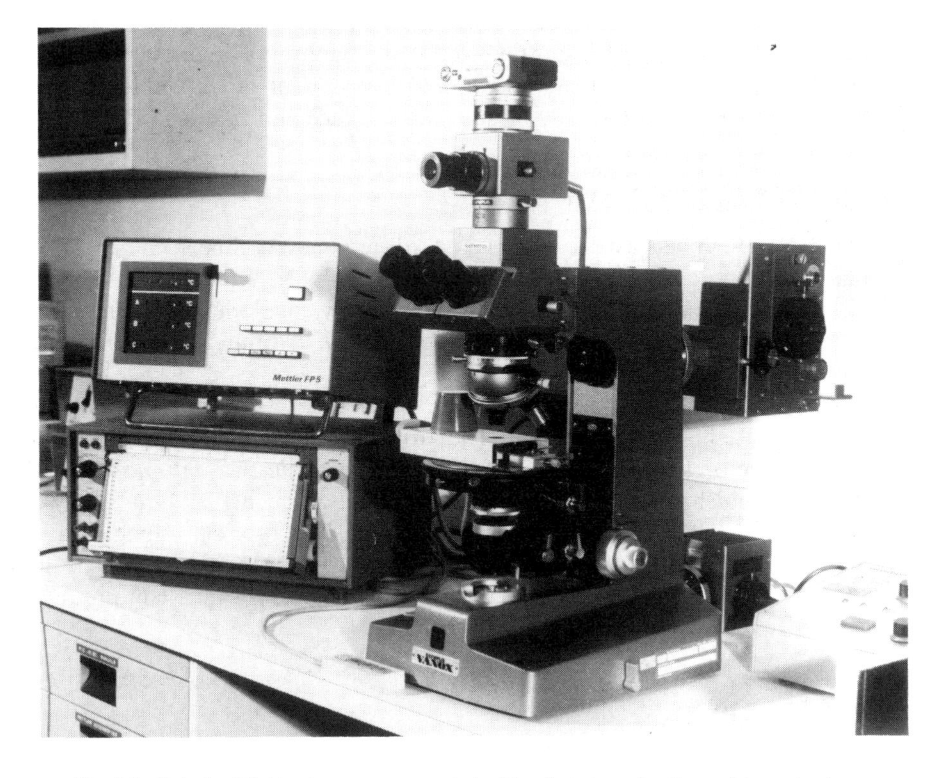

Fig. 3.7. Polarized light microscope mounted with a hot stage for thermal investigations.

sample handling can still be arduous; generally, small fibres must still be de-mounted from microscope slides, cleaned of mountant, and transferred to the microscope accessory. Often, to improve the sample area presented for measurement or simply to reduce the 'lens' effect caused by cylindrical fibres, the sample must be pressed. The greatest potential for this sampling accessory would seem to be with side-by-side bicomponent and biconstituent fibres whereby each component could be characterized *in situ*. Bartick (1987) and Tungol *et al.* (1989, 1990) have standardized these analytical protocols and constructed a computer searchable spectral library of reference fibres.

3.4 REFERENCE COLLECTIONS AND DATABASES

Collections of authentic samples are indispensable for reference purposes. They can be constructed by contributions from manufacturers, museums, plant tours, or obtained commercially. To be of real value, however, samples in any collection must be catalogued and characterized. With the widespread availability of personal computers and database software the handling of data generated from such investigations has been greatly facilitated. Two recent approaches have been taken:

- casework databases using samples from actual casework exhibits;
- standards databases using authentic collected samples.

Both are of forensic value. The casework approach provides information on textile materials in actual use at a particular time and is analogous to a population database. The other approach assists in an operational capacity of identifying questioned samples as well as in a support role of quality assurance and training.

3.5 FIBRE/TEXTILE DAMAGE

For routine textile damage assessment a standard stereomicroscope, ideally equipped with dual iris diaphragms to increase depth of focus, is routinely used, while on occasion a scanning electron microscope can provide valuable information [Hearle *et al.* 1989), Matsibora *et al.* (1990)]. See Hearle *et al.* (1989) for an in-depth treatment of this subject.

While the forensic specialist is often asked to 'match' a suspect weapon with damage found in a garment, this is rarely possible. The garment can be examined to characterize the age of damage present, for example, 'recent' or 'fresh' versus 'old' in nature; the proviso to distinguish recent from old being that the garment has not been worn extensively or laundered since the damage occurred. The type of damage may be characterized as cut, rip, or seam separation. The suspect weapon may also be used to produce test damage, simply to indicate whether or not it is capable of producing damage consistent with that in the garment.

3.6 CONCLUSION

Given the great diversity of textile fibres in commercial production and use, with their multiplicity of production variables such as diameter, cross-section shape, thermal properties, delustrants, and texturizing, combined with a huge selection of dyestuffs marketed, the possible combinations of characteristics are immense. Good discrimination between fibres of different manufacture is possible more often than not by relatively rapid and non-destructive microscopical methods. Given the good mobility of fibres with their often easy transfer from one object or place to another, this class of trace evidence continues to provide valuable information for the forensic specialist.

ACKNOWLEDGEMENTS

The author wishes to acknowledge the assistance of R. Van Gastel for the preparation of photographs, as well as Anne-Elizabeth Charland, Barry Gaudette, and Paul Roussy for their constructive comments.

APPENDIX 1. BASIC program for calculating modification ratio of trilobal fibres

```
10 REM ***********************MODRAT.BAS********************************
20 REM *                                                               *
30 REM *   CALCULATES THE MODIFICATION RATIO OF MAN-MADE FIBRES        *
40 REM *         BY S.A.LALONDE,G.R. CARROLL,E.J.CROCKER               *
50 REM *          R.C.M.P. CENTRAL FORENSIC LABORATORY                 *
60 REM *                     REV. 91/04/24                             *
70 REM *****************************************************************
80 CLS
90 PRINT
100 PRINT "                    MODIFICATION RATIO"
110 PRINT "                    ******************"
120 PRINT
130 PRINT "DO YOU NEED INSTRUCTIONS (Y/N) ";
140 INPUT A$
150 IF A$ =  "N" GOTO 380
155 IF A$ =  "n" GOTO 380
160 IF A$ =  "Y" GOTO 170
165 IF A$ <> "y" GOTO 130
170 PRINT
180 PRINT
190 PRINT "                       DIAMETER OF CIRCUMSCRIBED CIRCLE (D(C))"
200 PRINT "MODIFICATION RATIO = ---------------------------------------"
210 PRINT "                       DIAMETER OF INSCRIBED CIRCLE     (D(I))"
220 PRINT:PRINT
230 PRINT "FOR TRILOBAL FIBRES, ENTER THE FOLLOWING MEASUREMENTS:"
240 PRINT
250 PRINT TAB(10)"X(C): THE DISTANCE BETWEEN TWO TIPS OF THE FIBRE WHICH"
260 PRINT TAB(16)"FORM ITS BASE"
270 PRINT TAB(10)"Y(C): THE DISTANCE BETWEEN TWO TIPS OF THE FIBRE FROM"
280 PRINT TAB(16)"A BASAL TIP TO THE APICAL TIP"
290 PRINT TAB(10)"Z(C): THE REMAINING DISTANCE BETWEEN A BASAL"
300 PRINT TAB(16)"AND THE APICAL TIP"
```

```
310 PRINT TAB(10)"X(I): THE DISTANCE BETWEEN TWO ADJACENT TROUGHS WHICH"
320 PRINT TAB(16)"FORM ITS INSCRIBED BASE"
330 PRINT TAB(10)"Y(I): THE DISTANCE BETWEEN TWO TROUGHS FROM A BASAL"
340 PRINT TAB(16)"TROUGH TO THE APICAL TROUGH"
350 PRINT TAB(10)"Z(I): THE REMAINING DISTANCE BETWEEN TWO TROUGHS FROM"
360 PRINT TAB(16)"A BASAL TROUGH TO THE APICAL TROUGH"
370 PRINT
380 REM**** BEGINNING OF MAIN LOOP
390 PRINT
400 PRINT "ENTER SAMPLE IDENTIFICATION: ";
410 INPUT B$
420 PRINT
430 PRINT TAB(5);"X(C)";
440 INPUT X(1)
450 PRINT TAB(5);"Y(C)";
460 INPUT Y(1)
470 PRINT TAB(5);"Z(C)";
480 INPUT Z(1)
490 PRINT
500 PRINT TAB(5);"X(I)";
510 INPUT X(2)
520 PRINT TAB(5);"Y(I)";
530 INPUT Y(2)
540 PRINT TAB(5);"Z(I)";
550 INPUT Z(2)
560 FOR K = 1 TO 2
570 LET T = X(K)^2*Y(K)^2+X(K)^2*Z(K)^2+Y(K)^2*Z(K)^2
580 LET T = 2*T-Z(K)^4-Y(K)^4-X(K)^4
590 LET R(K) = X(K)*Y(K)*Z(K)*SQR (1/T)
600 NEXT K
610 PRINT
620 PRINT TAB(5)"MODIFICATION RATIO = ";R(1)/R(2)
630 PRINT
640 PRINT TAB(5)"RADIUS ONE = ";R(1)
650 PRINT TAB(5)"RADIUS TWO = ";R(2)
660 PRINT
670 PRINT "DO YOU WISH TO CONTINUE (Y/N) ";
680 INPUT C$
690 IF C$ = "N" GOTO 720
695 IF C$ = "n" GOTO 720
700 IF C$ = "Y" GOTO 710
705 IF C$ <>"y" GOTO 670
710 GOTO 390
720 END
```

APPENDIX 2. BASIC program for calculating birefringence from optical compensator data

```
10 REM***********************BIREF.BAS************************
20 REM                 BEREK COMPENSATOR CALCULATOR            *
30 REM                 (FOR FIBRE BIREFRINGENCE)               *
40 REM             GLENN R. CARROLL AND COLIN SAUNDERS         *
50 REM            R.C.M.P. CENTRAL FORENSIC LABABORATORY       *
60 REM                    VER 1.2     91.JUL.02                *
70 REM********************************************************
80 CLS
90  PRINT TAB(25) "OPTICAL COMPENSATOR CALCULATOR"
100 PRINT TAB(25) "******************************"
110 PRINT
120 PRINT "DO YOU NEED INSTRUCTIONS (Y/N)";
130 INPUT A$
140 IF A$ =  "N" GOTO 420
145 IF A$ =  "n" GOTO 420
150 IF A$ =  "Y" GOTO 160
155 IF A$ <> "y" GOTO 120
160 CLS
170 PRINT TAB(25) "INSTRUCTIONS
180 PRINT
190 PRINT "THIS PROGRAM CALCULATES FIBRE BIREFRINGENCE USING DATA OBTAINED"
200 PRINT "WITH A 3-ORDER OLYMPUS OR 10-ORDER LEITZ BEREK COMPENSATOR."
210 PRINT
220 PRINT "COMPENSATOR CONSTANTS USED ARE FOR DAYLIGHT (550 nm) SOURCES.
230 PRINT
240 PRINT "YOU WILL BE PROMPTED INITIALLY TO ENTER YOUR SAMPLE IDENTIFICATION.
250 PRINT "THEN YOU WILL BE ASKED TO INDICATE WHICH COMPENSATOR YOU USED,
260 PRINT "THE SIGN OF ELONGATION OF YOUR SAMPLE, THE ANGLES OF TILT WHICH YOU
270 PRINT "HAVE MEASURED, AND FINALLY THE FIBRE DIAMETER.
280 PRINT
290 PRINT "RETARDATION AND BIREFRINGENCE WILL THEN BE CALCULATED, FOLLOWED
300 PRINT "BY A LIST OF POSSIBLE FIBRE TYPES.
310 PRINT
320 PRINT "BEFORE USING FOR THE FIRST TIME, YOU MUST ENTER THE CONSTANTS
330 PRINT "FOR YOUR PARTICULAR COMPENSATOR. THESE ARE DENOTED IN THE
340 PRINT "PROGRAM BY THE VARIABLE C.
350 PRINT
360 REM****BEGINNING OF MAIN LOOP
370 PRINT
380 PRINT "ARE YOU READY TO PROCEED (Y/N) ";
390 INPUT Z$
```

```
400 IF Z$ =  "N" THEN GOTO 1500
405 IF Z$ =  "n" THEN GOTO 1500
410 IF Z$ =  "Y" THEN GOTO 420
415 IF Z$ <> "y" THEN GOT TO 380
420 CLS
430 PRINT "ENTER SAMPLE IDENTIFICATION: ";
440 INPUT B$
450 PRINT
460 PRINT "WHICH COMPENSATOR WAS USED (L = 10-ORDER LEITZ
470 PRINT "                           0 =  3-ORDER OLYMPUS);
480 INPUT C$
490 IF C$ = "L" THEN C = 2.04 :GOTO 520
495 IF C$ = "l" THEN C = 2.04 :GOTO 520
500 IF C$ = "O" THEN C = .876 :GOTO 520
505 IF C$ = "o" THEN C = .876 :GOTO 520
510 GOTO 460
520 PRINT
530 PRINT "ENTER SIGN OF ELONGATION (+,-,or 0): ";
540 INPUT E$
550 PRINT
560 PRINT "ENTER TILT ANGLE #1 (not required if ISOTROPIC): ";
570 INPUT T1
580 PRINT
590 PRINT "ENTER TILT ANGLE #2 (not required if ISOTROPIC): ";
600 INPUT T2
610 PRINT
620 PRINT "ENTER FIBRE DIAMETER IN µm: ":
630 INPUT D
640 PRINT
650 IF C$ = "L" THEN LET I = (T1 + T2)/2
655 IF C$ = "l" THEN LET I = (T1 + T2)/2
660 IF C$ = "O" THEN LET I = (T1 - T2)/2
665 IF C$ = "o" THEN LET I = (T1 - T2)/2
670 LET S = (SIN(I*3.1416/180))^2
680 LET F1 = S*(1+(.204*S))
690 LET F2 = 10000 * F1
700 LET R = C * F2
710 LET B = R/(D * 1000)
720 CLS
730 PRINT "OPTICAL COMPENSATOR RESULTS
740 PRINT "****************************"
750 PRINT
760 PRINT "SAMPLE IDENTIFICATION:";B$
770 PRINT
780 PRINT "TILT ANGLE #1:";T1
```

```
790 PRINT "TILT ANGLE #2:";T2
800 PRINT
810 PRINT "FIBRE DIAMETER:";D;"µm"
820 PRINT
830 PRINT "RETARDATION: ";R;" nm"
840 PRINT
850 IF E$="0" THEN PRINT "BIREFRINGENCE: ";E$:GOTO 870
860 PRINT "BIREFRINGENCE: ";E$;:PRINT USING "#.####";B
870 PRINT
880 PRINT "POSSIBLE FIBRE TYPE(S)"
890 PRINT "**********************"
900 REM****FIBRES WHICH ARE ISOTROPIC
910 IF E$ ="0" AND B=0! THEN G$(4)="GLASS":PRINT G$(4):FLAG=1:GOTO 1160
920 IF E$ = "+" GOTO 1010
930 REM**** FIBRES WITH NEGATIVE SIGN OF ELONGATION
940 IF B>=8.000001E-03 AND B<=.01 THEN G$(1) = "SARAN":PRINT G$(1):FLAG=1
950 IF B>=0! AND B<=.005 THEN G$(2) = "ACRYLIC":PRINT G$(2):FLAG=1
960 IF B>=0! AND B<=.005 THEN G$(6) = "MODACRYLIC":PRINT G$(6):FLAG=1
970 IF B>=0! AND B<=.002 THEN G$(3)= "NOVOLOID":PRINT G$(3):FLAG=1
980 IF B>=0! AND B<=.001 THEN H$(9)= "NYTRIL":PRINT H$(9):FLAG=1
990 GOTO 1160
1000 REM****FIBRES WITH POSITIVE SIGN OF ELONGATION
1010 IF B>=0! AND B<=.001 THEN G$(5)="TRIACETATE":PRINT G$(5):FLAG=1
1020 IF B>=0! AND B<=.005 THEN G$(6)="MODACRYLIC":PRINT G$(6):FLAG=1
1030 IF B>=.001 AND B<=.005 THEN G$(7)="ACETATE":PRINT G$(7):FLAG=1
1040 IF B>= 0! AND B<=.005 THEN G$(8)="VINYON":PRINT G$(8):FLAG=1
1050 IF B>=.002 AND B<=.046 THEN G$(9)="VISCOSE":PRINT G$(9):FLAG=1
1060 IF B>=.021 AND B<=.05 THEN G$(10)="MODAL":PRINT G$(10):FLAG=1
1070 IF B>=.021 AND B<=.037 THEN H$(1)="CUPRO":PRINT H$(1):FLAG=1
1080 IF B>=.041 AND B<=.051 THEN H$(2)="POLYNOSIC":PRINT H$(2):FLAG=1
1090 IF B>=.021 AND B<=.052 THEN H$(3)="OLEFIN":PRINT H$(3):FLAG=1
1100 IF B>=.04 AND B<=.05 THEN H$(4)="FLUOROFIBRE":PRINT H$(4):FLAG=1
1110 IF B>=.036 AND B<=.061 THEN H$(5)="NYLON":PRINT H$(5):FLAG=1
1120 IF B>=.12  AND B<=.4  THEN H$(6)="ARAMID":PRINT H$(6):FLAG=1
1130 IF B>=.044 AND B<=.056 THEN H$(10)="POLYESTER:Partially Oriented PET":PRINT H$(10):FLAG=1
1140 IF B>=.096 AND B<=.104 THEN H$(7)="POLYESTER:PCDT":PRINT H$(7):FLAG=1
1150 IF B>=9.399999E-02 AND B<=.191 THEN H$(8)="POLYESTER:NON-PCDT":PRINT H$(8):FLAG=1
1160 IF FLAG <>1 THEN PRINT "NO MATCHES FOUND"
1170 PRINT
1180 PRINT "DO YOU WANT A PRINTED REPORT (Y/N)";
1190 INPUT K$
1200 IF K$ =  "N" GOTO 1440
1205 IF K$ =  "n" GOTO 1440
1210 IF K$ =  "Y" GOTO 1220
1215 IF K$ <> "y" GOTO 1180
```

```
1220 REM****PRINTED REPORT GENERATOR
1230 LPRINT
1240 LPRINT "OPTICAL COMPENSATOR RESULTS FOR SAMPLE ";B$
1250 LPRINT "          ******************************"
1260 LPRINT
1270 LPRINT "TILT ANGLE #1: ";T1
1280 LPRINT "TILT ANGLE #2: ";T2
1290 LPRINT
1300 LPRINT "FIBRE DIAMETER: ";D;"µm"
1310 LPRINT
1320 LPRINT "RETARDATION: ";R;"nm"
1330 IF E$="0" THEN LPRINT "BIREFRINGENCE: ";E$:GOTO 1360
1340 LPRINT "BIREFRINGENCE: ";E$;
1350 LPRINT USING "#.####";B
1360 LPRINT
1370 LPRINT "POSSIBLE FIBRE TYPE(S)"
1380 LPRINT "************************"
1390 FOR I= 1 TO 10
1400 LPRINT G$(I);" ";H$(I);" ";
1410 NEXT I
1420 IF FLAG <>1 THEN LPRINT "NO MATCHES FOUND"
1430 LPRINT
1440 CLEAR
1450 PRINT "DO YOU WISH TO CONTINUE (Y/N)";
1460 INPUT J$
1470 IF J$ =  "N" GOTO 1500
1475 IF J$ =  "n" GOTO 1500
1480 IF J$ =  "Y" GOTO 1490
1485 IF J$ <> "y" GOTO 1450
1490 GOTO 420
1500 END
```

BIBLIOGRAPHY

AATCC Technical Manual (1975) Vol. 51. American Association of Textile Chemists and Colorists, Research Triangle Park, USA.

Bartick, E. G. (1987) Considerations for Fiber Sampling with Infrared Microspectroscopy. *The Design, Sample Handling, and Applications of Infrared Microscopes. ASTM STP 949*, Roush, P. B. (ed.), American Society for Testing and Materials, Philadelphia, USA, 64–73.

Bartick, E. G., Tungol, M. W., Carroll, G. R., Carnahan, E. J., Sprouse, J. F. (1990) A Combined Infrared Spectroscopic and Text Data Base for Forensic Fiber Identification. Presented at the 12th Meeting of the International Association of Forensic Sciences, Adelaide, Australia.

Carroll, G. R. (1985) The Use of Optical Compensators in Forensic Fibre Examination.

In: *Proceedings of the 10th Meeting of the International Association of Forensic Sciences, Oxford.* AC Associates, Solihull, UK.

Carroll, G. R., Lalonde, W. C., Gaudette, B. D., Hawley, S. L. and Hubert, R. S. (1989) A computerized database for forensic textile fibres. *Can Soc For Sci J* **21**, 1–10.

Catling, D. M. and Grayon, J. (1982) *Identification of Vegetable Fibres.* Chapman & Hall.

Cook, R. and Norton, D. (1982) An evaluation of mounting media for use in forensic textile examination. *J For Sci Soc* **22**, 57–63.

Crocker, E. J. (1986) Micrometric Determination of Fibre Cross-Section Modification (Aspect) Ratios. Presented at the 33rd Annual Meeting of the Canadian Society of Forensic Science, Niagara Falls, Canada.

Deadman, H. (1984a) Fiber evidence and the Wayne Williams trial (Part I). *FBI Law Enf Bull* **53**, 13–20.

Deadman, H. (1984b) Fiber evidence and the Wayne Williams trial (conclusion). *FBI Law Enf Bull* **53**, 10–19.

DeForest, P. R. (1982) Foundations of Forensic Microscopy. In: Saferstein, R. (ed.) *Forensic Science Handbook.* Prentice Hall, Englewood Cliffs, USA, 416–528.

Doyle, Sir A. C., 'The Adventure of Shoscombe Old Place' In: *The Complete Sherlock Holmes*, Vol. II, 1102, Doubleday & Co., Garden City, USA.

Gaudette, B. D. (1988) The Forensic Aspects of Textile Fiber Examination. In: Saferstein, R. (ed.) *Forensic Science Handbook*, Vol. II, Prentice Hall, Englewood Cliffs, USA, 209–272.

Grieve, M. C. (1983b) The role of fibers in forensic science examinations. *J For Sci* **28**, 877–887.

Grieve, M. C. (1983a) The use of melting point and refractive index determination to compare colourless polyester fibres. *For Sci Int* **22**, 31–48.

Grieve, M. C. (1990) Fibres and Their Examination in Forensic Science. In: Maehly, A. and Williams, R. L. (eds.) *Forensic Science Progress*, Vol. 4, Springer-Verlag, Berlin, Ger. 41–125.

Grieve, M. C. and Cabiness, L. R. (1985) The recognition and identification of modified acrylic fibres, *For Sci Int* **29**, 129–146.

Hartshorne, N. H. and Stuart, A. (1970) *Crystals and the Polarising Microscope.* 4th ed., Pitman Press, Bath, UK, 574.

Hearle, J. W. S., Lomas, B., Cooke, W. D., Duerden, I. J. (1989) *Fibre Failure and Wear of Materials.* Ellis Horwood, Chichester, UK.

Heyn, A. N. J. (1954), *Fiber Microscopy.* Interscience Publishers Inc., New York, USA, 341.

Identification of Textile Materials (1985), 7th ed. The Textile Institute, Manchester, UK.

Isenburg, I. (1967) *Pulp & Paper Microscopy.* Institute of Paper Chemistry, Appleton, USA, 222.

Laing, D. K., Cook, R., Hartshorne, A. W., Robinson, G. (1987) A fibre data collection for forensic scientists–collection and examination methods. *J For Sci* **32**, 364–369.

Macrae, R., Dudley, R. J. and Smalldon, K. W. (1979) The characterization of dyestuffs on wool fibers with special reference to microspectrophotometry. *J For Sci* **24**, 117–129.

Matsibora, N. P., Rakviashvili, R. N., and Nosov, M. P. (1990) Study of failure of tricotage fabrics made from various man-made yarns by the scanning electron microscope method, *Khimicheskie Volokna* **6**, 24–25.

Mauersberger, H. R. (ed.) (1954) Matthews' *Textile Fibers*, 6th ed. Wiley, New York, USA.

McCrone, W. C. (1974) Asbestos, *American Laboratory* **6**, 13–18.

McCrone, W. C. (1978) Identification of asbestos by polarized light microscopy. *National Bureau of Standards Special Publication No. 506*. National Bureau of Standards, Washington, USA, 235–248.

McCrone, W. C. (1979) Evaluation of asbestos in insulation. *American Laboratory* **11**, 19–31.

McCrone, W. C. (1985) Routine detection and identification of asbestos. *Microscope* **33**, 273–284.

Palenik, S. and Fitzsimons, C. (1990) Fiber Cross-Sections: Part I, *Microscope,* **38**, 187–195.

Petraco, N. (1985) The occurrence of trace evidence in one examiner's casework. *J For Sci* **30**, 485–493.

Scott, H. (1985) The persistance of fibres transferred during contact of automobile carpets and clothing fabrics. *Can Soc For Sci J* **18**, 185–199.

Strelis, I. and Kennedy, R. W. (1967) *Identification of North American Commercial Pulpwoods and Pulp Fibres.* University of Toronto Press, Toronto, Canada, 9.

Tungol, M. W., Montaser, A., and Bartick, E. G. (1989) FT-IR Microscopy for Forensic Fiber Analysis: The Results of Case Studies, *Proceedings of the International Society for Optical Engineering*, Vol. 1145 *Fourier Transform Spectroscopy*, 308–309.

Tungol, M. W., Bartick, E. G. and Montaser, A. (1990) The development of a spectral data base for the identification of fibers by infrared microscopy, *Applied spectroscopy* **44**, 543–549.

TAPPI Test Methods 1991 (1990) TAPPI Press, Atlanta, USA.

Wortmann, F. J. and Ams, W. (1987) Identifying Animal Fibers by their Cuticle Scale Height. Presented at the 11th Meeting of the International Association of Forensic Sciences, Vancouver, Canada.

4

Colour analysis by microspectrophotometry

James Dunlop, BSc MSc
Police Forensic Science Laboratory, Tayside Police Headquarters,
West Bell Street, Dundee, DD1 9JU, Scotland

4.1 INTRODUCTION

In the examination of textile fibres in a forensic context, the colour of the fibres in question is usually the most discriminating feature available. Microspectrophotometry, which uses a microscope and photometric measuring techniques, has proved to be an extremely valuable, if not indispensable, tool in comparing the colours of fibres. With its ability to provide colour information from very small areas of single fibres, microspectrophotometry can confirm suspected colour matches while removing the disadvantages of subjectivity which are associated with human perception of colour. Equally important is the ability of microspectrophotometry to immediately inform the user of spectral differences between two coloured fibres, in spite of a convincing visual match on the comparison microscope, a phenomenon known as *metamerism*. Commercially produced microspectrophotometers, of varying affordability, have been available for a number of years and have enhanced the forensic examination of fibres to a large extent. In addition, the use of microspectrophotometry, in combination with the appropriate software, has aided the establishment of an important fibre database, allowing the communication of precise colour information among users of the technique.

Microspectrophotometry is principally concerned with colour and its measurement, and a basic understanding of colour and how we perceive it is useful, if not essential, in acquiring an understanding of the technique as a whole.

4.2 COLOUR AND HOW WE PERCEIVE IT

If one were to present, say, four individuals with the same colour shade card and ask each of them to describe the colour as precisely as possible, it is unlikely that the answers would closely agree. Such an exercise would serve to illustrate that colour is in fact a subjective sensation. The sensation of colour, and our ability to distinguish colours, is dependent on the simultaneous presence of three factors, viz. the light source, the observed object, and the human factor, more specifically, the response of the eye.

4.2.1 Light

Daylight, the most natural form of light, is present in that region of the electromagnetic spectrum between 380 nanometres and 770 nanometres (1 nanometre = 10^{-9} metres). This is commonly called the visible region. The white light we known as daylight is, however, a mixture of numerous radiations whose wavelengths occupy the above range. When the complete mixture of radiations is distorted in some way, for example by placing an object in its path, some of the white light is 'lost' by absorption to the object. The remainder is reflected, and on striking the eye produces the sensation of colour.

The bands of radiation which go to make up natural white light, and which we see as different colours, are ordered in the visible region of the electromagnetic spectrum, with violet occurring at around 380 nm, followed by blue, green, yellow, orange, and red, ending at around 770 nm. One colour merges into the next in a virtually imperceptible way.

The light source, or illuminant, used to view an object can have a dramatic effect on the colour observed. The same object viewed with different illuminants will appear to change colour. An illuminant commonly in use in microspectrophotometry is Illuminant C, which represents average daylight with an overcast sky.

4.2.2 The observed object

The ability of different objects to absorb visible radiation selectively results in our being able to distinguish not only various colours but a vast array of associated shades. Two special situations can occur. Firstly, when the object absorbs almost all the light it appears black, and secondly, when almost all of the light is reflected, the object will appear white. It should be noted that total absorption or reflection of light is not possible in practice. Non-selectively reflecting objects range from black to white through a gradation of grey, which is known as the achromatic range.

The effect of selective absorption can be illustrated by considering two lengths of acrylic yarn, one of which has been coloured with a red dye, while the other has been dyed green. The red length of yarn appears red because the dye is absorbing much of the shorter wavelength region of the total light spectrum (the blue and green areas) and reflecting the longer wavelengths. Conversely, the green yarn is absorbing the longer wavelengths and reflecting those which are shorter.

4.2.3 The response of the eye

No single theory of colour vision has been able to explain all observations made by researchers in this field. One postulation, known as the *trichromatic theory*, is relatively simple. This theory proposes three different types of cone-shaped nerve endings, which are tightly packed together in a small pit in the centre of the retina, known as the fovea. (Other types of nerve endings, called rods, are present over most of the retina and play very little part in colour vision.) Some of the cones are 'red sensitive' and are activated by the longer wavelengths in the visible region, while others are 'green sensitive', responding to those wavelengths in the middle region of the visible spectrum. A third type of cone is 'blue sensitive' and is brought into play by short wavelength light.

The wavelength bands to which each of the cone types is sensitive have been estimated, with absorption peaks at 430, 540, and 570 nm for blue, green and red sensitive cones respectively. (Chamberlin & Chamberlin 1980).

When more than one type of cone responds, for example the green sensitive and red sensitive cones together, then, according to the trichromatic theory, the brain decodes this signal to produce a yellow sensation. On the simultaneous activation of all cones, the brain will interpret this message as white, although the spectrum is not complete.

Users of microspectrophotometry very often find it useful to translate an observed colour into mathematical form, assigning the colour a set of coordinates whereby it can be positioned quite precisely on a kind of colour map. For those concerned in the measurement of colour in this way, the trichromatic theory of colour vision fits the facts well and has subsequently been accepted on an international scale.

4.3 INSTRUMENTATION

More than thirty years ago the construction of a 'home made' microspectrophotometer for measuring the absorption spectrum of coloured fibres was reported (Amsler 1959). The instrument used a monochromator, or light filtering device, as the light source, a microscope, and a sensitive photometer. It was built because it was recognized that the conclusion of a 'match' between two coloured fibres based on microscopy alone is not really sufficient for the courts, and very often with short fibre lengths no characterization of the colourants by thin-layer chromatography is possible. Nowadays, most laboratories employing microspectrophotometry will have purchased a commercially available instrument.

Essentially, a microspectrophotometer system has as its component parts a microscope, a spectrophotometer, a control unit for the photometer, and a chart recorder or plotter (Fig. 4.1). Many users also incorporate as part of their system an analogue to digital converter to interface with a microcomputer. A widely used and relatively low cost system is the Nanospec10S made by the American company, Nanometrics Incorporated. A schematic diagram of this system is shown in Fig. 4.2.

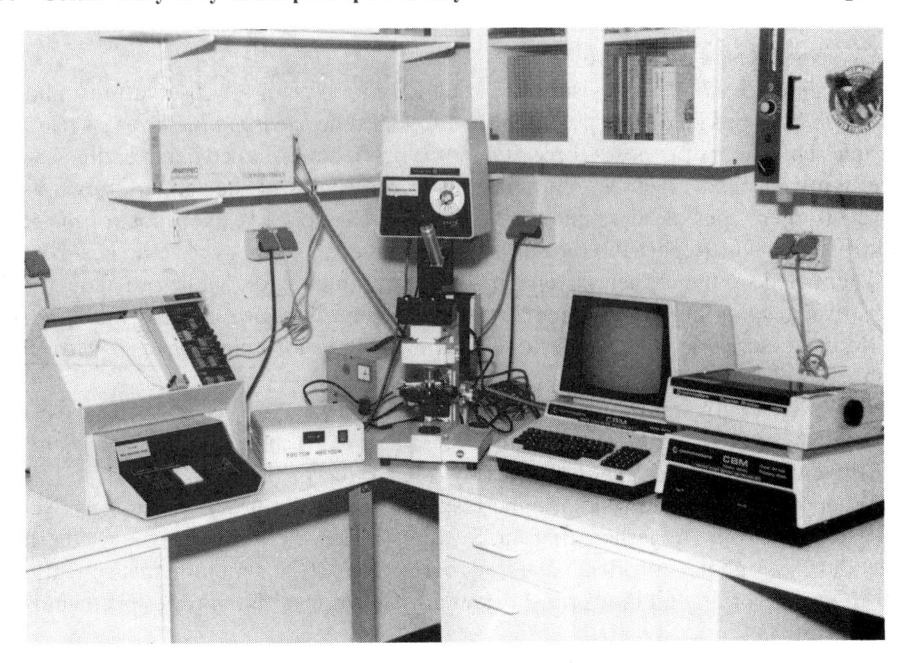

Fig. 4.1. The NANOSPEC 10S microspectrophotometer system

4.3.1 The microscope

Most optical microscopes equipped with a trinocular viewer and a camera port are suitable for use with the Nanometrics spectrophotometer. A voltage regulator must be used in conjunction with the light source for photometric measuring.

4.3.2 The spectrophotometer

The photometer head consists of a grating monochromator (in which the monochromatising effect is due to diffraction) and a photomultiplier detector. With glass optics, a wavelength range of 360–850 nm is available. Below 360 nm considerable extra expense is incurred, since quartz, fluorite, or reflecting objective lenses are required for the shorter wavelengths. Measurement in the visible range is accomplished with glass optics.

A single eyepiece viewer allows the fibre image on the entrance slit to be seen (Fig. 4.3). This slit is adjustable, and once the area to be measured has been selected, radiation from this area alone reaches the monochromator. The intermediate and exit slits are preset.

The detector is a photoelectric assembly using a gallium arsenide photocathode which has a significantly wider spectral response than other photocathodes. On irradiation of the photosensor an electric current is supplied to the amplifier. This converts and amplifies the input quantity into voltage which is supplied to the indicator. The indicator produces the measuring value which can be displayed on a scale (analogue) or in numbers (digital) and can be supplied to a recorder or plotter. A photo-intensity display meter is incorporated.

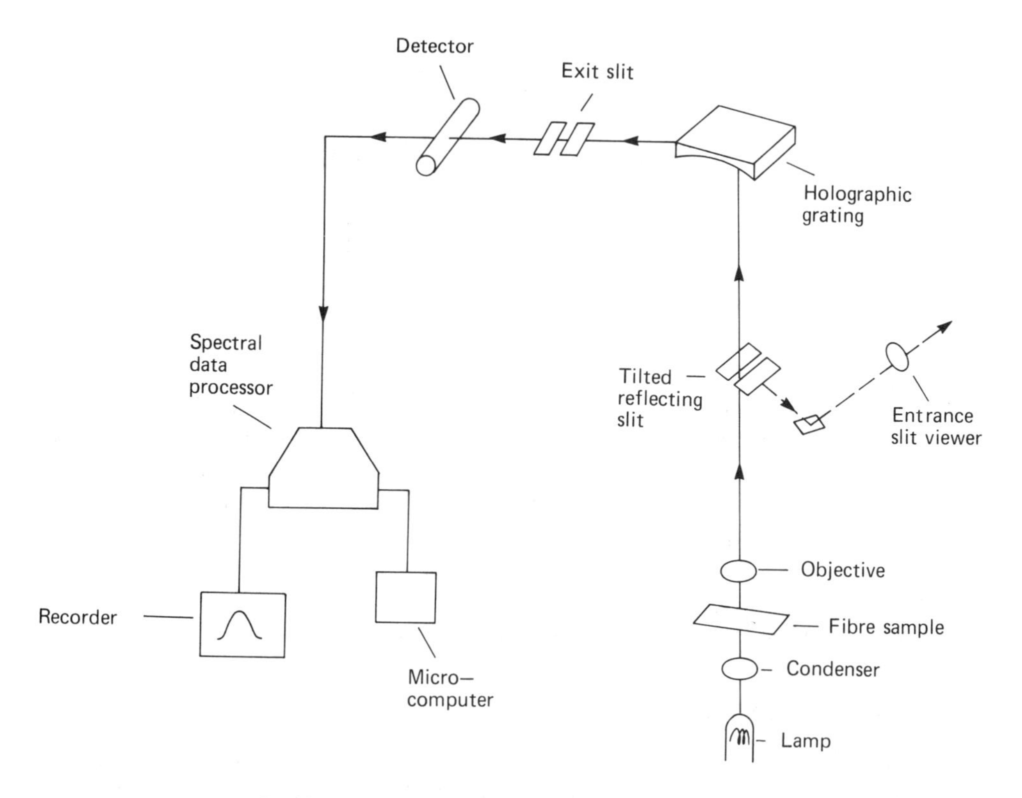

Fig. 4.2. Schematic diagram of NANOSPEC 10S system.

4.3.3 The photometer control unit

Nanometrics Inc. produce a unit which is in fact a powerful microcomputer, known as a spectral data processor. The unit is used to input the settings required for operation, for example, the wavelength range to be measured, full-scale settings, and scanning speed. A major feature of the spectral data processor is its memory storage capacity. Reference (background) measurements can be stored in the memory and are automatically subtracted when a coloured sample is being analysed to produce the pure spectrum of the sample. This effectively affords the instrument a double-beam capability, although, of course, it is a single-beam system. The background is normally run on an area near the fibre selected for the sample run.

4.3.4 The recorder/plotter

An ordinary paper chart recorder is sufficient for producing spectra for comparison. It is more convenient, however, to use an X-Y recorder. The incorporation of a printer/plotter can cut down the time taken to produce a correctly documented result as well as to simplify the fibre analysis and improve the clarity and presentation of the fibre spectra (Paterson & Cook 1983).

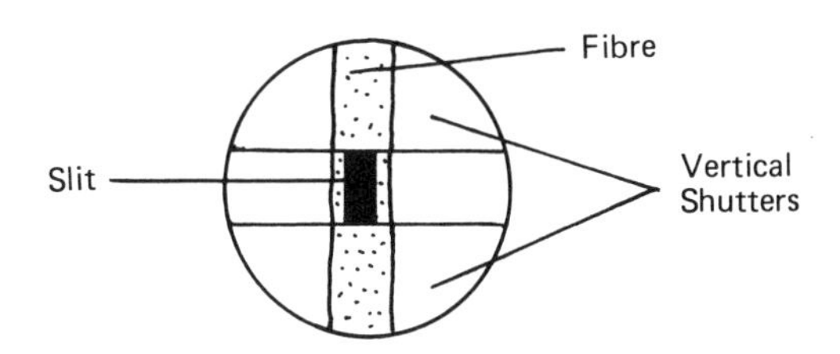

Fig. 4.3. The measuring slit (black rectangle) in position on the fibre sample.

The instrument described above, the Nanospec 10S, is but one of a number of commercially available instruments. Both Zeiss and Leica produce fully integrated systems, whilst Rofin, a newcomer to the marketplace, produce a spectrometer/data processing package which, like the Nanospec 10S, can be used with any suitable microscope. Although differing in detail from the Nanospec 10S, all the above systems must have the basic components of a spectrophotometer, a control unit (in more modern instruments a stand alone microcomputer) and a recorder with more sophisticated multi-colour laser printer/plotters now available.

4.4 EVALUATION OF THE TECHNIQUE

Forensic colour analysis of textile fibres was being carried out routinely in Europe, particularly in Switzerland, by the early 1970's. Halonbrenner & Meier (1973) report results achieved in casework using a Zeiss UMSP-1 instrument, which was designed to allow measurements to be made not only in the visible region but also in the ultraviolet range.

The discriminating power (DP) of the technique was not, however, assessed until several years later by Macrae *et al.* (1979) who used a Shimadzu MPS-50L spectrometer with a microscope attachment placed in the sample beam. Various coloured wools were examined. The results for blue wool fibres, for example, on comparing 153 pairs, showed the discriminating power of comparison microscopy (white light and UV) to be 0.68 (or 68% discriminated). The DP of microspectrophotometry, for the same pairs of blue wool fibres, was much higher at 0.99. In the same study it was shown that, in some instances, thin-layer chromatography could discriminate pairs of fibres which were spectrophotometric matches, and vice versa. These techniques are now generally accepted to be complementary.

This initial evaluation of the technique was followed by an evaluation of the Nanometrics instrument, the Nanospec 10S, and was carried out by Beattie *et al.* (1979) at the Home Office Central Research Establishment (now the Central Research and Support Establishment—HOCRSE). The wool samples used in the above study were again analysed, and very good spectral correlation was found. The DP of the Nanospec 10S was comparable to the initial findings. Having been shown to be a highly discriminating and non-destructive technique, able to produce spectra from

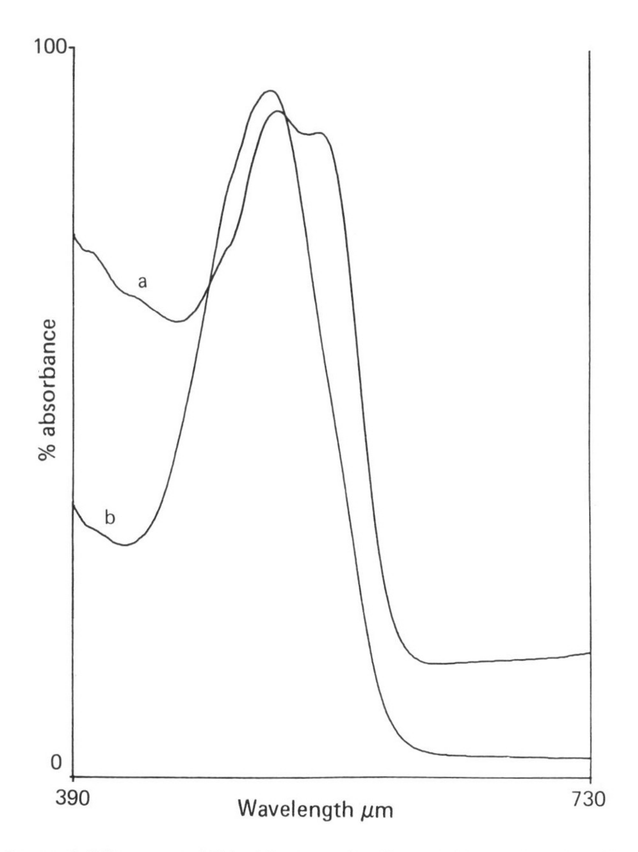

Fig. 4.4. Spectral differences exhibited by two visually matching red cotton fibres, a and b.

the shortest of fibres, use of the Nanometrics system quickly became widespread in the UK and elsewhere.

The discriminating power of microspectrophotometry, with respect to differentiating coloured cotton fibres, was illustrated by Grieve *et al.* (1988). With 46 samples of red cotton fibres, for example, every possible pairing was compared visually with the comparison microscope, and 320 pairs out of a possible 1035 could not be discriminated. After employing microspectrophotometry, however, only 10 pairs of red cotton fibres remained indistinguishable (Fig. 4.4).

4.5 SPECTRAL MEASUREMENT AND ANALYSIS

4.5.1 Instrument calibration

To ensure accuracy in spectral measurement, calibration is essential. The minimum essential calibration requires adjustment of the baseline to 100%*T*, that is, zero

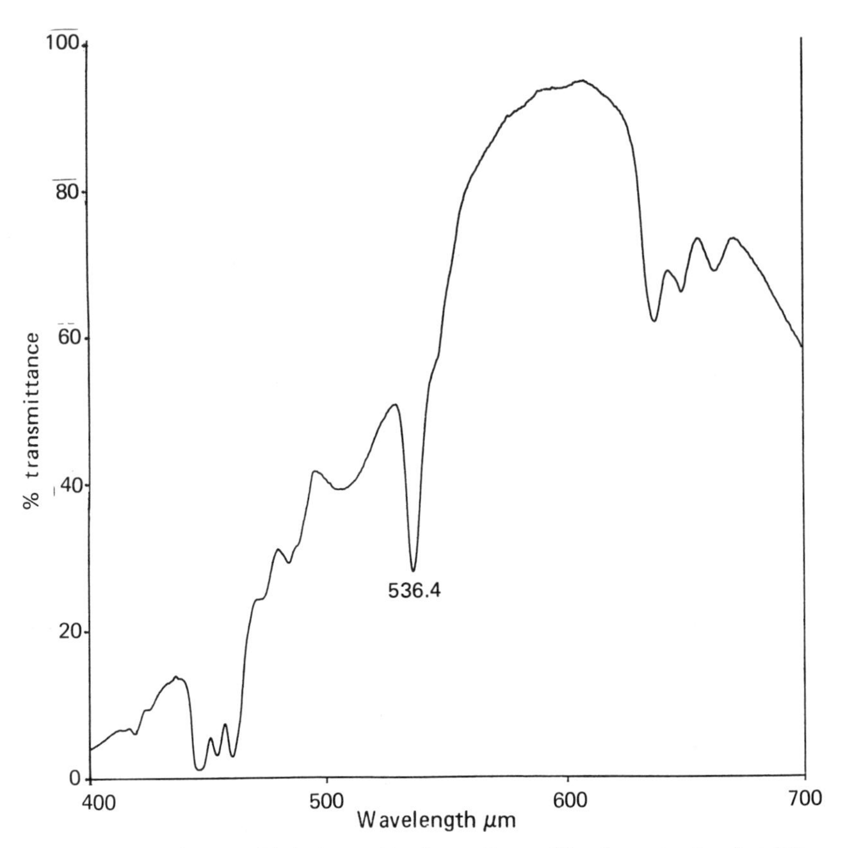

Fig. 4.5. Spectral curve of holmium oxide glass calibrant. The absorption band at 536 nm is very useful for wavelength calibration.

absorbance, and the use of glass filters, for example, holmium oxide glass or didymium glass. These materials have a number of strong and narrow absorption bands and are suitable for wavelength calibration (Fig. 4.5). Such calibrants should be safely stored since they change with time through exposure to light, wear, and surface film accumulation.

4.5.2 Spectral comparison

Normally there is a choice of recording absorbance or %transmittance curves ($\%T = 100T$). Absorbance spectra are usually the preferred choice when analysing the colour of fibre dyes. Very often fibres dyed with the same dye, or mixture of dyes, show different levels of dye uptake even along the length of a single fibre. The relationship between dye concentration and %transmittance is nonlinear, and problems may arise when attempting to compare spectral shapes recorded in transmittance. The equation which relates absorbance to transmittance is:

$$A = \log (100/T)$$

According to Beer's Law, absorbance (of a solution) is linearly related to its concentration or:

$$A = \alpha c$$

where A is absorbance, α is a constant, and c is the concentration. The linear nature of the absorbance/concentration relationship, that is to say, that absorbency is independent of concentration, makes it convenient to record absorbance spectra. This approach not only facilitates visual comparison of spectra in casework, but also has important implications if colour information is required in digital form (see section 4.6).

Although many spectral analyses require data to be taken from one point only on a curve, that is, the analytical wavelength, it is extremely important when comparing fibre colours by spectral means to consider the curve in its entirety. One of the easiest ways to do this is with the use of a light box. The spectrum of a 'suspect' fibre can be overlaid on spectra run as standards. Multiple spectra (a maximum of five is practical) from the standard sample can be run on a single chart sheet. Correlation between absorption maxima, 'shoulders', and the general shape of the curve must be established before a spectral match is concluded.

Often, especially with dyed natural fibres, intrasample spectral differences may be seen. These differences can take the form of absorption maxima shifts or peak ratio reversals (Grieve *et al.* 1990). When recording spectra from natural fibres (usually cotton or wool) a minimum of ten spectra from different fibres in the sample should be run. In such cases it may be decided that the spectrum from a 'suspect' fibre lies within the range exhibited by the replicate standard spectra. The above approach to spectral shape comparison is relatively crude. Macrae *et al.* (1979), in a study of coloured wool samples, explored more refined ways of comparing spectra. The approach taken was to convert spectra to unit areas. Normalized spectra were obtained by dividing the absorbance values by total band areas. Five so-called difference parameters were established.

(1) The sum of the absolute differences between corresponding data points at 10 nm intervals.
(2) The sum of the squares of the differences between corresponding data points.
(3) The sum of the absolute differences in gradients at corresponding data points.
(4) The maximum difference between the normalised spectra. (The longest vertical distance between two curves.)
(5) The maximum difference between the normalized cumulative distributions.

The wavelength of the absorption maximum (λ_{max}) was also used as a discriminating parameter. The results showed that, on attempting to discriminate between twelve visually similar red wools and eighteen blue wools, the most powerful discrimination was achieved by using a combination of sums of squares and λ_{max}.

More often than not the colorant under examination will be a mixture of two or more dyes as opposed to a single dye. The absorption spectrum from a coloured fibre, dyed with a multi-component mixture, is an overall definition of how the colour is made up from a number of components. When such multicomponent dyes are present there exists the potential for selective uptake of individual dyes, a source of spectral variation. Absorption spectra yield little information about the dye or dyes present on the fibre; rather they are an expression of colour only.

4.5.3 Wood's anomaly

At specific points on monochromator gratings, especially those which are holographically blazed, there occurs a polarization of diffracted energy. This causes enhancement of the signal at the wavelength of the anomaly. On the Nanospec10S gratings polarization occurs at 510 and 800 nm. The 510 nm polarization is known as *Wood's anomaly* (Fig. 4.6). The effect can be kept to a minimum if the degrees of polarization present in the background run and in the sample run are equal, and if the wavelength drive can be wholly synchronized from the background run to the sample run. With the Nanospec10S instrument a variation of up to 0.5 nm per run can be expected.

4.6 NUMERICAL COLOUR CODING

With time, analysts taking part in fibre examinations will accrue a large number of fibre standards. This represents an important source of data since such features as fibre type, presence or absence of delustrant, cross-sectional shape, and colour are recorded. The accumulation of this kind of information, and ready access to it, are necessary for evaluating the value of fibre evidence in any given case. Such a database may not be an exact reflection of the fibre population at large, but it should provide a reasonable approximation.

Of all parameters observed during fibre examinations in the laboratory, the colour of the fibre is the most difficult to record and communicate effectively. Visual descriptions of colour are very subjective, and simply recording absorption maxima alone will afford only limited colour information. A number of colour order systems have been developed. A system which assigns a code, (that is, precise mathematical values for any spectral colour encountered) would be of great value for storing and communicating colour information.

Paterson & Cook (1980) and Cook & Paterson (1981) classified colours initially by using dye extracts (solution spectrophotometry) and then applied the same classification system to coloured fibres by using microspectrophotometry. The method used was based on a widely accepted colour ordering system originally proposed by the Commission International de l'Eclairage of 1931 (CIE).

4.6.1 The CIE System

Colorimetry, the science of determining and specifying colours, is based on a number of principles put forward by Grassman (1853).

(1) The human eye can distinguish only three dimensions of colour.
(2) When one part of a two part mixture is changed, the colour of the mixture changes.
(3) Separate lights that appear of the same colour to the human eye will produce identical effects in mixtures, whatever their spectral composition.

As outlined in section 4.2, the colour of an object depends on three factors: the light source, the observed object, and the colour response of the eye. Before a

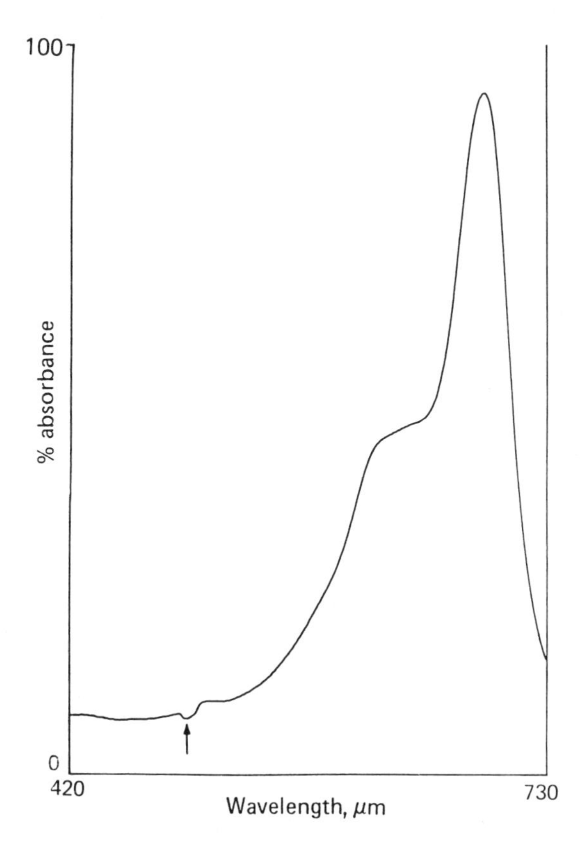

Fig. 4.6. The effect of Wood's Anomaly (arrowed) on the absorption spectrum from a blue–green cotton fibre.

mathematical description of colour can be arrived at, the light source and eye response factors have to be standardised. The CIE system required the development of such standards. Light sources, or illuminants, have been defined. The most commonly used illuminant in transmission work is Illuminant C which corresponds to simulated overcast sky daylight. Illuminant A is incandescent light, and Illuminant B is simulated noon sunlight.

The definition of a 'standard observer' was accomplished by getting a number of people to review numerous shades of colour. They were asked to reproduce or match a shade by means of mixing red, green, and blue lights. The human eye on stimulation by light reacts as it has three distinct colour receptors for red, blue, and green which have been given the symbols $\bar{x}$, $\bar{y}$ and $\bar{z}$ respectively. Enough information was collected in this way to define the values X, Y and Z, known as the tristimulus values, and x, y and z which are the chromaticity coordinates (Graham 1983). The chromaticity coordinates are calculated by normalising the tristimulus values, that is, by dividing each of them by their sums, as follows:

$$x = \frac{X}{X + Y + Z}$$

$$y = \frac{Y}{X + Y + Z}$$

and

$$z = \frac{Z}{X + Y + Z}$$

The sum of x, y and z is always 1, and subsequently only two coordinates are necessary for colour specification. The x and y values were chosen for this purpose. From these terms a chromaticity diagram can be constructed. All spectral colours between 380 and 770 nm may be located within the roughly triangular area of the diagram.

4.6.2 Complementary chromaticity coordinates (CCC)

The chromaticity coordinates, x and y, are given when the colour measurement is undertaken in %transmittance, and therefore they are dependent on dye concentration. This represents a drawback when a number of measurements are required from a standard fibre sample in which the dye concentration may vary. Operating in absorbance instead of transmittance produces the complementary chromaticity coordinates (CCC), x' and y'. Table 4.1 gives the CCC values computed for three different dyes. The CIE complementary chromaticity diagram is shown in Fig. 4.7.

Table 4.1. CCC values computed for 3 dyes. (illuminant C).

Dye	x'	y'
CIBACRON blue TRE	0.4321	0.4072
SOLOPHENYL navy blue RL	0.3690	0.3932
BASILEN-M red M 5-B	0.2321	0.4814

The CCC values for a number of coloured fibres, in a sample dyed with the same colorant, should now be effectively the same, irrespective of concentration, assuming that the colorant obeys Beer's Law. However, measuring the colour of dyed fibres, as opposed to solutions (for which the CIE system was devised), gives rise to complications. Such factors as selective uptake from multicomponent dyes and the inherent fibre colour can produce variations in the values of the coordinates (Laing *et al.* 1986). For single fibres, however, the coding method described above is extremely useful. Further investigation (Hartshorne & Laing, 1987) into the factors which may cause unexpected variations in CCC values showed that the presence or absence of delustrant and varying levels of delustrant can cause colour shifts between fibres dyed with the same colorant. Fibre shape was found to be of little significance.

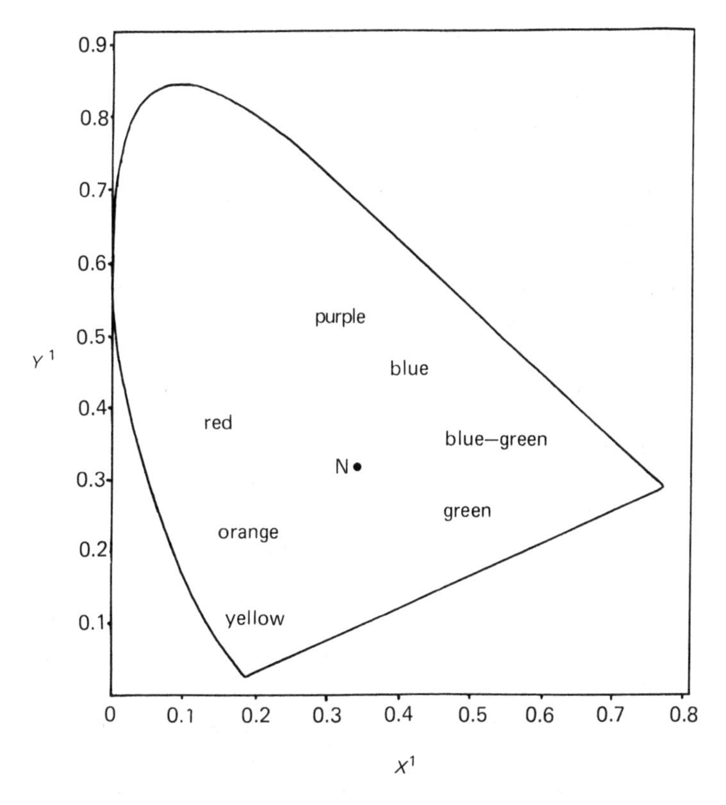

Fig. 4.7. The CIE Complementary Chromaticity Diagram (1931) showing the areas occupied by the main colours. N is the neutral point.

4.6.3 The CRSE fibres data base

To build up an extensive fibres database, forensic scientists in the UK took fibre samples from casework material over a period of several years. A standard questionnaire was filled in which on completion contained a substantial amount of characterising information about a fibre. This information included the fibre source, type and subtype, cross-sectional shape, delustrant status, and diameter. In addition to the questionnaire a sample of the relevant fibre was returned to HOCRSE. (Laing *et al.* 1987). Colour measurements were carried out on the samples by an outside contractor, using the Nanometrics system described in section 4.3 along with an interfaced microcomputer and suitable software. The computer program required the transcription of data on the questionnaire. Before measuring the fibre colour, the absorption spectrum of a purple glass calibrant was acquired, the resulting colour values were checked against standard values, and any correction factors required were calculated. All spectra measured subsequently were corrected against the standard values. Natural fibres require ten replicate measurements to take account of the variations in coordinate values expected from such samples. Synthetic fibres, however, give more reproducible results, and five determinations are considered sufficient. All the information was stored on floppy disks and eventually transferred to a Prime 550 computer. To date, the database contains information on approximately 18 000 fibres. These data are accessible to UK forensic scientists who can enter their own data, including the colour coordinates, with respect to a recovered 'suspect' fibre. They receive in return the number of fibres in the data base which match the characteristics of the 'suspect' fibre in question.

It was found that on plotting the replicate CCC values a roughly elliptical shape is formed, the so-called *error ellipse*. The size of the ellipses reflects the degree of variation within a sample; the larger the ellipse the greater the variation. A study on colour matching within the data base (Hartshorne & Laing, 1986) showed that the use of such ellipses required much computer space, and an alternative method using squares was found to be a simpler approach. The centre of the square is the point described by averaging five or ten CCC values, and the chosen size of the square is the average coordinates plus and minus 0.025. It was found that squares of this size tended to overestimate the number of colour matches for synthetic fibres.

Although information received in this way as to the number of fibres in the collection matching a 'suspect' fibre should be used with caution in a courtroom context, it does give the analyst a reasonable idea of the 'rarity' of the recovered fibre type.

4.7 SPECTRAL ANALYSIS USING UV RADIATION AND FLUORESCENCE

Increasing interest is being shown in extending the range of microspectrophotometry into the ultraviolet. Instruments such as the Zeiss UMSP 80 allows measurement from 240 nm. It is also capable of measuring fluorescence excitation spectra. The UMSP 80 employs a side monochromator, placed between the radiation source and the microscope. Fibres not discriminated in the visible region may well show differences in the UV range. Such differences are normally due to the presence or absence of optical brighteners or the application of different brightening agents. Undyed acrylic fibres produced by the same manufacturer have also been discriminated (Adolf 1989). Nanometrics Inc. have recently introduced a new microspectrophotometer, the Nanospec DUV (Deep Ultra Violet) with a spectral range of 190–900 nm.

Comparisons of fluorescent dyes and optical brighteners are commonly carried out on a modified comparison microscope. An assessment of the value and the practicability of measuring the emission (fluorescence) spectra of such additives has shown promising results (Hartshorne & Laing 1988).

The option of objectively measuring the emission characteristics of a coloured fibre as well as its absorption in the visible range will provide the analyst with a further valuable technique for the comparison of fibres.

BIBLIOGRAPHY

Adolf, P. (1989) Personal communication.

Amsler, H. (1959) Die Mikro-Spektralphotometrie, ein wichtiges Hilfsmittel fuer den Farbvergleich kleinster corpora delicti. *Archiv fuer Kriminologie* **124**, 85–94.

Beattie, I. B., Dudley, R. J., Laing, D. K. and Smalldon, K. W. (1979) An Evaluation of the Nanometrics Incorporated Microspectrophotometer the 'Nanospec 10S'. Personal communication.

Chamberlin, G. J. and Chamberlin, D. G. (1980) *Colour–its measurement, computation and application.* Heyden, London, 4–8.

Cook, R. and Paterson, M. D. (1981) Colour Evaluation for Computer Storage. *Proceedings of the Australian Seventh International Symposium on the Forensic Sciences.*

Graham, L. A. (1983) Color Order Systems, Color Specification, and Universal Color Language. In: Celikiz, G. and Kuehni, R. G. (eds.) *Color technology in the textile industry.* American Association of Textile Chemists and Colorists, Research Triangle Park, North Carolina, 135–152.

Grassman, H. (1853) Zur Theorie der Farbenmischung. *Annalen der Physik und Chemie,* **89**, 69–84.

Grieve, M. C., Dunlop, J. and Haddock, P. (1988) An assessment of the value of blue, red and black cotton fibres as target fibers in forensic science investigations. *J For Sci* **33**, 1332–1334.

Grieve, M. C., Dunlop, J. and Haddock, P. (1990) An investigation of known blue, red, and black dyes used in the coloration of cotton fibers. *J For Sci* **35**, 301–315.

Halonbrenner, R. and Meier, J. (1973) Mikrospektralphotometrische Untersuchungen an Textilfasern. *Kriminalistik*, August 1973.

Hartshorne, A. W. and Laing, D. K. (1987) The definition of colour for single textile fibres by microspectrophotometry. *For Sci Int* **34**, 107–129.

Hartshorne, A. W. and Laing, D. K. (1988) Color matching within a fiber data collection. *J For Sci* **33**, 1345–1354.

Hartshorne, A. W. and Laing, D. K. (1988) The microspectrofluorimetry of fluorescent dyes and brighteners on single textile fibres. Personal communication.

Laing, D. K., Hartshorne, A. W. and Harwood, R. J. (1986) Colour measurements on single textile fibres. *For Sci Int* **30**, 65–77.

Laing, D. K., Hartshorne, A. W., Cook, R. and Robinson, G. (1987) A fiber data collection for forensic scientists—collection and examination methods. *J For Sci* **32**, 364–369.

Macrae, R., Dudley, R. J. and Smalldon, K. W. (1979) The characterization of dyestuffs on wool fibers with special reference to microspectrophotometry. *J For Sci* **24**, 117–129.

Paterson, M. D. and Cook, R. (1980) The production of colour coordinates from microgram quantities of textile fibres. Part 1 *For Sci Int* **15**, 249–258.

Paterson, M. D. and Cook, R. (1983) The use of a printer plotter in the microspectrophotometry of textile fibres. Personal communication.

5

Chromatographic analysis for fibre dyes

Peter C. White, PhD, CChem, MRSC
Forensic Science Unit, University of Strathclyde,
204 George Street, Glasgow, G1 1XW, Scotland
and
Ian R. Tebbett, BPharm, PhD, MIBiol, CChem, MRSC, MRPharmS
Department of Pharmacodynamics,
University of Illinois at Chicago,
Box 6998, Chicago, IL 60680,
USA

5.1 INTRODUCTION

Fibres used for the manufacture of clothing and furnishings may be synthetic or nonsynthetic, or even mixtures of both types. In their natural state, these fibres are opaque, but are usually coloured for commercial purposes, by the application of a dye or a mixture of dyes. The colorants used to dye fibres can provide additional evidence in forensic casework for the comparison of two fibres from different sources. This chapter therefore describes the examination of fibre dyes in the forensic context.

5.2 BASIC THEORY OF COLOUR AND COLOURANTS

A dye is a compound that is fixed on a substance in a more or less permanent state and which evokes the visual sensation of that colour. The dye absorbs light rays on a selective basis, causing the fabric to reflect those rays that are not absorbed. Since normal daylight consists of light rays in the visible spectrum, that is, in a wavelength range 380–750 nm, the colour observed by the eye will depend on the absorbance/reflectance properties of the fibre and dye. If the material completely reflects all of the light impinging on it, it appears white. If it completely absorbs all the light, it appears black. If the material absorbs a constant fraction of the light across the spectrum, it appears grey. White, black, and grey are called *achromatic* colours. It is only when the material absorbs light of only certain wavelengths that it appears coloured, and the colours that we observe are referred to as *chromatic* colours. If an object absorbs light of only a single wavelength, the eye sees a mixture of all the other colours. These are known as the *complementary* colours. The relationship between wavelength, the colour absorbed by an object, and the colour seen by the eye is shown in Table 5.1.

Table 5.1. Relationship between the wavelength of colour absorbed by a material and the colour seen by the eye

Wavelength range Absorbed (nm)	Colour absorbed	Colour seen by eye
380–430	Violet	Yellow-green
430–480	Blue	Yellow
480–490	Green-blue	Orange
490–500	Blue-green	Red
500–560	Green	Purple
560–580	Yellow-green	Violet
580–590	Yellow	Blue
590–610	Orange	Green-blue
610–750	Red	Blue-green

Note: that solids which appear green are actually characterized by absorption at two wavelengths typically, 380–430 nm (blue) and 610–750 nm (yellow).

Synthetic colourants used for the dyeing of fibres are organic, and the colour of the dye is related to its chemical structure. In Table 5.2 the chemical structures of five organic compounds are depicted. Of these, only the last two are capable of behaving as dyes. All of these molecules absorb radiation, but in the case of compounds 1–3, radiation is absorbed in the ultraviolet region of the spectrum, that is, below 380 nm. The compounds therefore appear colourless. However, with compounds 4 and 5, radiation is absorbed not only in the ultraviolet region but also in the visible region of the electromagnetic spectrum. These compounds therefore give a UV/visible absorbance spectrum with maximum absorbance in the visible region (λmax) as indicated, and appear coloured. For any compound to absorb visible radiation it must contain at least one *chromophore*, and most dyes contain an *auxochrome*. A chromophore is that part of the molecule that absorbs light and is characterized by areas of high electron density. These are associated with unsaturated moieties such as: $—CH_2{=}CH_2—$, $—N{=}N—$ and benzene rings.

Table 5.2. The colour of organic molecules

1	Ethanol	COLOURLESS	CH_3CH_2OH
2	Benzene sulphonic acid	COLOURLESS	
3	β-naphthol	COLOURLESS	
4	Acid orange 7	ORANGE λ max = 480 nm	
5	Acid red 88	RED λ max = 512 nm	

$ = $SO_3^+H^-$

If several chromophores are in close proximity (separated by no more than one single bond), then the degree of conjugation is increased and absorption of light occurs at a longer wavelength. With examples 4 and 5, there is sufficient conjugation to ensure that absorption occurs in the visible region of the spectrum and the compound appears coloured.

An auxochrome is a functional group which when added to a molecule changes the intensity of the colour but does not greatly influence the maximum absorbance value.

Auxochromic groups include —COOH, —SO_3H, —NH_2, and —$N(CH_3)_2$, which play a major role in fibre dye chemistry. Apart from changing the depth of a colour, they can also greatly influence the solubility of the dye and hence its ability to bind to a fibre.

5.3 CLASSIFICATION OF FIBRE DYES

5.3.1 Introduction

Dyes can be classified in several ways on the basis of their hue, chemical class, method of application, and type of fibres to which they are applied. The method of application of the dye and the type of fibre to which the dye is applied, is largely dependent on the relative solubility of the dye in water. The chemical class of a dye can be determined by reference to the *Colour Index* (Society of Dyers & Colourists 1976). The index was first published in 1956 by a joint committee of the Society of Dyers and Colorists and the American Association of Textile Chemists and Colorists. It lists dyestuffs under several classifications, and is an invaluable source of information for analysts attempting to identify dyes or to develop new techniques. The colour index now consists of five volumes. An example of how a dye is classified is given below.

STRUCTURE

OH

HO_3S—(naphthalene)—N=N—(naphthalene)

SO_3Na

COLOUR INDEX NUMBER—CI 14720

All dyes have been given a colour index number (CI). This is always a five digit number and is related to its chemical structure. The dye in this particular example is a monoazo dye. Some acidic dyes have been produced as different salts, and their colour index numbers would be represented as XXXXX: 1, XXXXX: 2, etc.

GENERIC NAME—Acid red 14

This name describes the class of dye and provides an indication of its use. Some dyes can have more than one generic name, indicating that it has other applications. In this example, the other generic names for the dye are Food Red 3 and Mordant Blue 79.

CLASSICAL NAME—Carmosine

This is the trivial name given to a dye, often by the manufacturer. Hence with a particular dye which has been produced by several companies several classical names may exist.

In addition to providing these various classifications, the colour index is also an excellent source for obtaining information about manufacturers, dye solubilities, and their method of preparation. The manufacture of dyes is complex, generally requiring a number of reactions and intermediates to produce the desired shade. This may involve the use of a combination of several dyes. There are hundreds of dyes at present available to the textile industry, and identification of the dye is possible only after considerable effort and by reference to an extensive collection of standards. For this reason, forensic laboratories generally do not attempt to identify the dye; instead, the similarity or dissimilarity between the known and unknown fibre dye components is assessed, based on their physical and chemical properties. In this respect, a great deal of emphasis is placed on methods which separate dyes into classes based on their solubility and/or their chromatographic properties.

In this regard, the fibre dye classification scheme is of great importance. Apart from being able to classify dyes according to the method used to dye fibres, this scheme can also be used to identify the different fibre types that can be dyed with a particular class of dye. This method of classifying dyes is used extensively by the fibre dye industry and by analysts carrying out colorant analysis of fibres in forensic science laboratories. Table 5.3 presents a synopsis of the named dye classes, the fibre types that can be dyed with each class of dye and the methods used to dye fibres.

Table 5.3. Common fibre/dye combinations

Dye class	Fibre	Method of dyeing
ACID	Wool	Acidic conditions
	Silk	
	Nylon	
AZOIC	Viscose,	Diazonium coupling
	Cotton	
BASIC	Acrylic	Acidic conditions
	Modified acrylic	
DIRECT	Cotton	Electrolytic
	Viscose	
DISPERSE	Polyester	Aqueous dispersion
	(Nylon)	
	(Acrylic)	
MORDANT	Wool	Inorganic mordant, for
	(Nylon)	example Cr_3^+
	(Mod acrylic)	
PREMETALLIZED	Wool	Metal complex
	Cotton	
	Nylon	
REACTIVE	Cotton	Chemical reaction—covalent
	Wool	bonding
	Nylon	
SULPHUR	Wool	Reduction of disulphide dyes
	Cotton	followed by oxidation
VAT	Wool	Aqueous dispersion
	Cotton	

5.3.2 Chemical classification

Dyes can be classified either by their chemical structure or by the dyeing methods and areas of application. Fibre dyes can be roughly divided into two groups: *water soluble* and *water insoluble*. The water soluble dyes can be further classified as being *cationic* or *anionic*. The anionic water soluble dyes have at least one salt forming group, usually a sulphonic acid (SO_3H) or a carboxylic acid (COOH).

Classes of dye within this group are: *acid, acid–mordant, direct, reactive, solubilized vat, condensed sulphur* and *leuco sulphur* dyes. The only group of dyes which are water soluble and cationic are the basic dyes. The dyeing process for acid dyes involves a chemical reaction between the fibre molecules and the acid dye in an acid bath. The fibres attract the acid dye, and an associative bond is established. Obviously, acid dyes can be used only on fabrics resistant to acid damage, such as acrylic, nylon, spandex, polypropylene, and proteinaceous fibres such as wool and silk. Most acid dyes contain one or more SO_3Na or SO_3H groups. A few acid dyes contain carboxylic acids. Some other dyes can acquire acid dye properties by the introduction of sulphonic groups.

Acid mordant dyes are acid dyes whose fixation on the fibre is improved by metallic mordants. Two metals used as mordants are chromium and cobalt. These metals render the dye insoluble on the fibre polymer. The two types of acid mordant dyes are 1 : 1 metal–dye complexes, which always contain sulphonic groups, and 1 : 2 metal dye complexes which may or may not contain sulphonic groups.

Direct dyes are the largest and most commercially significant group of dyestuffs. They are so called because they are added directly to the fabric, unlike mordant and vat dyes, which require an auxiliary agent to precipitate the dye on the fabric. The dye is dissolved in water together with sodium chloride or sodium sulphate, which is added to control the absorption rate of the dye by the fibre. The amount of dye absorbed depends upon the size of the dye molecules and the size of the pore opening in the fibre. Direct dyes exhibit poor colour steadfastness, which is why finishing chemicals must be added. Most direct dyes are azo dyes having one or more azo groups (—N=N—) which form bridges between organic residues.

Reactive dyes are dyes which react chemically with the fibre molecule by forming covalent bonds with the functional groups of fibres in the dyeing process. Reactive dyes are structurally similar to acid dyes with the important difference that they contain a nucleophilic group which results in greater stability. They are susceptible to damage from chlorine and are mainly used for cellulose and proteinaceous fibres, but also for polyamides such as nylon.

Solubilized vat dyes are derived from parent vat dyes and are reduced to the water soluble form by sulphation of the leuco compound. The aqueous solution may be oxidized back to the parent dye by treatment with sodium nitrite and acid. They are used for pale shades on cotton and rayon.

Cationic or **basic dyes** are ammonium, sulphonium or oxonium salts to which acetic acid is added to improve the solubility. At present, basic dyes are mainly used on acrylic fibres. Basic dye properties can be accomplished in other dye classes by the introduction of amino or dialkylamino groups followed by conversion into an ammonium salt.

Water insoluble dyes are categorized as *disperse*, *vat*, *pigment*, and *sulphur dyes*. **Solvent dyes**, so called because they are applied to the yarn in an organic solvent, also fall into this category. They are, however, rarely used today. Although pigments are dyestuffs, they are not true dyes and are included in this category because of their insolubility in water.

Disperse dyes are multiphase dispersions of pigments, resins, and water. The dye has a low solubility in water and readily passes out of solution into the fibre. Disperse dyes are applied to cellulose acetate and polyester fibres as aqueous dispersions. Within the class of disperse dyes most are azo dyes and a few are anthraquinones. Rather than being dissolved in the dye bath they are dispersed and attach themselves to the fibre. They are commonly used in combination with vat or direct dyes.

Vat dyes are so called because they were originally applied to yarns or fabrics in large vats. Although vat dyes are insoluble in water they can be transformed by reduction (vatting) into a compound (leuco form) soluble in aqueous alkali, and dyed in this form. These leuco compounds fix onto the fibre by oxidation in the air without the use of a mordant. Air oxidation will reform the original dye on the fibres. The leuco form of vat dyes can be reduced to a water soluble form by conversion to the leuco esters. They exhibit differences in colourfastness because of differences in chemical structure. Their insolubility in water makes them colourfast, and they are used for all cellulose fibres.

Pigments are not considered to be dyes. Pigment colours have no affinity for fibres, therefore they are attached to fibres or fabrics by means of some type of adhesive or bonding agent. This creates a fabric whose colour is relatively permanent but whose durability is dependent on the binding agent. Pigments can be organic or inorganic such as delustrants, and are soluble only in a limited number of solvents, making their analysis difficult.

Sulphur dyes are compounds formed by treating aromatic amines, and phenols with sulphur. To apply these dyes to fabrics, the compounds must be reduced with sodium sulphide in an alkaline medium. Most sulphur dyes are navy blue or black and are used largely for the dyeing of cellulose fibres.

Metal complex dyes can either be soluble or insoluble in water, depending on the dye class to which they are attached. They contain a metal ion combined with a dye particle which contains free electron pairs (ligand). A metal such as aluminium, chromium, or nickel is incorporated into the fibre, rendering it dyeable with mordant dyes. Chelates are formed when the ligands occupy many sites on the metal ion. The metal complexes of azo compounds are also formed with direct and reactive dyes. These dyes are expensive and are available in only a limited number of shades.

Although there are several other dyestuffs commercially available, those described above are the ones most commonly seen in forensic casework. Because of the small sample size, little attempt has been made to apply classical textile fibre dye analysis to the identification of the chemical components.

5.3.3 Common fibre/dye combinations

The chemical differences of fibres results in some fibre/dye combinations being more favourable than others. These combinations are summarized in Table 5.3, for cotton, wool, nylon, polyester, and acrylic fibres.

Acid dyes are the most important group of dyes for wool and other proteinaceous fibres. The complex structure of wool can accept a wide range of dyestuffs, whereas the acid dye affinity for nylon fibres is primarily due to the terminal amino group of the nylon. The use of acid mordant dyes for wool is possible because of the basic side chains of wool which can combine with chromic acids. Polyester fibres lack the presence of ionizable sites and are therefore usually dyed with a disperse dye.

Polyester is hydrophobic in nature, and its tightly packed molecular structure, together with its lack of reactive groups, render it unreceptive to dye molecules. Certain disperse dyes can be applied, but under normal conditions, adsorption is slow and only pale shades are obtained. The disperse dyes are based on anthraquinone and diazo compounds.

The most important dye classes for acrylic fibres are basic dyes, whilst cotton fibres are usually dyed by either direct or reactive dyes. It is thought that the direct dye attaches to the cellulose molecule via hydrogen bonding with the hydroxyl group.

It is evident from these classification schemes that many dyes have been developed for the coloration of fibres because of the various fibre types and classes of dye produced. Therefore, apart from the problem of sample size, the analyst must also consider extraction methods since, as shown above, a variety of dyeing methods are used. Furthermore, with so many dyes having been produced it is important in casework that the analytical technique is capable of a high degree of discrimination between colourants.

5.4 EXTRACTION OF FIBRE DYES

Before any chromatographic comparison of fibre dyes can be achieved, the dye must first be extracted from the fibre. In the most commonly used method, the fibre is first placed, together with a small volume of extracting solvent, in a glass capillary tube that has been sealed at one end. The other end of the tube is then heat sealed and the tube heated at about 100°C. This process extracts the dye mixture from the fibre into the solvent. The solvent can then be removed with a fine capillary tube or syringe and used for either thin layer (TLC) or high performance liquid chromatographic (HPLC) analysis. An alternative to the use of capillary tubes is simply to place the fibre in a capped tapered glass vial. Again, solvent is added and the sample is heated in a heating block to extract the dye.

For TLC or HPLC analyses to be successful, the extraction of dyes from the fibre must be as efficient as possible. In addition, the choice of a suitable TLC solvent or HPLC mobile phase is best achieved once the nature of the dye is known. Dye extraction schemes have therefore been developed to (a) identify the nature of the dye, and (b) to optimize the extraction of that dye from the particular fibre. The choice of solvent used to extract the dye from the fibre is largely dependent on the type of dye present, since different dye types will have differing solvent solubilities. Identification of the type of fibre (nylon, wool, cotton, etc.) can give an indication as to the nature of the dye present and therefore the optimum solvent for its extraction. This can often be achieved by microscopic examination or by infrared spectrophotometry. In addition, dye extraction schemes have been developed which allow a fibre to be sequentially extracted with different solvents in order to determine the most efficient extraction procedure and at the same time to give information regarding

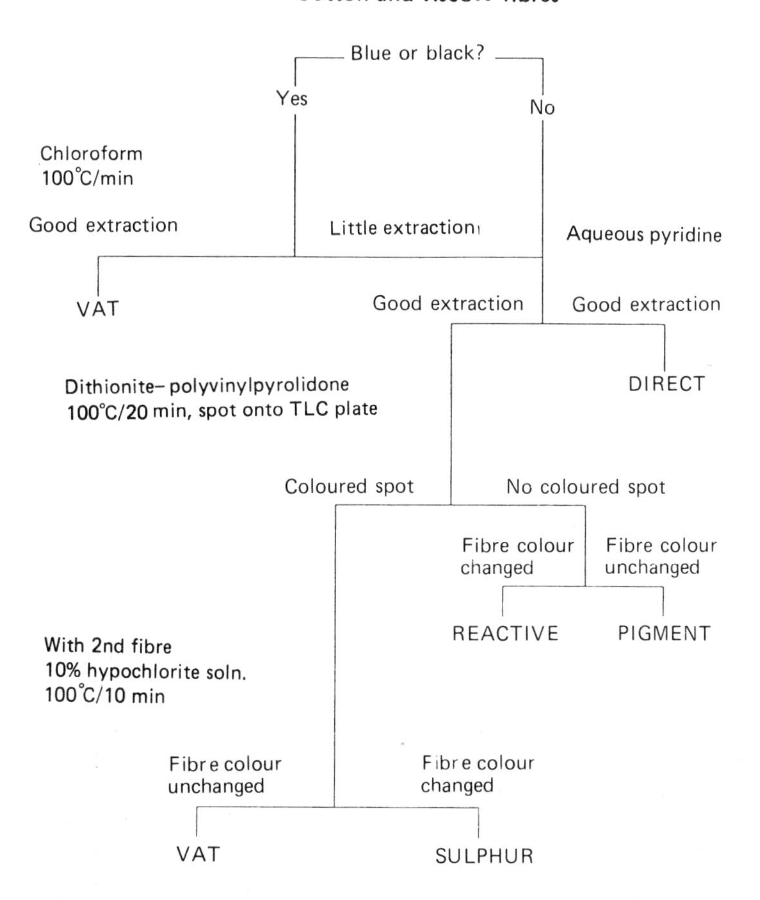

Fig. 5.1. Classification and extraction of dyestuffs from different fibre types. (Courtesy of the Metropolitan Police Forensic Science Laboratory, London, England.)

the nature of the dye present.

A number of studies have been carried out on the extraction and classification of dyestuffs from wool, nylon, polyacrylonitrile and polyester, polypropylene, and cellulose acetate fibres. Macrae and Smalldon's work with wool fibres (Macrae & Smalldon 1979) developed a three stage extraction procedure for the major dye classes found on wool. The fibres were initially extracted with pyridine/water (4:3 v/v) either at room temperature or at 90°C. If this solvent resulted in an efficient extraction of the dye, the presence of an acid dye was indicated. If the extraction failed, the fibres were pretreated with 2% oxalic acid for 20 minutes, dried, and then again extracted with pyridine/water. If after this process the extraction was improved, the presence of a chrome or metal complex dye is suggested. If both procedures failed to extract the dye, a reactive dye was indicated, and extraction with 1, 2-diaminoethane was attempted.

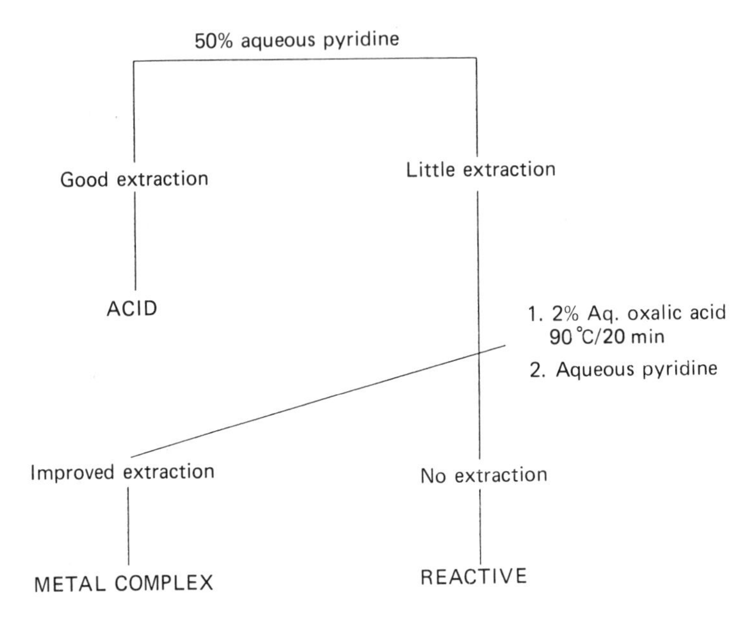

Fig. 5.2. Classification and extraction of dyestuffs from different fibre types. (Courtesy of the Metropolitan Police Forensic Science Laboratory, London, England.)

Resua (1980), produced information on woollen and synthetic fibres. His extraction solvents were essentially the same as those previously described by Macrae and Smalldon, but extracts were then applied to silica gel thin layer chromatographic plates and developed in a series of solvent systems. Two systems were used for acid/base dyes and two additional systems for disperse dyes. The chromatographic behaviour of the dye in a solvent referred to as the screening solvent system (SSS), namely chloroform : methanol : acetic acid (70 : 20 : 10), indicated the class of dye and the subsequent TLC solvent system to be used. Reactive, direct, and vat dyes were listed as immobile in the screening solvent. This piece or work was later criticized in a review article by Fong (1989) because Resua was considered to have used unrealistically large samples for his tests. Nonetheless, the idea of a systematic approach to dye extraction and optimal choice of TLC solvent system was firmly established. Although classification schemes have been described for cellulosic fibres such as cotton and viscose, most rely on a combination of solvent extraction and TLC.

Probably the most comprehensive and widely used scheme for the classification and extraction of fibre dyes is that employed by the Metropolitan Police Forensic Science Laboratory, London, outlined in Fig. 5.1—5.6. A recent report by Cheng *et al.* (1991), describes a scheme for the extraction of direct, vat, azoic, and sulphur dyes from cotton fibres. This work describes the sequential use of five extraction solvents on a single cotton fibre. The efficiency of each extraction was scored on a scale of 1 to 5, and this value was used to determine the nature of the dye. Unfortunately the

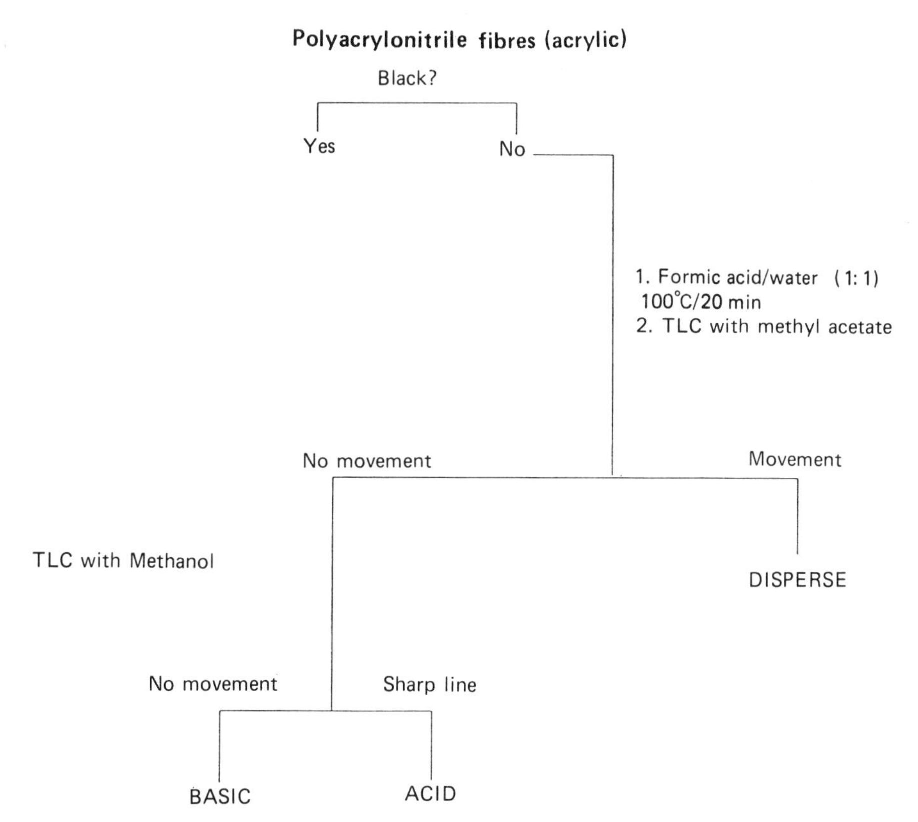

Fig. 5.3. Classification and extraction of dyestuffs from different fibre types. (Courtesy of the Metropolitan Police Forensic Science Laboratory, London, England.)

subjectivity of this approach means that dyes are difficult to classify. However, further work along these lines, perhaps coupled with HPLC analysis could prove fruitful in the identification of these particular dyes.

5.5 DYE ANALYSIS

5.5.1 Introduction

Owing to shedding of fabrics, single strands of fibres are transferred through contact with other items. A survey has shown that the lengths of single strands of fibres transferred were variable, with the highest percentage of fibres falling within a length of 2–5 mm (Kidd & Robertson 1982, Robertson *et al.* 1982). This information is invaluable since it is possible to estimate that the quantity of dye present on this length of fibre is approximately 5 ng. The minimum suggested length of fibre needed for a successful thin layer chromatographic analysis is shown in Table 5.4 (Gaudette 1988). Therefore, even if this amount of dye can be extracted quantitatively from a strand of fibre, only a limited number of analytical techniques can be considered for the examination of fibre dyes.

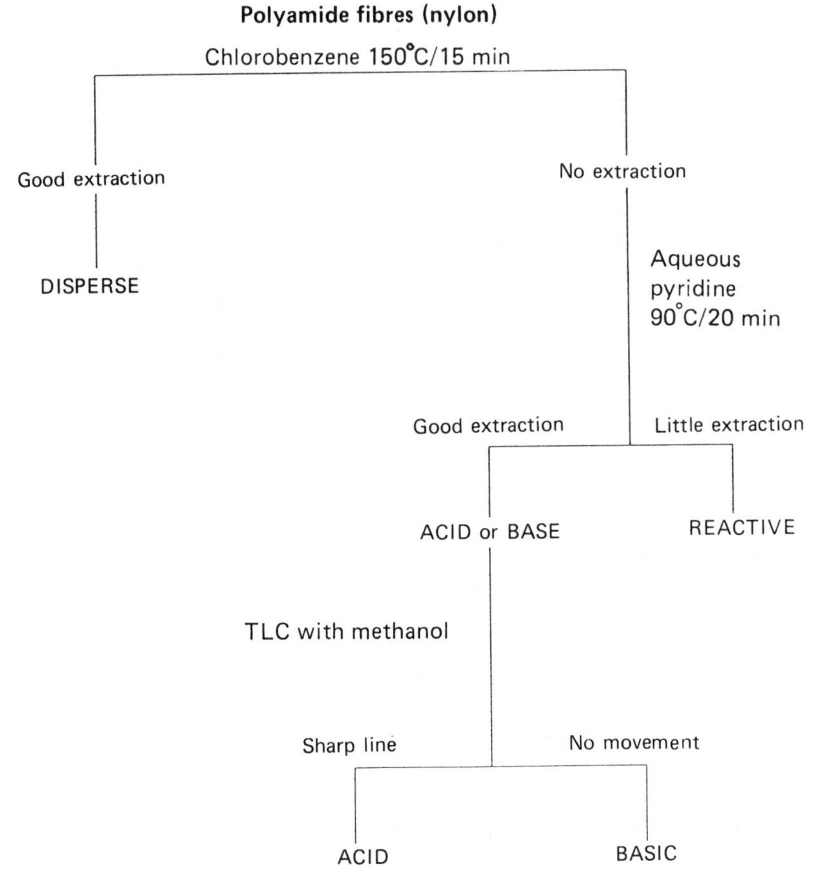

Fig. 5.4. Classification and extraction of dyestuffs from different fibre types. (Courtesy of the Metropolitan Police Forensic Science Laboratory, London, England.)

There is a further limitation that can govern the choice of analytical techniques, It is not uncommon for the colouration of a fibre to be produced by using a mixture of dyes or an impure dye. Hence analytical techniques that can provide a separation between the components are preferable, and these have the potential of improving dye discrimination and identification. Unfortunately, the desire for separation coupled with the detection sensitivity required severely limits the number of techniques that could possibly be used.

The classical structural elucidation techniques of infrared (IR) and mass spectrometry (MS) have to date proved unsuccessful even when used to analyse pure dye samples. Problems due to the inadequate sensitivity of characterization of these techniques have been experienced. In many laboratories attempts have been made to improve IR results by using Fourier transform techniques, but this approach has not been highly successful.

Table 5.4. Minimum suggested fibre length needed for successful TLC

Fibre type	Colour	Shade	Total length or fibre (mm)
Cotton	Black		10
	Red	Dark	8
		Medium	15
	Green	Dark	8
		Medium	12
		Light	150
	Blue	Dark	5
		Medium	50
		Light	160
	Yellow	Dark	80
		Light	200
Wool	Black		2
	Red	Dark	2
		Medium	2
		Light	2
	Green	Dark	2
		Medium	4
		Light	20
	Blue	Dark	3
		Medium	4
		Light	6
	Yellow	Dark	6
		Medium	14
Acrylic	Black		5
	Red	Dark	2
		Medium	2
		Light	100
	Green	Dark	8
		Medium	8
	Blue	Dark	4
		Medium	2
		Light	20
	Yellow	Dark	12
		Medium	20
		Light	150

Table 5.4.—*Continued*

Nylon	Black		2
	Red	Dark	3
		Medium	4
		Light	60
	Green	Dark	4
		Medium	4
		Light	16
	Blue	Dark	2
		Medium	6
		Light	20
	Yellow	Dark	14
		Medium	20
		Light	100
Polyester	Black		1
	Red	Dark	2
		Medium	2
		Light	10
	Green	Dark	6
		Medium	8
		Light	20
	Blue	Dark	6
		Medium	6
		Light	30
	Yellow	Dark	6
		Medium	10

(after Gandette, 1988)

Thin layer chromatography (TLC) is one analytical technique that is capable of separating mixtures of dyes, and it has been used extensively for the analysis of fibre dyes. Owing to the spectral properties of dye molecules and their high absorptivities, the detection levels required can generally be achieved by the human eye. There are, however, certain aspects which make this technique less highly desirable. For example, the precision of Rf data is not ideal, the colour and intensities of the separated components are assessed subjectively, and because of fading of colours there are occasionally problems in trying to keep permanent records of the analytical results. To some extent the introduction of instrumentation which can record spectra of the components directly from a TLC plate has improved this situation considerably.

High-performance liquid chromatography (HPLC) would appear to offer a viable approach for the analysis of fibre dyes, since the separation and detection power of this technique has been established for many years. The interest in this technique in forensic laboratories for the analysis of dyes, and in particular fibre dyes, has, however, been very limited. This has probably been due to chromatographic difficulties and to the lack of a detection system that can provide more than simple retention time data.

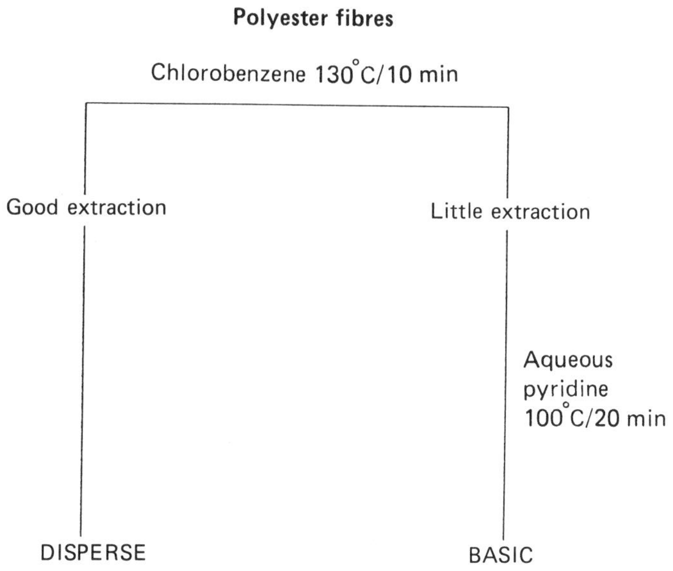

Polyester fibres

Chlorobenzene 130°C/10 min

Good extraction

Little extraction

Aqueous pyridine 100°C/20 min

DISPERSE

BASIC

Fig. 5.5. Classification and extraction of dyestuffs from different fibre types. (Courtesy of the Metropolitan Police Forensic Science Laboratory, London, England.)

Within the past decade multiwavelength UV/visible detection systems have been developed. These detectors usually permit spectral data to be recorded for solutes as they are eluted from an HPLC column. This has resulted in a keener interest in the use of HPLC for the analysis of dyes.

TLC and HPLC are the only techniques that can meet some of the main requirements of the forensic scientist who is required to undertake the analysis of fibre dyes. However, it must be appreciated that improvements in these techniques are still being sought.

5.5.2 Thin layer chromatography

Thin layer chromatography (TLC) has been applied to the separation of dyes extracted from textile fibres for many years and, within the last decade or so, for the comparison of fibres transferred during the commission of a crime. The advantages of TLC are that it is a rapid, simple, and inexpensive technique which can be applied to the examination of most dye types. It does, however, lack the sensitivity and resolution of some of the more sophisticated analytical techniques such as high performance liquid chromatography. Nonetheless, thin layer chromatography is an important tool for the initial chemical comparison of two dyestuffs, and it is widely employed by crime laboratories in the investigation of fibre evidence.

As described above, the dyes are extracted by placing the fibres in glass capillary tubes, and extracting them at around 100°C with an appropriate solvent. After

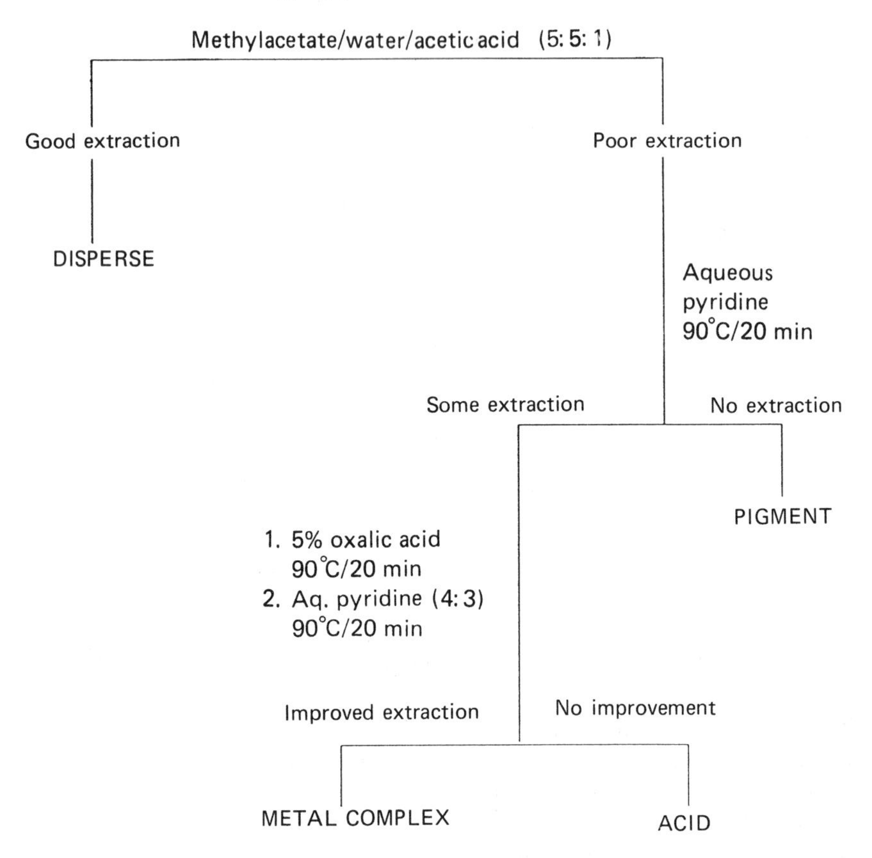

Fig. 5.6. Classification and extraction of dyestuffs from different fibre types. (Courtesy of the Metropolitan Police Forensic Science Laboratory, London, England.)

adequate extraction of the dye from a fibre, the top of the capillary tube is broken off, and the extract is applied to a thin layer chromatographic plate. The TLC plate typically consists of silica gel on an inert aluminium, plastic, or glass support. It is common to use so called high performance thin layer chromatographic (HPTLC) plates. These differ from normal TLC plates in that the particle size of the silica is smaller. This results in a high surface area which leads to better resolution of the dye components. Several extracts can be applied to the same plate to allow comparison of different dyes or dyes from different sources. The chromatographic plate is then placed in a chamber containing eluent which consists of a mixture of organic solvents and/or aqueous buffers. The eluent rises up through the TLC plate by capillary attraction and causes separation of the dyestuffs on the TLC plate. This separation is the result of the relative affinity of the dye for the silica plate compared to its affinity for the eluent. Generally, less polar compounds will travel the farthest up the plate, while polar compounds will remain near the bottom. The original dye will thus be separated into its individual components which can be seen as a series of spots on the TLC plate by viewing in normal daylight or under an ultraviolet lamp. The distance that a particular spot travels on the chromatographic plate may be described by its *Rf* value, defined as the distance from the point of application to the centre of the spot of interest divided by the distance travelled by the eluent. These *Rf* values do, however, vary greatly with temperature and slight differences in the composition of the eluent, and should be considered purely as an indication of the relative behaviour of different compounds under a given set of conditions. In the forensic examination of dyestuffs, however, TLC is generally used to compare the component dyes extracted from two fibres in order to determine whether or not the fibres could have originated from a common source.

As with extraction solvents, eluent systems for thin layer chromatography have been optimized over the years for the separation of different dye types. These eluent systems are listed in Table 5.5.

Table 5.5. Eluents in order of preference for the common dye class/fibre type combinations

Fibre type	Dye class	Eluent number
Wool	Acid or metallised	1 2
Cotton and viscose	Direct or reactive	1 4 3
Polyacrylonitrile	Basic	11 12 1
Polyester	Disperse	6 7 8 5
Polyamide	Acid	9 10

Eluent No.	Eluent composition	Proportions (V/V)
1	*n*-Butanol, Acetone, Water, Ammonia	5:5:1:2
2	Pyridine, Amyl Alcohol, 10% Ammonia	4:3:3
3	*n*-Butanol, Ethanol, Ammonia, Pyridine, Water	8:3:4:4:3
4	Methanol, Amyl Alcohol, Water	5:5:2
5	Toluene, Pyridine	4:1
6	Chloroform, Ethyl Acetate, Ethanol	7:2:1
7	*n*-Hexane, Ethyl Acetate, Acetone	5:4:1
8	Toluene, Methanol, Acetone	20:2:1
9	*n*-Butanol, Acetic Acid, Water (upper phase)	2:1:5
10	*n*-Butanol, Ethanol, Ammonia, Pyridine	4:1:3:2
11	Chloroform, Methyl Ethyl Ketone, Acetic Acid, Formic Acid	8:6:1:1
12	*n*-Butanol, Acetic Acid, Water (upper phase)	4:1:5

Thin layer comparison of the extracted dyes of two fibre samples under separate chromatographic conditions, such as different eluent systems, can be performed when adequate sample is available. It has been argued that such an approach can be used to increase the discriminatory power of TLC. In other words, the more tests which can be performed showing similarities in the behaviour of two dyes, the greater the certainty that they are of a common origin. Studies to evaluate the relative discriminating power of two different TLC systems have been reported by Resua *et al.* (1981), and later by Golding & Kokot (1989). Solvent systems that produce unrelated *Rf* information, that is, reflecting separations achieved on the basis of different properties of the dye molecule, were considered to offer the greatest degree of discrimination. While this approach has some merit in increasing the value placed on the results of TLC comparisons of fibre dyes, few eluent systems have in fact been described which achieve this goal of increasing discrimination between similar dye types. Golding and Kokot indicated that this could be best achieved by examining the behaviour of the dye on both a silica gel plate and a reversed phase plate, separation on the reversed phase system being a factor of the dye's interaction with a hydrophobic long chain hydrocarbon stationary phase.

In fact, the limited amount of increased information obtained by performing a second thin layer chromatographic analysis on an extracted dye suggests that it is probably more appropriate to use a completely different technique to support the initial TLC finding. In particular, recent work with high performance liquid chromatographic analysis of fibre dyes suggests that this technique offers greater discriminatory power, resolution, and sensitivity than the TLC systems at present employed.

5.5.3 High performance liquid chromatography (HPLC)

5.5.3.1 Introduction

HPLC was introduced as an analytical technique about 25 years ago, and today it is used extensively for the qualitative and quantitative analysis of drugs, metabolites, proteins, polymers, and natural products. A schematic diagram of the instrument is shown in Fig. 5.7, and a typical instrument shown in Fig. 5.8. The mobile phase is drawn from the solvent reservoir and pumped through the analytical column at a rate of 1–2 ml/min. This is achieved by means of a pump operating typically at around 1 to 2000 p.s.i. The sample, dissolved in a suitable solvent, is applied to the column via a valve (injection port), and the mobile phase takes it through the column. The column is packed with fine particles of an adsorbent material, usually consisting of modified silica. This acts in the same way as a TLC plate to separate the individual components of the sample. These components are then sequentially eluted from the column and pass into a detector. Although several different HPLC detection systems are available, for fibre dye analysis an ultraviolet or visible spectrophotometer is generally employed. By use of a computerized data handling system the dye components can be visualized as a series of peaks on a chart, and their relative concentrations determined. Although HPLC has become the method of choice for a wide range of analyses, it is interesting to note that this technique has not gained the same degree of use in the field of colour chemistry.

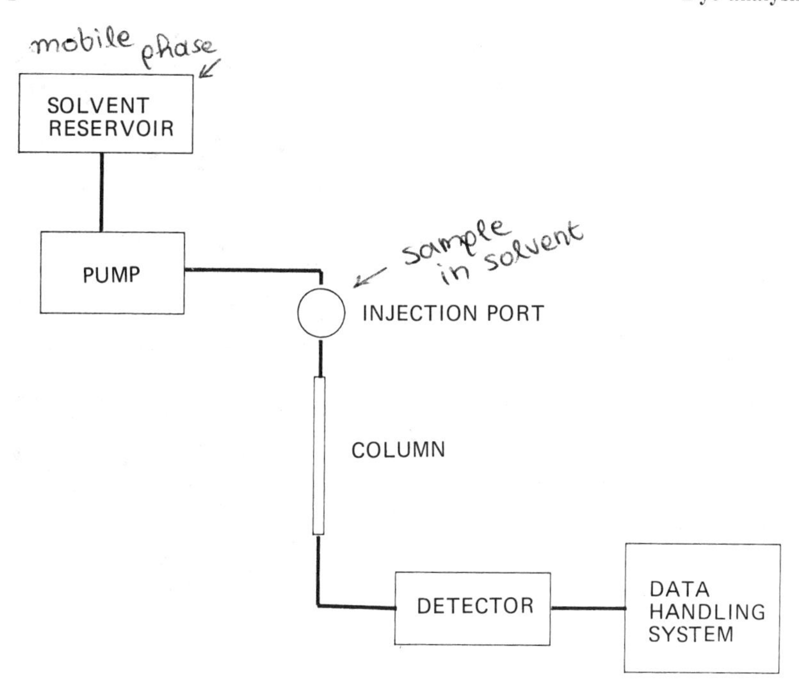

Fig. 5.7. Schematic diagram of the working components of an HPLC.

As detailed earlier in this chapter, dyes can be either ionic (acidic or basic) or neutral, and in theory HPLC is the ideal separation technique for these types of molecule. Furthermore, a visible wavelength spectrophotometric detector can be used to monitor dye separations with relative ease. Therefore, the reasons for the small number of reported uses of HPLC for the analysis of dyes are not obvious, but may possibly be attributed to dye manufacturers attempting to preserve commercial confidentiality, or simply for historical reasons. Dye manufacturers have for many years relied upon basic physical and chemical tests, for example, solubility, lightfastness, spot tests, paper chromatography, TLC, spectroscopy, and microscopy for the quality control of their products. Stricter regulations are now being introduced which require the use of more sophisticated analytical techniques and HPLC is starting to be used to assess purity and to identify impurities in food, cosmetics and even fibre dyes.

Within forensic science laboratories a similar picture can be observed, with HPLC being used extensively as an analytical technique, but only recently being considered for the analysis of dyes. This attitude is quite surprising, especially since in theory HPLC can offer many advantages over thin layer chromatography for dye analysis.

An interesting point to note is that fibre dye analyses have been performed in most forensic science laboratories by biologists who have therefore inherited the responsibility of completing fibre dye analyses. It is now evident that any further progress in this subject requires the skills of the analytical chemist. Certainly, knowledge of chemical structures, extraction procedures, and separation and detection techniques are essential when attempting to develop suitable HPLC methods.

Fig. 5.8. Typical HPLC instrumentation.

From the literature it is evident that there are some practical difficulties in developing HPLC methods which are suitable for casework samples. In this chapter we shall review the developments that have occurred, identify some of the existing problems, and indicate, where possible, if these can be resolved. Since major areas of development can be identified—separation, detection, and extraction—separate discussions on these topics will be presented.

5.5.3.2 HPLC separation of dyes

It was outlined earlier (Table 5.3) that a number of classes of dye are used to colour fibres, although chemically these dyes can be divided into only three main groups: acidic, basic, and neutral. Furthermore, within each class of dye there can be many dyes of widely differing chemical structures. Therefore the chromatographer is faced with some major analytical problems.

In the first instance it is very difficult to separate the three dye types on a single chromatographic system. Therefore, it is necessary to employ a separate system for each of the three types of dye. Secondly, the chemical nature of the fibre dye is in most instances unknown and must be determined before any analyses to ensure correct selection of the chromatographic system. Fortunately, in many cases the latter situation can be resolved by using an extraction classification scheme on control fibres.

Once it has been possible to identify the chemical nature of the dye, the next problem is to achieve satisfactory chromatography of any dyes extracted from a fibre. Some chromatographic systems have been reported for the separation of acidic, basic, and neutral dyes used to colour fibres. These are presented in Table 5.6.

Table 5.6. HPLC separation systems developed for the analyses of acidic, basic and neutral dyes

Dye type	Column	Eluent	Reference
Neutral (disperse)	250 × 3.2 mm 10 μm Lichrosorb Si-60	Hexane/ethyl acetate gradient	West (1981)
	125 × 3.2 mm 5 μm Hypersil ODS	Acetonitrile/water citric acid Isocratic	Wheals *et al.* (1985)
Basic	160 × 4.5 mm 5 μm Spherisorb S5W	Water/methanol/ammonia gradient	Griffin *et al.* (1988)
	250 × 4.5 mm 5 μm Spherisorb S5W	Ammonium acetate/ methanol Isocratic	Griffin *et al.* (1989)
Acidic	40 × 1 mm 5 μm Hypersoil ODS	Methanol/water/TMAC[a] gradient	Laing *et al.* (1988)
Various	200 × 4.4 mm 5 μm Spherisorb S5W	Dichloroethane/ethanol/ formamide Isocratic	Tebbett *et al.* (1990)

[a] Tetramethylammonium chloride.

A critical examination of these publications reveals that two references, (Wheals *et al.* 1985, Griffin *et al.* 1989), meet the criteria required by the forensic scientist for the analysis of fibre dye extracts. The remaining references are unsuitable because they are incompatible with the detection techniques employed (Griffin *et al.* 1988, Laing *et al.* 1988), and/or have been evaluated only with quantities of dyes which could be extracted from a fibre if greater than 5 mm in length (West 1981, Laing *et al.* 1988).

Interestingly, although both *gradient* and *isocratic* (single eluent) elution systems have been investigated, the latter, despite having less resolving power, has been shown to be the most successful. A comparison of these techniques also illustrates that if gradient elution is used, then other significant practical problems are introduced. These include a considerable increase in analysis time, poorer precision of retention time data, and, most disturbing of all, the creation of a major detection problem when using multiwavelength detection procedures. These detection problems have been identified in the work reported by Griffin *et al.* (1989) and Laing *et al.* (1988). Because of these problems the former group elected to develop and use an isocratic system (Griffin *et al.* 1989).

To chromatograph acidic, basic, or neutral dyes on silica or modified silica column packing materials, several modes of separation can be used for each type of dye. These include, *normal phase*, *reversed-phase*, *ion exchange*, *ion suppression* and *ion-pair*

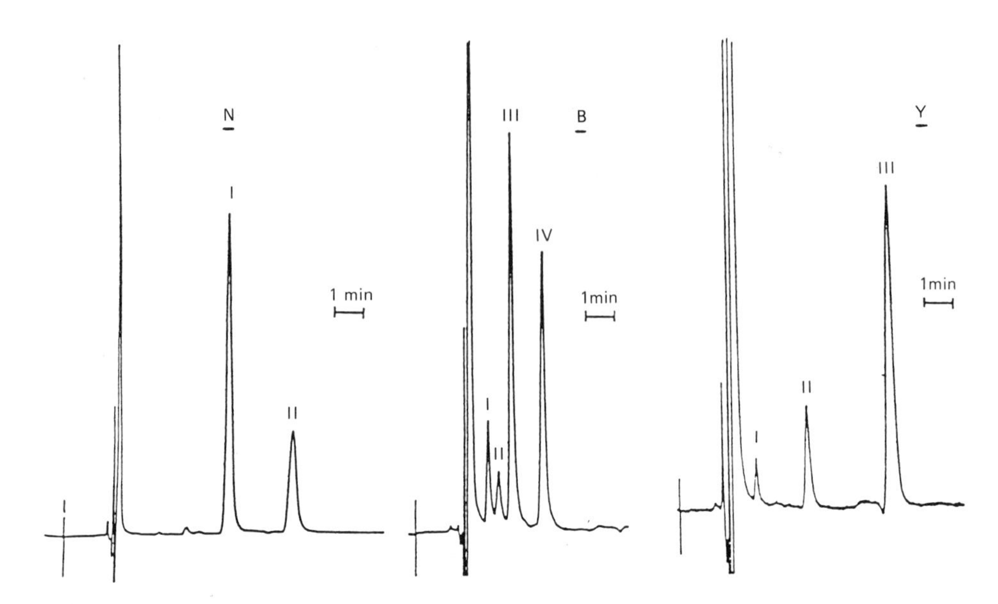

Fig. 5.9. Chromatograms of disperse dyes extracted from 5 mm lengths of a single fibre using a reversed-phase system, N = ICI Dispersol Navy C-4R (5% shade), B = ICI dispersol Black D-B (9.5% shade) and Y = ICI Dispersol Yellow C-36 (1.5% shade) monitored at wavelength of 600, 500 and 400 mm respectively and 5 mAUFS.

chromatography. The final choice of packing material and mode of separation for any particular type of dye can in general be reached only after extensive tests with a large number of colorants. In these tests analysis time, chromatographic efficiency, and resolution should all be studied in detail as they are important parameters. Two different chromatographic systems have been developed for the analysis of neutral (disperse) dyes (West 1981, Wheals *et al.* 1985). In the study conducted by West (1981), retention is obtained under normal phase chromatographic conditions, whereas Wheals *et al.* (1985) elected to use a reversed-phase system. Control of retention and peak shape reproducibility is extremely difficult to maintain with a normal phase system because the introduction of solutes or even water can change the activity of the silica surface. Therefore, the use of this type of system is not recommended, especially if there are alternatives. Tebbett *et al.*, (1990) also reported using a normal phase system for fibre dye analysis, but this was only for some exploratory work to assess detection techniques and was not intended for casework samples.

Apart from improved reliability achieved with the reversed-phase system for the analysis of the disperse dyes it was also found that, in comparison with these normal phase systems, a considerable gain in chromatographic efficiency could be obtained (Wheals *et al.* 1985). A total of 57 dyes (ICI Dispersol range) extracted from polyester fibres were investigated in this study, and some typical results obtained from 5 mm lengths of single strands of fibres are shown in Fig. 5.9.

These chromatograms illustrate clearly that fibre dye extracts can contain several components. With the dyes investigated in this study, 38% of the samples analysed were found to contain two or more major components (a major component was arbitrarily defined as any peak with a height greater than 10% of the major peak), and most of the extracts characterized by one major peak also contained minor peaks in the chromatogram. The ability of any system that can offer this degree of selectivity is clearly an advantage, especially if comparing suspect and control extracts. Therefore it would appear that the reversed-phase system is the most favourable for the separation of neutral dyes.

Basic dyes are commonly used for the coloration of acrylic fibres. They have been studied extensively by Griffin *et al.* (see Table 5.6). The chromatographic system employed for this type of dye is based on the ion-exchange properties of silica (Griffin *et al.* 1989). At the stated eluent pH value of 9.76, both the dye and the silica will be partially ionized, hence retention of the dyes occur.

A high proportion of the fibre dyes encountered by Griffin *et al.* were black, therefore efforts were concentrated upon this particular colour. Since a black coloration on a fibre is usually obtained by employing an admixture of blue, red, and yellow dyes, the major requirement was for an HPLC system that could provide adequate resolution between these components.

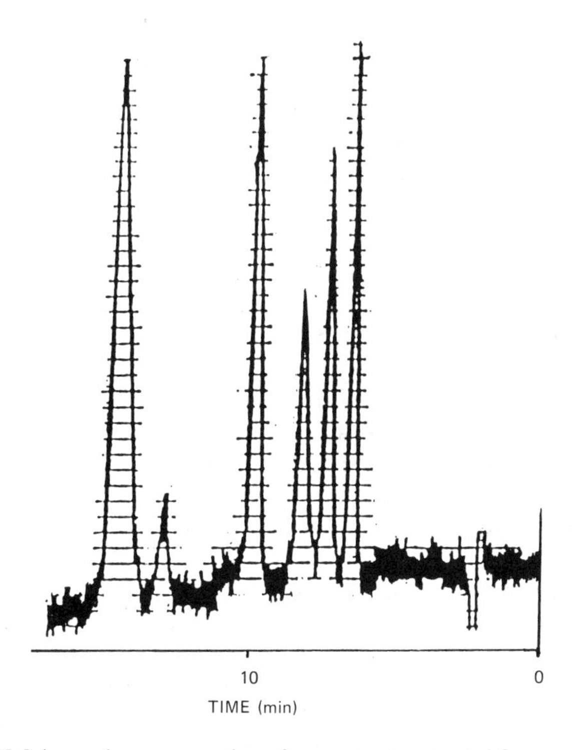

Fig. 5.10. HPLC ion-exchange separation of components extracted from a black acrylic fibre. Separation monitored at 500 nm. With permission of R. M. E. Griffin *et al.* 1989).

The results obtained from the analysis of a black fibre dye extract, illustrating the performance of the ion-exchange system, are presented in Fig. 5.10. From this chromatogram and other results presented it is apparent that good chromatographic efficiency and adequate separation between components can be achieved with this system. An interesting point noted by the authors was that if compared against TLC, better sample discrimination could be achieved with their HPLC system.

Chromatographers are aware that the ion-exchange properties of silica can vary from batch to batch because of the apparent difficulty in controlling its surface activity during manufacture. Therefore if this mode of separation is employed for the analysis of basic dyes, long-term variations in retention time data and even changes in elution sequences might be experienced. Analysts should be aware of these problems, especially if contemplating the compilation and use of a reference collection of data for the identification or comparison of colorants.

An alternative method for the separation and analysis of basic solutes is the established technique of *reversed-phase ion-pair* (RP–IP) chromatography. Under these conditions the silica is modified, typically with octadecyl groups, to reduce surface activity. The use of RP–IP can bring additional benefits if this approach is employed, since the ion-pairing reagent contained in the eluent forms ion-pairs with the solutes of interest, and these chromatograph as neutral molecules. RP–IP has been used successfully for the forensic examination of basic dyes in ballpoint pen ink formulations (White & Wheals 1984), but no similar method has been published for the analyses of basic fibre dyes. However, some preliminary studies have shown that this could be a viable alternative to the ion-exchange system (White & Bennett unpublished results).

Forensic applications of HPLC for the analysis of *acidic dyes* have been limited mainly to the food colorants found in heroin (Clark & Miller 1978), and illicit drugs and tablets (Joyce & Sanger 1979, Joyce & Humphreys 1982, Joyce *et al.* 1982). Laing *et al.* (1988) reported the use of this technique for the analyses of acidic fibre dyes. In our view, this approach can be discounted because the system fails to meet the important criteria required by the forensic scientist. This lack of an HPLC method for acidic fibre dyes but success with food colorants, indicates that there are some practical problems.

Unlike the basic dyes, or even with the small group of food colorants or dyes, acidic fibre dyes are composed of several classes of dye, for example, acid, azoic, direct, mordant, premetallised, reactive, sulphur, and vat. Therefore the chromatographer is faced with the problem of trying to develop a system that can accommodate a diversity of chemical structures found either between classes or even within a selected class of this type of dye.

After an extensive investigation conducted by White and Harbin (White & Harbin 1989) it was concluded that many of the problems experienced with the analysis of acidic dyes could be attributed to undesirable properties of silica based packing materials. This prompted their evaluation of a polymeric packing material polystyrene divinylbenzene (PSDVB).

PSDVB packing materials have started to be used only recently in HPLC because they can be produced commercially with a particle size of 5 μm and the desired rigidity. Potentially, packing materials of this type offer many advantages over silica, because they are chemically inert and stable over a pH range of 1–14.

To assess this packing material and to optimize eluent conditions a selection of 52 dyes with a wide range of chemical structures were analysed. To obtain retention of these dyes it was necessary to include an ion-pairing agent tetrabutylammonium hydrogen sulphate (TBAH) in the eluent. On the basis of chromatographic efficiency and analysis time the eluent composition selected finally consisted of acetonitrile/water (50:50) containing citric acid (0.7 g/l) and TBAH (3.369 g:0.01 M) adjusted to pH 9.0 with concentrated ammonia.

All 52 dyes could be chromatographed successfully with this system. Some typical chromatograms obtained are illustrated in Fig. 5.11A—G. The PSDVB packing material would appear to offer considerable advantages over silica and to resolve some major chromatographic problems. The precision of retention time data are reported to be excellent, and these studies indicate that retention of the dyes occurs predominantly via an ion-pair reversed-phase mechanism. Some successful analyses of acidic dyes in casework samples have been accomplished with this PSDVB system (White unpublished results, White & Catterick 1990), but these have not included any analyses of fibre dye extracts. The reasons for this are mainly due to extraction and not chromatographic problems, a point that is to be considered later in this chapter.

5.5.3.3 Detection and monitoring of dye separations

Traditionally dyes have been monitored in HPLC eluates with a visible wavelength spectrophotometric detector. With a good quality fixed wavelength detector it is possible to achieve minimum detection levels of less than 1 ng of a single dye component. However, it is important to appreciate that detection limits will be dependent upon several factors, including absorptivity of the dye, detector noise level, and the number of components in a dye extract.

Unfortunately, although these fixed wavelength detectors can generally offer the desired sensitivity for fibre dye analysis, all qualitative analyses are based solely upon retention time data. Another problem is that these detectors can usually operate only at a single wavelength during an analysis, hence a considerable amount of analytical information can be lost if a dye contains several coloured components.

The method reported for the analysis of basic fibre dyes actually favours the use of a single wavelength detector (Griffin *et al.* 1989). To obtain sample discrimination the dye extract is analysed at three selected wavelengths. This confirms the sensitivity of this type of detector since only one third of the extract is analysed at each wavelength. A major disadvantage of this approach is that the sample analysis time is increased considerably.

About ten years ago there was a major development in detector technology with the introduction of the multiwavelength detector. Since this enabled several wavelengths to be monitored simultaneously and permitted spectra to be recorded of components as they eluted from an HPLC column, many chromatographers found that it was possible to overcome some of the problems outlined above.

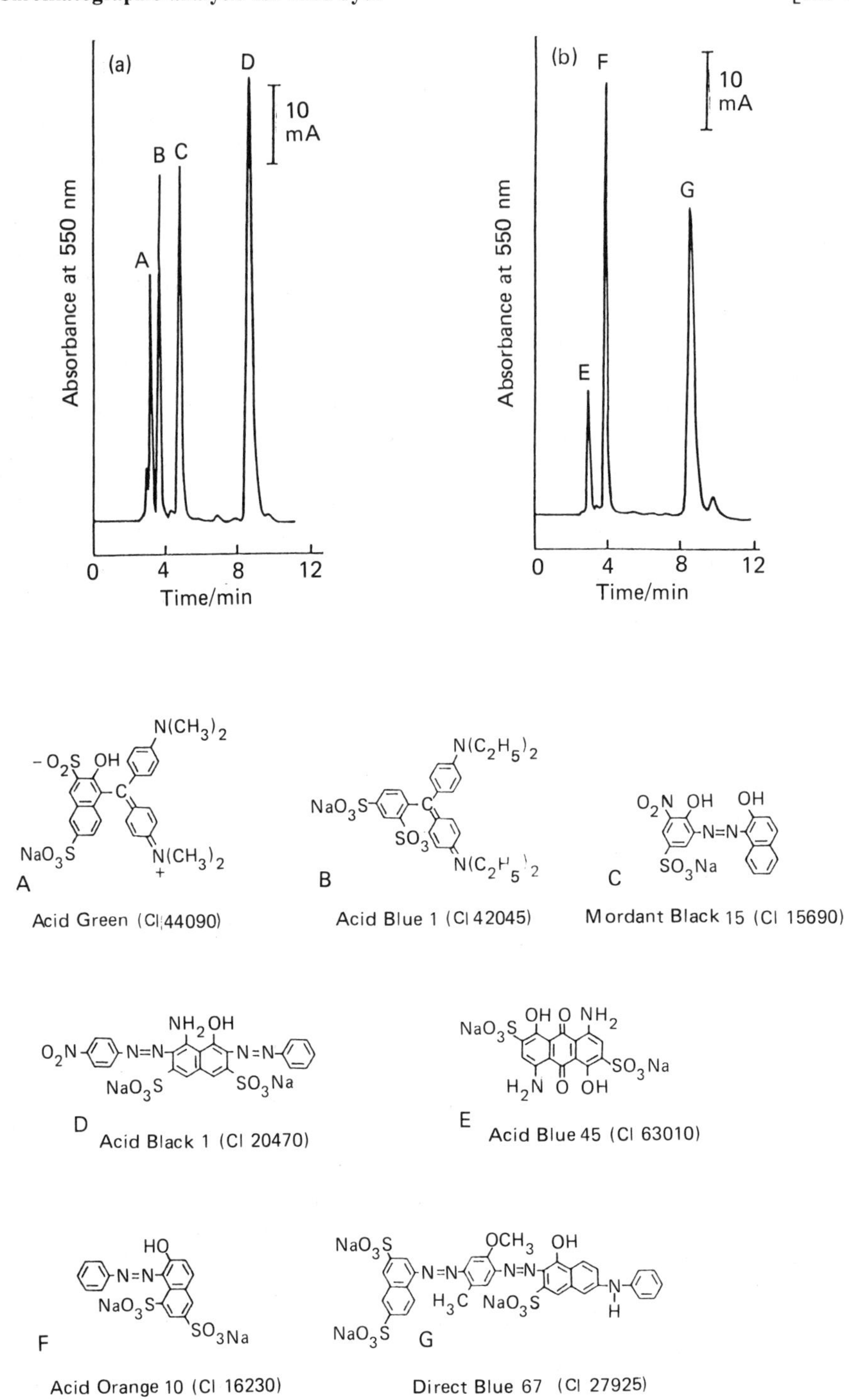

Fig. 5.11. HPLC analysis of some acidic dyes chromatographed on a PSDVB packing material dynamically modified with TBAH.

Several groups of analysts believed that these detectors could also improve the qualitative examination of fibre dyes, but unfortunately, various studies generated conflicting results. To clarify this situation, important considerations including detector design, sensitivity, and routines used to evaluate the analytical data will now be examined in greater detail.

Detectors

Several manufacturers now produce a multiwavelength detector, but not all of these are suitable for the analyses of dyes. Some detectors are reported to be capable of operating over a UV/visible wavelength range, but require separate lamp sources for the UV and visible wavelength ranges. Since many dyes produce a significant UV spectra and some of the best multiwavelength detection techniques that have been developed require full UV/visible spectra, detectors that operate with only one source are more highly favoured.

There are now several designs of detector available, with the difference between them being the method used for the dispersion and capture of the radiant energy after the light has passed through the flow cell. The oldest and possibly most common design is based on a fixed grating/diode array system. Newer designs include a fixed grating with a movable single diode or rotating grating and fixed diode array.

The newer designs have been developed to improve sensitivity. In assessing any of these detectors an important factor to consider is sensitivity and hence the signal/noise performance. This performance test should be compared against a fixed wavelength filter detector since the latter is generally accepted as one of the most sensitive. Ideally this should also be carried out for a range of wavelengths, and if diode array detectors are tested the optimum bandwidth (that is, the number of diodes bunched together) should be determined (Wheals *et al.* 1985).

Spectral data throughout a chromatographic analysis can be generated with most multiwavelength detectors, but the number of wavelengths that can be monitored simultaneously can very between instruments. All manufacturers offer a variety of routines for evaluating data, but, as will be illustrated, many are unacceptable because they fail to meet the demands of analysts wishing to perform fibre dye examinations.

Sensitivity

Sensitivity is a major factor that has to be considered in fibre dye analysis because of the small sample size encountered. From published results it would appear that there is some conflict as to whether or not the multiwavelength detector can offer the desired sensitivity. Admittedly, sensitivity can vary between different commercial detectors, but since most can achieve the sensitivity desired there must be some other contributory factors. From an examination of the relevant publications the reported lack of sensitivity can possibly be attributed to the design of the chromatographic system and/or the type of data evaluation routine used.

All chromatographic processes cause the analyte to be diluted. One of the most effective ways to minimize this effect in HPLC is to reduce the column volume. In the disperse dye study carried out by Wheals *et al.* (1985), comparisons of column volumes and relative responses for six different column dimensions were performed.

It was shown that by reducing column dimensions from 250 × 4.9 mm to 125 × 3.2 mm this effectively increased sensitivity by a factor of four.

The advantages of using smaller column dimensions have been verified because, if using a 125 × 3.2 mm column, detection levels approaching 200 pg (White & Catterick 1987) have been reported for some disperse dyes. A similar reduction in column volume, if applied to the analysis of the basic dyes studied by Griffin *et al.* (1989), could possibly improve detection levels. Furthermore, this might permit a multi-wavelength detector to be used in place of the complex single wavelength detection procedure required for improving sample discrimination.

Any further attempts to reduce column volumes by using smaller diameter or even microbore columns are not favoured, since these can introduce additional practical problems, for example, column packing reproducibility and column overload effects.

It is evident that isocratic elution systems are preferable to gradient elution systems not only in the separation processes but also in detection procedures. Two independent groups confirm that severe refractive index effects occur with gradient elution systems, and this is irrespective of the column diameter (Griffin *et al.* 1989, Laing *et al.* 1988). These refractive index changes result in poor baseline stability and effectively increase the background noise level of the detection system.

Returning to the other factor related to the data evaluation routine used, it must be appreciated that some routines will give satisfactory results only if sample absorbance measurements are relatively high. In some instances these thresholds are higher than the minimum detector level of the instrument, therefore this lack of sensitivity is not due to the detector but to the routine employed. There has been a move toward the development of evaluation routines that will provide results from data obtained from very small sample sizes. This is covered in further detail in the following section.

Data evaluation routines

Several multiwavelength detectors have the advantage of being able to monitor several wavelengths simultaneously. An example of the facility being used for the analysis of fibre dye extracts is illustrated in Fig. 5.12. This process ensures that extracts containing eluted dyes cannot go undetected because of an incorrect choice of monitoring wavelength, as could be the case when using a single wavelength instrument. Furthermore, the additional information not only improves sample discrimination but indicates that the black fibre dye could contain the same dye as that used on the navy blue fibre and also an orange/red dye ($RT = 8$ min) and yellow dye ($RT = 3.5$ min).

A major advantage of most multiwavelength detectors is that all the chromatographic and spectral data obtained in a single analysis can be collected and then accessed and evaluated following the completion of a run. Several routines have been developed commercially for the evaluation of these stored data, including; absorbance/wavelength/time (3-D) plots, isoabsorbance plots, normal and derivative spectra, and absorbance ratio chromatograms (George & Maute 1982).

For qualitative analyses, routines for generating and comparing spectra have been employed extensively. Nevertheless, it must be emphasized that the quality of any

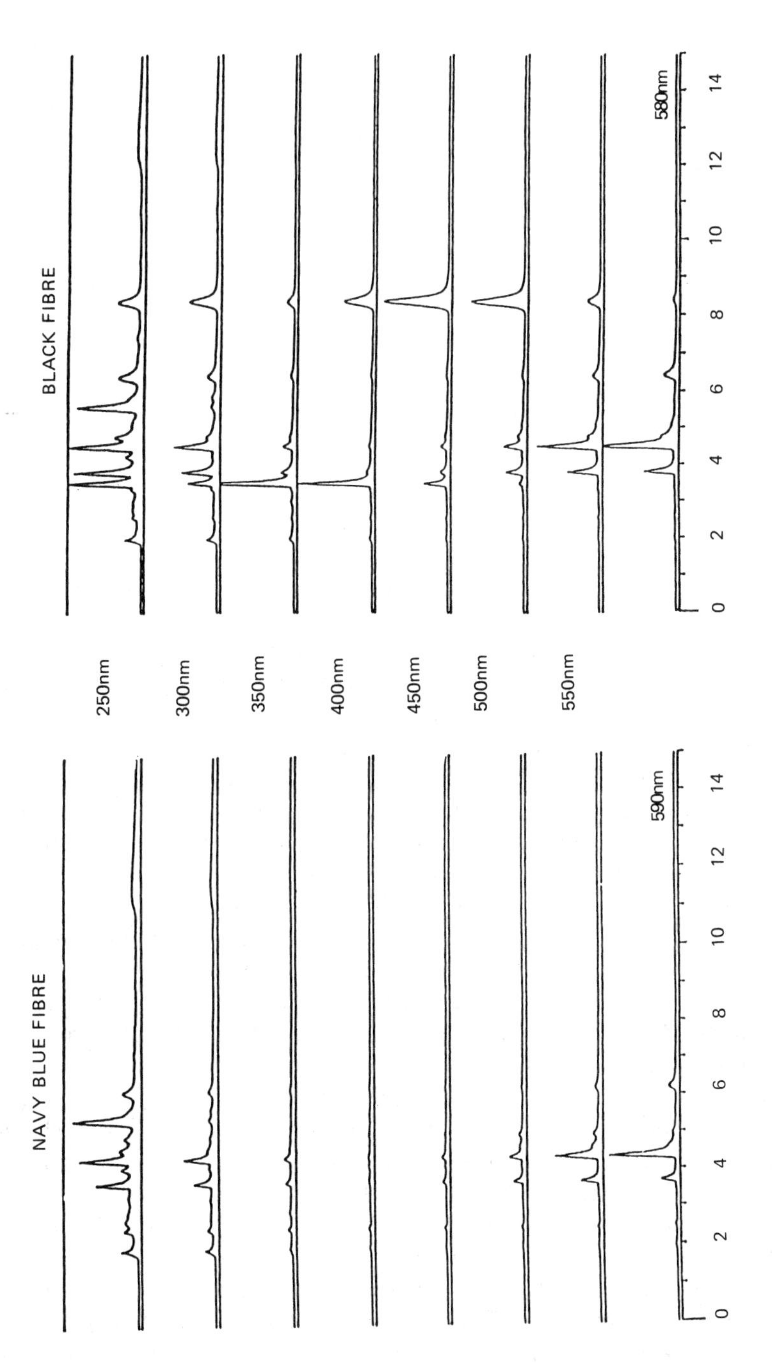

Fig. 5.12. Multiwavelength monitoring of dyes extracted from individual 5 mm lengths of two different single strands of polyester fibres. These chromatograms illustrate the additional information that can be gained by monitoring at different wavelengths (both UV and visible), thereby improving sample discrimination.

spectrum produced will be dependent upon sample concentration and the optical resolution of the detector. With the majority of the multiwavelength detectors the optical resolution is usually of the order of 4 nm, which is poor if compared with a high performance spectrophotometer (the optical resolution of these instruments is typically <1 nm).

In casework, the quantity of dye extracted from a fibre will display an absorbance of typically less than 10 mAU at its λ_{max} value. For some detectors there are difficulties in achieving satisfactory spectra with sample absorbtivities as low as this, therefore spectral comparisons cannot be made. Beware! Some manufacturers use curve smoothing routines to improve the appearance of spectra. This practice is not favoured since it can lead to spectral distortions.

To confirm the identity of a dye or be able to discriminate between dyes, spectral comparisons can be made either visually or with the help of curve matching routines. In general, the correct analytical conclusions can be reached only if measurable differences can be observed between spectra. If spectra are similar, it is advisable to use only the values generated by the curve matching routine. However, since there appears to be no background study to verify the precision of these values or what effect sample concentration might have on them, it follows that there must be considerable caution in expressing any conclusions from the results obtained. An important point to remember is that dyes can have similar retention time and spectra, therefore the forensic scientist must be aware of these limitations in using these spectral routines for the comparison of fibre dyes.

Before the introduction of the first commercial multiwavelength detector one group of analysts had been concentrating on trying to improve the HPLC characterization of solutes. Their studies resulted initially in the development of a technique called *absorbance ratioing*, which is based on the use of a combination of absorbance ratio and retention time data (White & Catterick 1983).

Absorbance ratios can be determined either from chromatographic data obtained to produce a multiwavelength plot or from spectral data of an eluting component as shown in Fig. 5.13. By selecting one wavelength as the reference wavelength, viz, 450 nm, absorbance ratios can be calculated. In the example shown these ratios would be $A450/A400$; $A450/A350$ and $A450/A250$ where Ax is the absorbance of the component at a wavelength of x nm.

An extensive evaluation of this technique has been completed on a commercial multiwavelength detector, and with this particular instrument it was shown that absorbance ratios could be generated with a long-term precision of less than 2% Relative Standard Deviation (RSD) provided that samples displayed an absorbance of greater than 1.0 mAU (White & Catterick 1987). A comparison of this technique against the other commercial data handling routines mentioned previously indicates that absorbance ratioing offers some significant advantages. These are of considerable benefit in forensic examination. They include the ability to analyse smaller sample sizes, faster acquisition, and simpler presentation of results. Another very important aspect is that since results can be evaluated statistically there is an increased confidence in the reporting of analytical results. This technique, although originally used for the forensic examination of drug samples, has been applied subsequently to casework

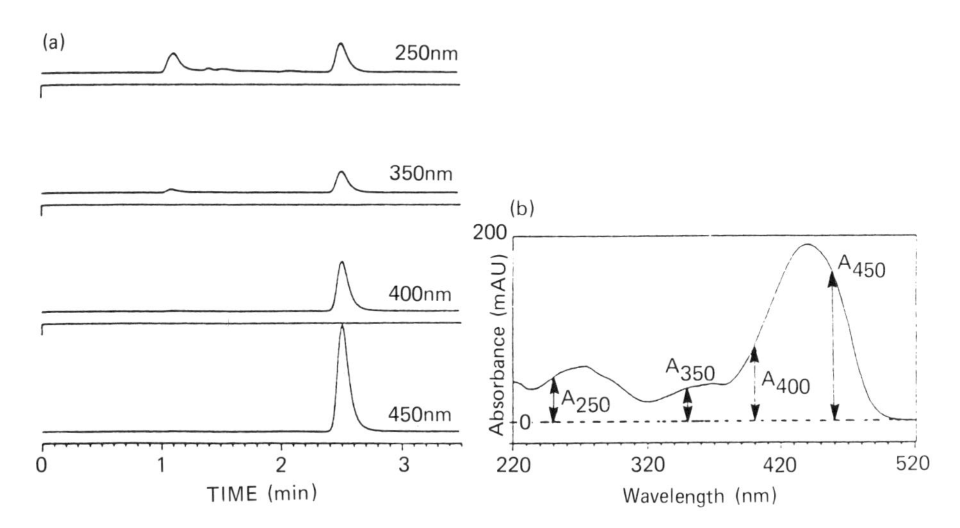

Fig. 5.13. (a) Multiwavelength plot and (b) UV-visible spectrum of the dye component eluted with a retention time of 2.5 min.

where dye analysis has been required. Successful applications have included the identification and discrimination of ink (White & Wheals 1984), food (White & Harbin 1989), and fibre dyes (Wheals *et al.* 1985, White & Harbin 1989, White & Catterick 1990, White & Catterick 1987).

Having realized the potential of this technique, but still concerned about the extremely large number of dyes that are commercially available, and hence the certainty of correct identification, this group continued to investigate other techniques that could be used with a multiwavelength detector. This has resulted in the development and use of two further techniques, *peak purity parameters* (PPP) and *complementary chromaticity coordinates* (CCC). Both techniques require spectral data for example, absorbance and wavelength values, and they provide similar advantages to absorbance ratioing since these data are reduced to a considerably smaller number of numerical values. The PPP technique was introduced by Varian when they marketed their diode array multiwavelength detector, and it generates a value which is the absorbance weighted mean wavelength of a spectrum for a specified wavelength range (Alfredson & Sheehan 1985, Alfredson *et al.* 1987). As illustrated in Fig. 5.14 the PPP value for any selected wavelength range, for example, $(x_1 - x_n$ nm) is calculated by using the following algorithm:

$$\text{PPP} = \frac{A_1^2 x_1 + A_2^2 x_2 + \ldots A_n^2 x_n}{\Sigma_1^n A^2}$$

where A is the absorbance measured by each diode and x is the monitoring wavelength of the diode (Varian Associates 1988).

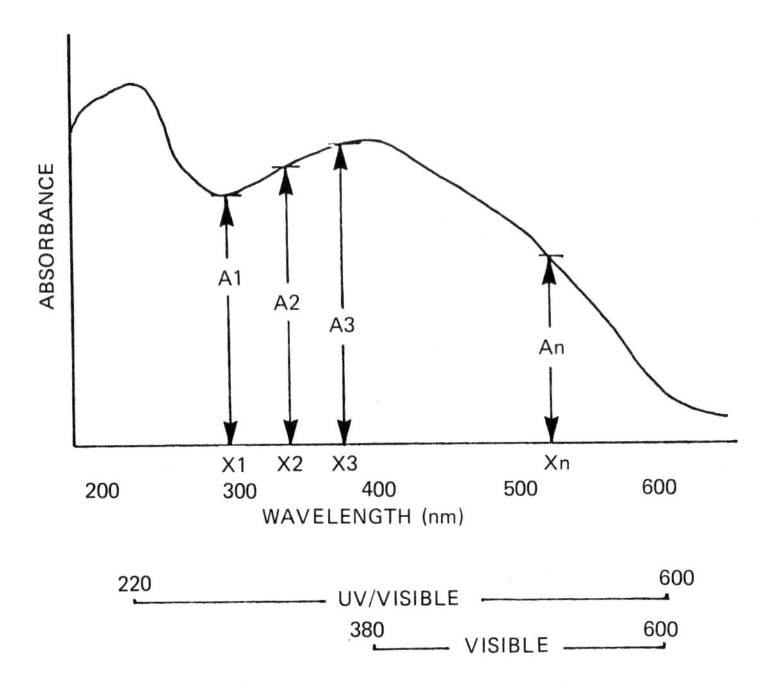

Fig. 5.14. Spectral data required for the determination of PPP values and typical wavelength ranges used for the analysis of dyes.

Originally, this technique was introduced for the analysis of UV absorbing spectra, but in a recent study PPP values were produced from UV/visible spectra recorded for some dye samples (White & Catterick 1990). Two wavelength ranges were recommended: 200–600 nm (UV/visible) and 380–600 nm (visible). The precision, effect of sample size, and sample discrimination were investigated for each of these wavelength ranges, and proved that a precision of less than 1% RSD could be achieved with 25 ng dye samples.

Chromaticity coordinates have been used by spectroscopists for the comparison of colorants, and more recently complementary chromaticity coordinates (CCC) have been introduced for the analysis of fibre dyes separated on a TLC plate (Paterson & Cook 1980). The latter technique, although suitable for the analysis of dyes in HPLC eluates, has attracted attention only very recently, since the data required to calculate these CCC values can be generated rapidly with a multiwavelength detector. This approach has been evaluated (White & Catterick 1990), and the calculation of the CCC values was based upon a routine described previously by Paterson and Cook (1980).

Owing to the upper wavelength limit of the multiwavelength, detector CCC values could be determined over only a restricted wavelength range of 380–600 nm, and a wavelength interval of 8 nm was selected for this study. To assess this routine, CCC values were calculated from the same set of analytical data that had been used to investigate the PPP technique (White & Catterick 1990). This provided a basis for comparing directly the performance of the CCC and PPP techniques, and, with the dyes selected, slightly poorer precision was recorded for the CCC technique. However, a slight improvement in sample discrimination could be achieved if the CCC values were employed.

Clearly, all of these numerical spectral techniques should be considered by the forensic scientist, especially if examining fibre dye extracts, because they enhance the characterization of dyes and increase confidence in reporting results. Since each of these techniques measures different parameters from UV/visible spectra, it has been suggested that a combination of all three routines should be used. Evidence to support this suggestion has been published in a case reported recently which required the identity of a dye to be confirmed (White & Catterick 1990).

Furthermore, as a result of a comprehensive study in which small quantities of dye samples were analysed, it has been shown that structural information can be derived by a combined use of these absorbance rationing, PPP, and CCC routines (White PhD thesis). If it is considered that no other analytical technique can currently provide this type of information, these results certainly highlight the potential of using HPLC, multiwavelength detection, and numerical spectral techniques for the qualitative analysis of fibre dyes.

5.5.3.4 Extraction procedures

Whilst an HPLC method might provide the desired chromatographic efficiency, resolution, and sensitivity with dye standards, the successful analysis of a fibre dye extract can be very dependent upon the choice of extractants and extraction procedure. Ideally, an extraction procedure should be capable of removing all dyes from a fibre quantitatively, reproducibly, quickly, and without degradation of the dye or, one hopes, the fibre. Furthermore, any extraction used must be compatible with the HPLC separation systems or detection techniques employed.

Experience gained from the TLC analyses of fibre dyes has to some extent been useful in the development of extraction procedures for the HPLC examination of fibre dyes. Some modifications have been required. A summary of the published procedures quoted for the HPLC of acidic, basic, and neutral groups of dyes is given in Table 5.7.

To be able to compare suspect and control fibre dye extracts accurately it is essential that extraction procedures provide reproducible results. Fortunately, an advantage of HPLC is that it can be used both qualitatively and quantitatively to assess results, hence the reproducibility of an extraction procedure can be determined. However, some care is required in the interpretation of any reproducibility study because of the use of multicomponent dyes, and dye content can vary along a length of a fibre or between fibres.

The extraction reproducibility for the DMF method used to extract disperse dyes from polyester fibre has been investigated and some interesting results have been reported (Wheals *et al.* 1985). It was found that with multicomponent dyes the variation in peak height of individual components ranged from 8.2 to 18.2% RSD, but if ratios of peak heights were compared the RSD ranged from 1.1 to 6.9%.

Table 5.7. Summary of published extraction procedures used for the HPLC analyses of fibre dyes

Dye group reference	Extraction solvent	Extraction conditions volume (μl)	temp (°C)	time (min)
Acidic (Laing *et al.* 1988)	Pyridine/water (4 : 3)	10	90	20
(Tebbett *et al.* 1990)	*Pyridine/water* (4 : 3)	100	100	20
Basic (Griffin *et al.* 1989)	Formic acid/water (1 : 1)	4	100	15
(Tebbett *et al.* 1991)	Formic acid/water (1 : 1)	100	100	20
Neutral (Wheals *et al.* 1985)	DMF[a]/Acetonitrile (1 : 1)	4	120	2
(Wheals *et al.* 1985)	Chlorobenzene	5	100	15

[a]DMF = dimethylformamide (250 ml) modified by addition of 2,6-di-tert-butyl-4-methyl phenol (1.25 g) and citric acid (1 g) to prevent oxidation.

This information is invaluable because (a) it implied that the extraction procedure was not preferentially dissolving one component from the dyed fibres, and (b) it provided a method for evaluating the extraction procedure. Unfortunately, similar studies have not been reported for acidic or basic dyes, and also the variation of dye content along a length of fibre has not been investigated.

Every effort should be made to ensure that an extraction procedure is highly efficient, because of the very small sample sizes encountered. With the published methods which have been shown to meet forensic requirements (Wheals *et al.* 1985, Griffin, *et al.* 1989), several precautions have been taken to maximise extraction efficiency. These have included the choice of extractants, the use of minimal sample manipulation, and small extraction volumes. Injection volumes also have an effect on band-broadening in the chromatographic system, and hence sensitivity. Therefore it is prudent to use small injection loop volumes, preferably not greater than 5 μl.

The choice of extractants is very important since these can affect both extraction efficiency and analysis time. Furthermore, any extractants used must be compatible with the HPLC eluent, and with the separation and detection techniques employed. It is worthwhile referring to the published results on the analysis of disperse dyes (Wheals *et al.* 1985), since the difficulties related above were encountered and had to be addressed in the development of the HPLC method.

Two alternative extraction procedures were evaluated in this study, one based on the use of chlorobenzene and the other dimethylformamide (DMF)/acetonitrile. Whilst both appear to work successfully, some advantages and disadvantages have been noted with both of the extractants, and these had further practical implications.

It has been claimed by Kissa (1979), that some disperse dyes undergo oxidation upon extraction, and with hot organic solvents the use of an antioxidant and an organic acid was recommended. To avoid sample degradation the DMF/acetonitrile extractant was modified with an antioxidant, 2.6-di-tert-butyl-4-methylphenol, and citric acid. A comparison of the DMF and chlorobenzene extractants indicated that, with the fibre dye extracts analysed, no evidence of sample degradation could be detected with either of the DMF or chlorobenzene extraction methods. Since this oxidation problem could still not be ruled out, the authors considered it prudent to use the DMF/acetonitrile extractant system. However, direct analysis of this extractant was impossible because of incompatibility with the HPLC system. This was overcome by diluting the extract (4 μl) with 6 μl of the eluent without a substantial loss in detector sensitivity. Unfortunately, the inclusion of the antioxidant precludes UV monitoring because of its substantial UV absorbance. Furthermore, it is retained, and when monitored in the visible wavelength region it gave a chromatographic baseline deflection, although this can act as an internal standard for the comparison of retention time data.

With the chlorobenzene extraction method this required longer extraction periods and more sample manipulation because of the need to evaporate the solvent. This last stage was found to be very important because if any residue of chlorobenzene was still present, this solvent produced chromatographic interferences. However, once all residual chlorobenzene was removed full UV/visible data could be obtained.

Clearly, these examples identify the problems of developing extraction procedures for the HPLC analysis of fibre dyes, but they do appear to be resolved for the neutral dyes and a selection of basic dyes. When attempting to repeat these methods it is advisable to ensure that the extraction procedure is compatible with the HPLC system. Any change in the column packing materials, especially those based on silica, can lead to a variation in retention of dyes and/or extractants, hence producing a chromatographic or detection problem. With the introduction of polymeric packing materials these might offer different selectivities for both samples and extractants. Additionally, because of the inert properties of this type of material and stability over a pH range of 1–14, there is the possibility that these packing materials may permit a wider range of extractants to be used.

It has already been identified that the forensic scientist does not currently have an HPLC method that is suitable for the analysis of acidic fibre dyes. Unlike neutral or basic dyes, the acidic group of colorants used to dye fibres includes several classes of dye as outlined earlier in Table 5.3. With the complexity of dye structures and some dyes being chemically bound to the fibre, this presents not only chromatographic problems but also extraction difficulties. Whilst some success has been achieved in resolving the former problem it is evident that the lack of a suitable HPLC system is due mainly to extraction difficulties.

Ideally, a single extraction procedure for all the acidic dyes is required, but it is evident from TLC studies that this cannot be achieved. For example, pyridine/water (4 : 3) will extract several classes of acidic dyes but is ineffective for others, for example, mordant, premetallized, and reactive dyes. Extraction for some of these dyes with this particular extractant fails because they are chemically bound to the fibre.

To overcome this extraction problem one possible solution is to use an extractant which is capable of breaking the dye-fibre chemical bond. This approach is being tested with some dyed woollen fibres. Several extractants including thioglycollic acid and some enzymes, for example Subtilisin carlsberg, under a variety of extraction conditions have already been investigated (Rudd & White unpublished results). To date, the results have not been very encouraging because of poor and variable extraction efficiency and problems with compatability between the extractants and the HPLC system. From the difficulties faced it is evident that a considerable amount of effort will be required to solve this particular problem. Hence, it may be a considerable time before an HPLC system suitable for the forensic examination of acidic fibre dyes can be introduced.

5.6 CONCLUSIONS

During the last decade or so, procedures for the extraction of dyes from single fibres, for chromatographic analysis, have been developed and optimized. Similarly, thin layer chromatographic systems have been described, and continue to be 'fine tuned' for the separation of the dyestuffs commonly encountered in the forensic laboratory. Thin layer chromatography has become the accepted method of choice for the comparison of two or more fibre dyes.

High performance liquid chromatography, however, has great potential for improving the sensitivity and resolution of TLC in the examination of fibre dyes. Although in its relatively early stages of development, HPLC has been shown to be a viable technique for the forensic examination of some classes of fibre dyes. Systems have already been described for the analysis of basic and neutral dyes, but currently none is available for acidic dye analysis. It should be emphasized that HPLC is not yet an option as a technique for the routine examination of fibre dyes in an operational forensic laboratory. Further developmental work is required to circumvent some of the problems outlined above. Nontheless, a great deal of progress has been made over the last few years, aided by the introduction of polymeric column packing materials and the increased sensitivity of multiwavelength detection systems. These factors, coupled with mathematical approaches such as absorbance ratioing, peak purity parameters and complementary chromaticity coordinates, will soon make HPLC a strong candidate for inclusion into the trace evidence section of the forensic laboratory.

HPLC offers the advantage of being able to examine smaller sample sizes, with a high degree of precision, and it can be evaluated statistically. These factors will improve sample identification and/or discrimination, and therefore confidence in reporting results.

Finally, now that the potential of HPLC for the forensic examination of fibre dyes has been illustrated, we expect to see a gradual increase in the acceptance and further development of this application over the next few years.

BIBLIOGRAPHY

Alfredson, T. & Sheehan, T. (1985) *Am Lab* **17**, 40.

Alfredson, T., Sheehan, T., Lenert, T., Aamodt, S. & Correia, L. (1987) *J Chromatogr* **385**, 213.

Cheng, J., Wanogho, S. O., Watson, N. D. & Caddy, B. (1991) *J For Sci Soc* **31**, 31.

Clark, A. B. & Miller, M. S. (1978) *J For Sci* **23**, 21.

Fong, W. (1989) *J For Sci* **34**, 295.

Gaudette, B. D. (1988) In: Saferstein, R. (ed.), *Forensic science handbook*, Vol. II. Prentice Hall, New Jersey, 209–272.

George, S. A. & Maute, A. (1982) *Chromatographia* **15**, 419.

Golding, G. M. & Kokot, K. (1989) *J For Sci* **34**, 1156.

Griffin, R. M. E., Kee, T. G. & Adams, R. W. (1988) *J Chromatogr* **445**, 441.

Griffin, R. M. E., Kee, T. G. & Elliot, L. (1989) *NIFSL Technical Note* No. 711.

Joyce, J. R. & Sanger, D. G. (1979) *J For Sci Soc* **19**, 203.

Joyce, J. R. & Humphreys, I. J. (1982) *J For Sci Soc* **22**, 253.

Joyce, J. R., Sanger, D. G. & Humphreys, I. J. (1982) *J For Sci Soc* **22**, 337.

Kidd, C. B. M. & Robertson, J. (1982) *J For Sci Soc* **22**, 301.

Kissa, E. (1979) *Text Res J* **46**, 245.

Laing, D. K., Gill, R., Blacklaws, C. & Bickley, H. M. (1988) *J Chromatogr* **442**, 187.

Macrae, R. & Smalldon, K. (1979) *J For Sci* **24**, 109.

Paterson, M. D. & Cook, R. (1980) *For Sci Int* **15**, 249.

Resua, R. (1980) *J For Sci* **25**, 168.

Resua, R., DeForest, P. R. & Harris, H. (1981) *J For Sci* **26**, 515.

Robertson, J., Kidd, C. B. M. & Parkinson, H. M. P. (1982) *J For Sci Soc* **22**, 353.

Rudd, S. & White, P. C. Unpublished results.

Society of Dyers and Colourists, (1976) *Colour Index*. Chorley and Pickersgill Ltd, Leeds.

Tebbett, I. R., Wielbo, D. & Strong, K. (1990) Presented at the 12th meeting of the International Association of Forensic Sciences, Adelaide, Australia.

Varian Associates (1988) *Publication No. 03-914252-00*, Varian, Sunnyvale, CA.

West, J. C. (1981) *J Chromatogr* **208**, 47.

Wheals, B. B., White, P. C. & Patterson, M. D. (1985) *J Chromatogr* **350**, 202.

White, P. C. (1991) PhD thesis, Brunel University.

White, P. C. Unpublished results.

Whist, P. C. & Bennett, A. Unpublished results.

White, P. C. & Catterick, T. (1983) *J Chromatogr* **280**, 376.

White, P. C. & Wheals, B. B. (1984) *J Chromatogr* **303**, 211.

White, P. C. & Catterick, T. (1987) *J Chromatogr* **402**, 135.

White, P. C. & Harbin, A.-M. (1989) *Analyst* **114**, 877.

White, P. C. & Catterick, T. (1990) *Analyst* **115**, 919.

6

The application of infrared microspectroscopy to the analysis of single fibres

K. P. Kirkbride, BSc, PhD
State Forensic Science, Forensic Science Centre, 21 Divett Place, Adelaide, South Australia 5000, Australia

6.1 INTRODUCTION

Forensic scientists are often called upon to analyse fibres in an effort to assist police with their investigations or to help courts decide whether a suspect can be associated with a crime. In the case when it appears that fibres have been transferred during a crime, for example between victim and offender, the analytical task becomes two-fold. One part involves comparative tests, in which the transferred fibres are compared to those taken from a putative source. Secondly, an identification step is performed, that is, a test which aims to establish the classes to which the transferred fibres belong (such as polyesters, nylons, cotton). This step is important for two reasons. When the fibres have been identified their likely behaviour during the crime, and afterwards, can be considered. Secondly, the extent to which the fibres are distributed in the community can be estimated once they are identified. Therefore, only after an examination involving both comparative and identification steps is a forensic scientist in a position to comment as to the significance of fibres transferred during a crime.

Infrared spectroscopy is a powerful technique which not only allows fibres to be compared on the basis of their characteristic infrared spectra but also identified because the data produced by the spectrometer are rich in diagnostic information.

It is the aim of this chapter to discuss the application of infrared spectroscopy to the analysis of fibres in terms of scope, limitations, and practical aspects. This

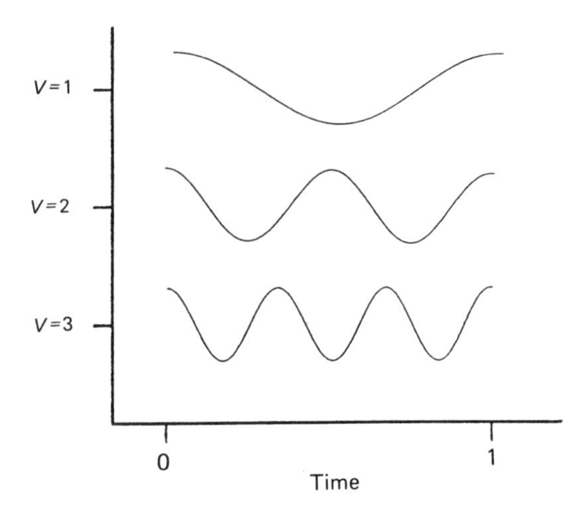

Fig. 6.1. Waveforms of frequency 1, 2, and 3 cycles per second. The horizontal axis represents 1 second.

discussion, necessarily based on physics and chemistry, must start off with an introduction to these topics if it is to be comprehensible to persons not specializing in the field; the remainder of the chapter is structured with this view in mind.

6.2 RADIATION, MATTER AND SPECTROSCOPY

6.2.1 Radiation

Light, X-rays, radiowaves, and microwaves are physical phenomena with which most persons are acquainted. What might not be so familiar is the fact that each of these phenomena is a manifestation of one physical entity called *electromagnetic radiation*. The difference between microwaves and X-rays, and any other portion of the spectrum of electromagnetic radiation, is the amount of energy that the radiation carries. Early in this century it was shown that the energy of radiation is dependent upon its frequency, and this idea is expressed in equation (6.1), which is called Planck's function after its discoverer.

$$\text{Energy} = \text{Planck's constant} \times \text{Frequency}$$

or

$$E = h \times \nu \tag{6.1}$$

What then is the frequency of radiation? All forms of electromagnetic radiation can be visualized as waves, similar to the type which can be set up in a piece of string tied securely at one end. The frequency of the wave set up in the string is the number of times it vibrates in one unit of time; if the string vibrates once in one second it has a frequency of one, twice in one second a frequency of two, three times in one

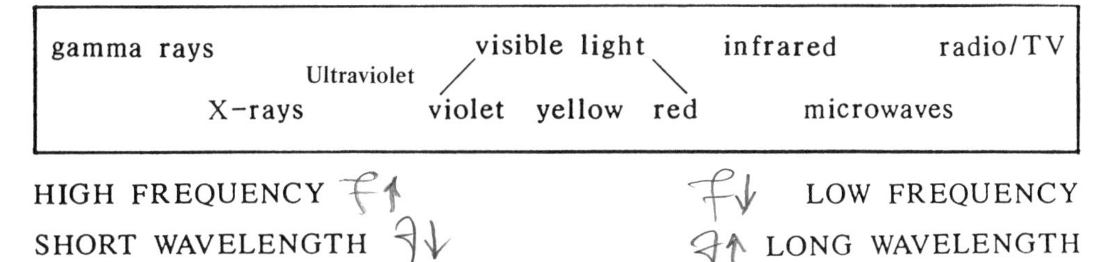

Fig. 6.2. The electromagnetic spectrum.

second a frequency of three, and so on. As one might expect, more energy is required to vibrate a string twice per second than to vibrate it once, and this situation is exactly that described by Planck's equation.

A more subtle relation also comes from this important equation; it is shown in Fig. 6.1, which represents hypothetical pieces of string in which have been set up waves of frequency one, two, and three; the horizontal axis depicts time, in this case one second. As can be seen, the waveforms in the three cases vary, in particular, the distance between crests in the waves gets shorter as the frequency increases from one to two to three. The distance between the crests is termed the wavelength, and it is the other important parameter (together with frequency) used to characterize electromagnetic radiation.

Fig. 6.2 shows how common forms of radiation fit into the electromagnetic spectrum in terms of their frequency, wavelength, and energy.

6.2.2 Matter

All forms of matter, the liquids, solids, and gases that surround us, are composed of molecules, and, in turn, molecules are built up from atoms. There are only ninety-two different types of naturally occurring atoms, and any molecule represents only one of the enormous number of combinations of them. Let us take pure water as an example and its well known molecular formula H_20. This chemical code tells us that the water molecule contains two atoms of hydrogen and one atom of oxygen. If we were to look at water with an extremely powerful microscope we would firstly see that it is composed of small units—these are water molecules. If the magnification of the microscope was made greater, we would then see that each one of the individual molecules was composed of three building blocks—the two hydrogen atoms arranged in a V formation around the oxygen atom. Formally, this molecule is described as an oxygen atom bonded to two hydrogen atoms, and the two bonds are described as oxygen to hydrogen.

A bond between atoms is caused by the interactions of their constituent parts or subatomic particles; these are negatively charged electrons and the positively charged nuclei of the atoms. Electrons situated between adjacent atoms experience an attractive

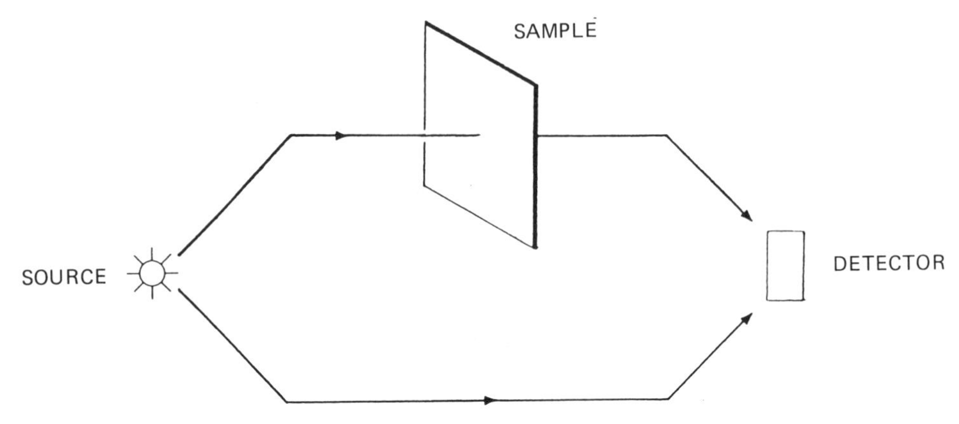

Fig. 6.3. Schematic diagram of a hypothetical spectrometer.

force toward the positively charged atomic nuclei, and this causes the atoms to stick together or bond. Certain bonds are formed when a larger number of electrons gather between nuclei; these are termed double and triple bonds. The bonds described above for water are termed single bonds.

The molecules present in synthetic fibres are usually composed of combinations of only a few types of atoms; namely carbon, hydrogen, oxygen, nitrogen, and, rarely, sulphur, chlorine, and fluorine. Again, if we could look at fibres under a very powerful microscope, we would see that their constituent molecules are long chains of repeated units—much like the fibres they make up. In section 6.4 the structure of fibres is considered in more detail.

Chemists do not look at a substance under a powerful microscope to ascertain the structure of its molecules, nor to see if more than one type of molecule is present. Indirect methods have to be relied upon—one of the most important is spectroscopy.

6.2.3 Spectroscopy

Spectroscopy is the technique whereby electromagnetic radiation is allowed to interact with matter in the hope that the interaction will yield some information about the nature of the matter. Usually the data yielded by spectroscopy do not tell the chemist how many of each type of atom is present in a molecule. Instead, the information relates to the types of atoms present and the bonds between them.

Fig 6.3 shows a hypothetical spectrometer. A beam of radiation is split into two; one is passed directly into a light measuring device (a detector) and the other is passed through a sample and then on to the detector. Unless the sample is completely transparent, it will absorb some of the radiation; it is the detector's job to sense the fact that the beam which has passed through the sample is attenuated compared to the other beam. A simple analogy is to consider one's eye as a detector, some pieces of glass as samples, and a light bulb as a source of electromagnetic radiation. Consider first looking at the bulb, and then looking at it through a piece of smoked glass. The

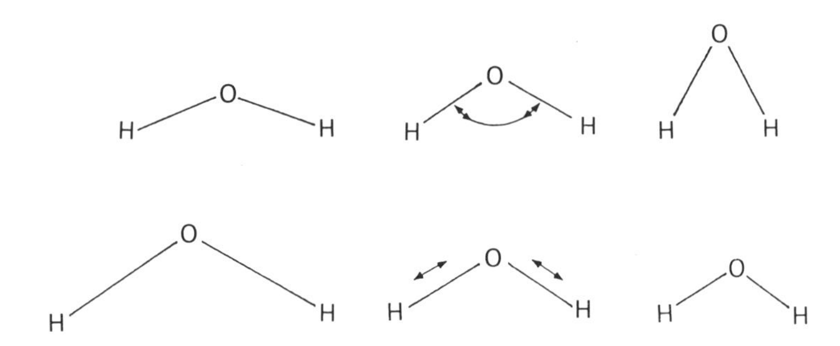

Fig. 6.4. The effect of infrared radiation on water. Top—bond bending. Bottom—bond oscillation.

eye senses that the light which has passed through the smoked glass is of lesser intensity than the unimpeded light.

Spectroscopy would be of little use if its only application was in measuring relative levels of transparency. Consider now the case of looking at the light through pieces of coloured glass, say a red piece then a blue piece. The eye can still sense that the light from the source is diminished by the pieces of glass, but, more importantly, it can also sense that the two pieces of glass are transmitting radiation of different frequencies, that is to say, light of different colours. This is the real power of spectroscopy, being able to distinguish between samples—and ultimately to identify them in terms of the atoms and bonds they contain—by virtue of the electromagnetic radiation that they absorb and transmit.

Although the basis of spectroscopy is the absorption of radiation by matter, the exact mechanism by which it is absorbed depends upon the type of radiation used as the probe. Absorption of low energy radiation, such as microwaves, causes only low energy changes to occur in matter, such as inducing molecules to tumble and spin. High energy radiation, such as far ultraviolet, gives far more energy to matter, enough to perturb its subatomic structure, and in some cases cause it to breakdown (this is why plastics often degrade in sunlight).

In infrared spectroscopy, matter absorbs radiation of medium energy, and in so doing its constituent molecules gain *vibrational energy*. Fig. 6.4 shows the effect of infrared radiation on water molecules. The energy from the radiation is captured by the molecules and causes the oxygen–hydrogen bonds either to stretch and contract (oscillate) or bend. The restrictions of quantum mechanical theory, the 'rules' by which physical processes occur at the atomic level, do not allow molecules to absorb energy in a random or imprecise fashion. If the energy required to bring about the bond oscillation is X, then X is the amount of energy that the radiation must supply to the molecule. If the energy requirement is strictly controlled, then, from Planck's function, so is the frequency of the radiation which can be absorbed.

Consider the situation when a sample of water, for example, is placed into the beam of an infrared spectrometer. The source emits radiation composed of a range of frequencies, but only radiation matched in energy to bond vibrations will be absorbed by the sample. The remaining energy will be transmitted by it.

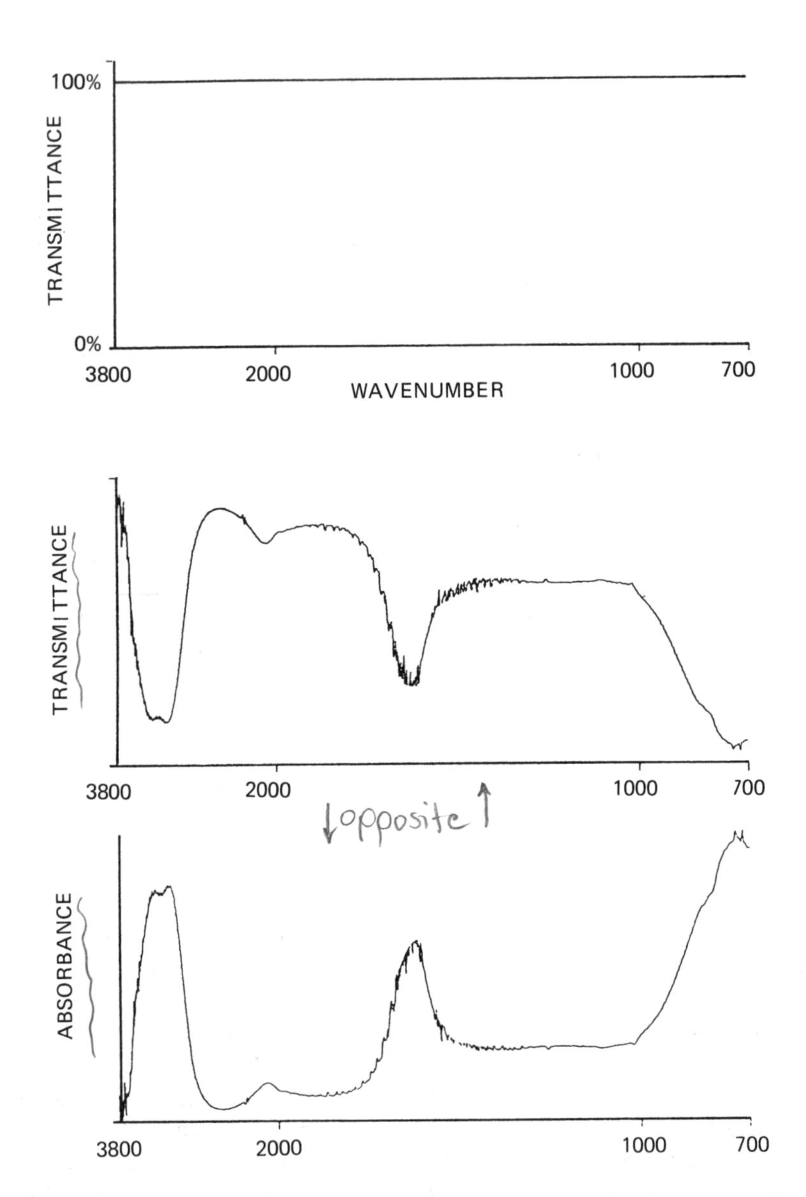

Fig. 6.5. Top—unattenuated spectrometer beam. Middle—infrared transmittance spectrum of water. Bottom—infrared absorbance spectrum of water.

This concept is described in Fig. 6.5. The top graph shows the response of the detector to an infrared source that is not attenuated by any matter plotted as a function of frequency. Each frequency emitted by the source is received at full intensity.

In the middle graph can be seen the response when the infrared beam is passed through a sample of water. Over certain frequency regions the intensity of the radiation received by the detector is low; these regions correspond to the frequencies absorbed

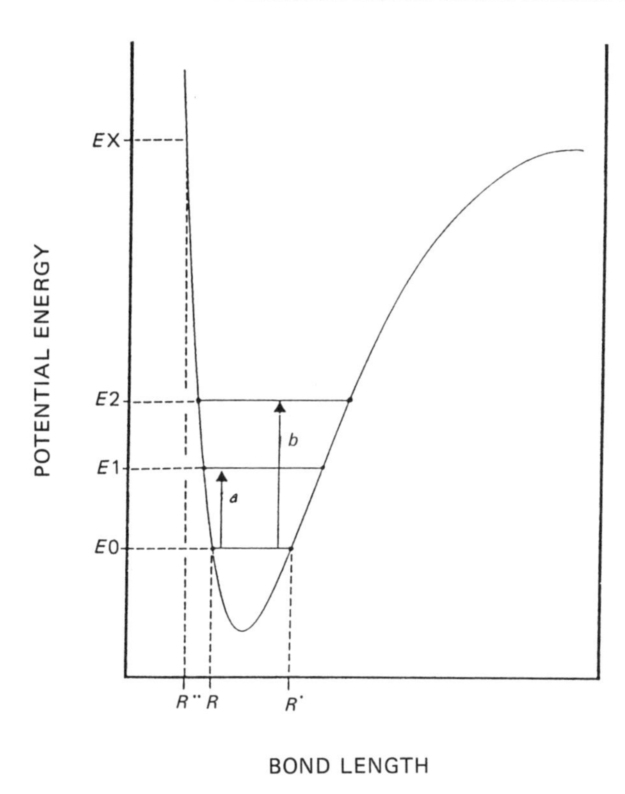

Fig. 6.6. Potential energy of a vibrating bond expressed as a function of bond length.

by water. The two vibrational motions shown in Fig. 6.4 give rise to the peaks located near the centre of the graph and at its left-hand side.

The middle graph is called the infrared spectrum of water, and the position and intensity of the absorptions (confusingly, these troughs are commonly known as peaks) are characteristic of water. The horizontal scale of the graph is frequency expressed in units called *wavenumbers* (also called reciprocal centimetres or cm^{-1}); these units are commonly used by infrared spectroscopists. The vertical scales range from 100% transmission of the beam (that is, no absorption of it) to 0% transmission (or complete absorption of it).

It is also possible to construct the vertical axes to display absorbance of radiation versus frequency, and the resulting spectrum, displayed in the bottom graph, is termed an absorbance spectrum. Infrared spectroscopists vary in their predilictions for presenting data, so both absorbance and transmittance spectra appear in the literature.

6.3 THE INTERACTION BETWEEN INFRARED RADIATION AND MATTER

The changes brought about when a molecule absorbs infrared radiation are described in Fig. 6.6. At room temperature, the atoms within a molecule are in a constant state

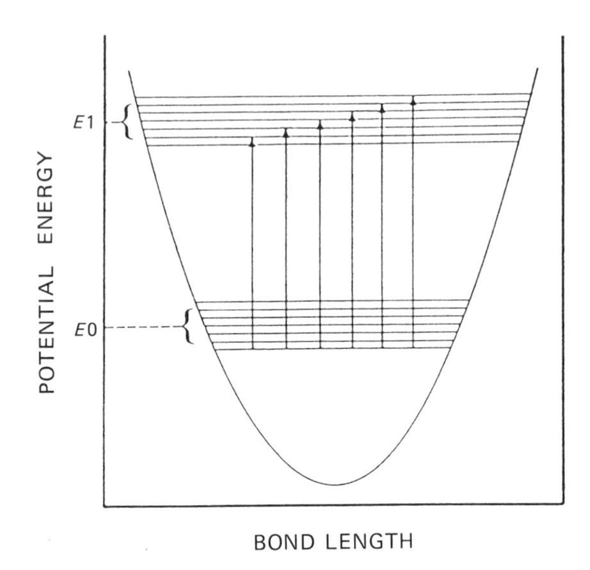

Fig. 6.7. Enlarged view of Fig. 6.6.

of motion, undergoing vibrations and tumbling. Even at absolute zero, the coldest temperature attainable, quantum mechanics shows that molecular atoms move relative to each other.

The curved line in Fig. 6.6 is the potential energy of a molecular bond as it undergoes lengthening and contractions (that is oscillates); the horizontal axis depicts the distance between the atoms or the bond length.

The restrictions of quantum mechanical theory allow bonds to possess only discrete or quantized amounts of energy, and the lowest amount is termed the zero point energy. In Fig. 6.6, the zero point energy is shown as $E0$ and corresponds to the situation when the molecular bond oscillates between a length of R and R''.

If the bond gains enough energy from radiation, then it might be 'promoted' to a higher vibrational state, such as that depicted by the horizontal line at $E1$. It turns out that for all molecules, the right amount of energy required to bring about promotion can be supplied by infrared radiation. If the bond was to absorb enough energy to promote it to energy level Ex, then it would oscillate between a length of R' and infinity; in other words, the bond would dissociate or fall apart.

A fundamental transition has taken place when the bond gains enough energy to promote it to level $E1$. Most peaks in infrared spectra are due to this phenomenon. Peaks of low intensity, called *overtones*, are sometimes observed; these arise when a bond absorbs enough energy to promote it to the second excited state of energy $E2$. As can be seen from Fig. 6.6, the energy required to promote a bond to its second excited state is almost twice that of the fundamental transition. For this reason (using Planck's function) overtone bands are observed in spectra at twice the frequency of the fundamental peak.

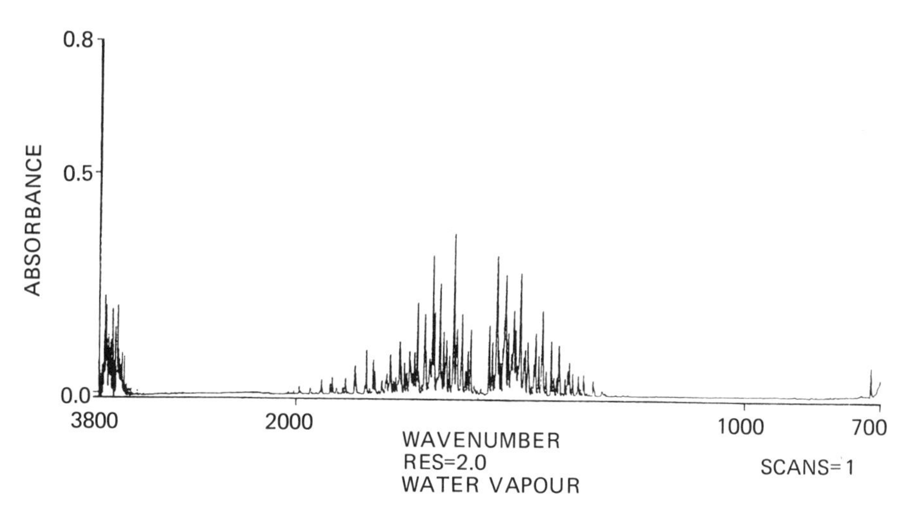

Fig. 6.8. Infrared spectrum of water in the vapour phase cf Fig. 6.5 in which is shown the infrared spectrum of liquid phase water.

Exactly the same argument applies when bonds absorb radiation and increase the energy of other vibrational modes, such as the bending described for water in Fig. 6.5.

Fig. 6.7 depicts in greater detail the first two levels E0 and $E1$ in Fig. 6.6. In this diagram the horizontal lines depict many *rotational energy levels* superimposed on the two vibrational states. The rotational states arise because molecules are constantly spinning, in the same manner as a child's top. It is another restriction of quantum mechanics that a change in vibrational state must be accompanied by a change in rotational state. Many transitions, all very close in energy (frequency) and all fundamental, are therefore possible; these are depicted by the vertical arrows.

In theory, then, the peaks as shown in Fig. 6.5 should resemble those shown in Fig. 6.8. In practice, however, these transitions are visible only in high resolution spectra of gases; in the condensed phase, intermolecular collisions act to broaden out the rotational fine structure, giving rise to broad, unresolved absorptions as shown in Fig. 6.5. Rotational fine structure might therefore appear to be of academic interest only, but it does intrude into routine analyses performed on Fourier transform spectrometers (this type of instrument will be discussed later). Incomplete compensation for water vapour leads to residues of the fine structure centred at 1595 cm^{-1} appearing in spectra.

The absorption of infrared radiation by matter leads to an increase in its vibrational energy, but what is the mechanism by which a wave interacts with the atomic particles? As previously stated, all molecules are constantly undergoing vibrations, and all molecules consist of an arrangement of charged particles distributed unevenly through space. If we consider a water molecule, the distribution of charged particles in it is such that the oxygen atom carries a small negative charge (designated −), while the two hydrogen atoms each carry a small positive charge (designated +) as illustrated in Fig. 6.9.

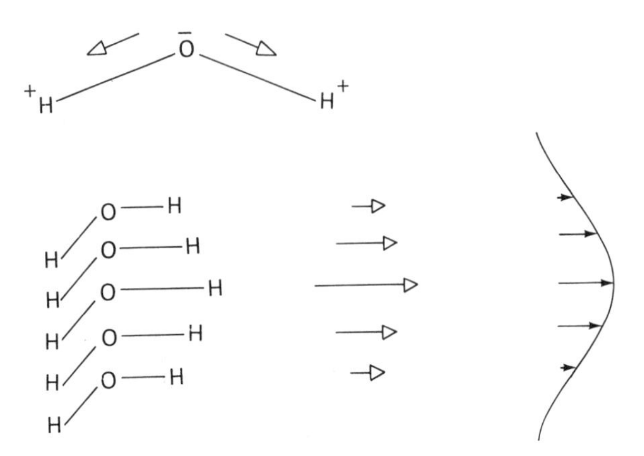

Fig. 6.9. The interaction between infrared radiation and water. Top—dipole moments associated with water. Bottom left—oscillation of one hydrogen–oxygen bond. Centre—dipole moment associated with bond oscillation. Right—electric field associated with radiation.

Each of the two oxygen–hydrogen bonds therefore have associated with it an electric field, the strength and direction of which are represented by the arrows (strictly speaking these arrows represent vectors called the dipole moments of the OH bonds). During the oscillation of the O—H bonds, the magnitude of the dipole moments will oscillate at the same frequency as the bond vibration (see Fig. 6.9).

Radiation also has an electric field associated with it (it also has an associated magnetic field, hence the name electromagnetic radiation), and this field oscillates at the frequency of the radiation (see Fig. 6.9).

Transferral of energy from radiation to matter can therefore be viewed as a resonant interaction between oscillating electric fields; if the frequency of the electric field of the incident radiation closely matches the frequency of the oscillation of the bond's dipole moment, then a resonant interaction can take place and the bond will absorb energy. Viewed simply, this process is analogous to two persons operating a large wood saw; efficient transferral of energy can take place only when the two operators work at the same frequency or in unison.

Infrared spectroscopy would be almost useless if all molecular bonds vibrated at the same frequency. Fortunately they vibrate over a wide and somewhat predictable range. An obvious corollary is that infrared spectroscopy is not a useful technique for the analysis of nonpolar substances (that is, those molecules which do not have small charges localized on some of their atoms), and might not be able to distinguish between molecules constructed of similar arrangements of similar bonds.

How then does one take the information contained within an infrared spectrum of a substance and convert it into a useful description of its structure? The frequency at which a bond oscillates depends upon its strength and the masses of the atoms at its termini. Pictured crudely, a bond might be thought of as a spring holding together

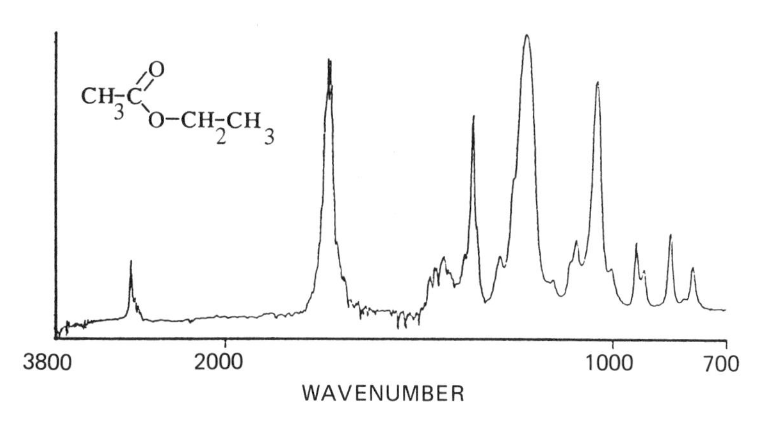

Fig. 6.10. Infrared absorbance spectrum of ethyl acetate.

two particles. For a given spring (bond) strength, the natural oscillation frequency is inversely dependant upon the masses of the particles (atoms)—heavier atoms result in a lower frequency. Similarly, for different springs between two given particles, the frequency of oscillation is proportional to the spring strength—a strong spring will oscillate at a higher frequency, a weaker one at a lower frequency.

Atomic masses are constant and are known accurately. Oxygen atoms and hydrogen atoms, for example, have specific masses. Bond strengths between two given atoms vary only between relatively narrow limits. As a consequence, the absorptions for bonds between oxygen and hydrogen atoms, for example, fall within a narrow and specific frequency range.

Literature dealing with infrared spectroscopy is full of tables listing the frequency at which particular groups of atoms absorb infrared radiation; the works by Colthup *et al.* (1964), Weast (1975), Avram & Mateescu (1972), and Bellamy (1975) are good sources of such data. Table 6.1 is an extract from the literature, listing the absorption frequencies of groups present in common fibres. A standard chemical shorthand is used in this Table: bold face letters represent atoms, with C representing carbon, N for nitrogen, O for oxygen, and H for hydrogen. The dashes between atoms represent the types of bonds holding them together: — for single bonds, = for double bonds, and ≡ for triple bonds.

From Table 6.1 in conjunction with Fig. 6.10, the spectrum for a liquid called ethyl acetate, one can see how the features of a spectrum arise. Hydrogen is the lightest atom known, and bonds containing it give peaks at the high frequency end of the spectrum. The peak near 3000 cm^{-1} in Fig. 6.10 is due to oscillations of the C—H bond. Absorbances near 3000 cm^{-1} are often found in spectra because most substances analyzed contain carbon and hydrogen.

Table 6.1.

Group of atoms	Infrared absorption frequency	Fibres in which this group is present
O—H	3200–3600	Cellulose based
N—H	3100–3500	Nylons, protein fibres
C—H	2800–3000	All except mineral fibres and 'Teflon' based fibres
C≡N	2200–2300	Acrylics, modacrylics
C=O	1650–1850	Modacrylics nylons, polyesters, cellulose acetate
C—N	1600–1650	Nylons, protein fibres
C—H	1350–1450	All expect mineral fibres and 'Teflon' based fibres
C—O	1000–1300	Polyesters, nylons, cellulose based fibres, modacrylics
C—F	1000–1400	'Teflon' based fibres
C—H	700–800	All expect mineral fibres and 'Teflon' based fibres
C—C1	600–800	Vinyl chloride based fibres

In the ethyl acetate molecule are three types of bond between carbon and oxygen, one double and two single. Oxygen atoms weigh much more than hydrogen atoms, therefore C—O bonds must give peaks at a frequency lower than C—H bonds. The carbon-oxygen double bond is much stronger than either of the two single bonds, therefore its peak is found at a higher frequency (1740 cm^{-1}). Although the two carbon-oxygen single bonds might appear to be equivalent, the infrared spectrum shows them not to be. One of the carbon-oxygen bonds is strengthened by its close proximity to the double bond therefore its peak is found at a frequency (1250 cm^{-1}) slightly higher than that for the remote bond (1150 cm^{-1}).

So far, only peaks arising from bond oscillations have been considered. As shown in Fig. 6.4, bonds also bend, but the energy required to bring about a change in this vibrational mode is comparatively low. Therefore, while the peaks associated with C—H oscillations in ethyl acetate are found near 3000 cm^{-1}, those associated with C—H bending modes are found at 1375 and 1450 cm^{-1}.

Peaks at the low frequency end of the spectrum (below 1000 cm^{-1} or the low end of the so-called fingerprint region), in addition to arising from deformations of bonds between heavy atoms, are due to movements of the molecule as a whole, and low energy bending movements. To visualize these motions, think of the entire molecule, not just single bonds in isolation, as being analogous to a loose concertina, able to expand, contract, and deform into waves.

Tiny peaks at 3470, 2470, and 2090 cm^{-1}, visible only after magnification of the spectrum of ethyl acetate, are overtones of the large peaks at 1740, 1240, and 1050 cm^{-1}; their appearance in this spectrum is characteristically small.

This example shows the wealth of information contained within an infrared spectrum. How does one go about interpreting, *a priori*, the spectrum of an unfamiliar substance? In general, the high frequency region (above 1500 cm^{-1}) yields information as to the gross structure of the molecule—does it contain C—H, C=O, C≡N, O—H, or N—H bonds, for example. Many molecules contain bonds of those types, so the chemists uses the fingerprint region (below 1500 cm^{-1}) in an effort to differentiate between the many possible structures. This latter step is accomplished by comparing standard spectra with the questioned spectrum. This technique leaves a lot of scope for intuition, and, to assist, many computer programs exist which enable comparison of the questioned spectrum with standards.

A facet of infrared spectroscopy, in addition to its use as a qualitative technique, is that it is quantitative. If we have a mixture of two substances, X and Y, then its spectrum is the sum of the individual spectra of X and Y (assuming there is no chemical interaction between the substances). Also, if the amount of X in the mixture is increased, the effect on the spectrum of the mixture is that the peaks due to X will increase.

The foregoing is succinctly expressed in a mathematical relation called *Beer's law* (Equation 6.2). This law shows that the absorbance of a peak (A, its height or intensity) is the product of a mathematical constant (E, called the *extinction coefficient*), the concentration of the molecule responsible for the peak (C), and the length of the path the radiation takes through the substance (L).

$$A = E \times C \times L \tag{6.2}$$

One direct result of Beer's law is that if the thickness of a sample is doubled, then the size of the peaks in its spectrum will also double. For example, peaks in the spectrum of a sheet of polystyrene 20 μm thick will have absorbance values twice as large as corresponding peaks in the spectrum of a 10 μm thick film. Similarly, a solution of a substance at a concentration of 2% will yield a spectrum with peaks twice as intense as those in a spectrum acquired from the same substance at a level of 1%. Beer's law eventually breaks down when very thick films or concentrated solutions are analysed.

The concept of extinction coefficients is more difficult to grasp. Bearing in mind the mechanism of the interaction of infrared radiation with matter, it is reasonable to find that different types of bond with different strengths and extents of polarization absorb different amounts of radiation. A large extinction coefficient for a bond implies that it strongly absorbs radiation and will be associated with an intense peak. It is common in infrared spectroscopy to express an extinction coefficient qualitatively, thus descriptive words like strong, medium, and weak abound. Peaks arising from bond oscillations are usually strong, particularly those arising from C=O bonds. Overtone peaks are usually very weak, while peaks due to bond bendings and entire molecule deformations range from medium to weak, respectively. Fig. 6.10 shows the typical range of peak sizes found in infrared spectra.

Taking the mixture of X and Y again as an example, Beer's law tells us that even if the peaks due to X are larger than those due to Y, the amount of X present is not necessarily greater. It might be the case that the concentration of Y is higher, but the extinction coefficients of its absorbances are small. It also follows that peak absorbance values should be universal constants—the absorbance values of peaks in the spectrum of the 20 μm sheet of pure polystyrene produced by one spectrometer should, in theory, be very close to those produced by another. In practice, this ideal situation might not be achieved, owing to instrumental variations and non-reproducibilities in sample preparation.

Infrared spectroscopy is a powerful tool because it enables us to characterize the chemical structure of a substance. For a substance X, the frequencies of its peaks, and their relative intensities, are characteristic. As a means of identification, infrared spectroscopy allows us to compare the spectral data of substance X with those of known standard substances. With the advent of high-speed computers capable of being attached to spectrometers, and copious digitized libraries of standard spectra, such spectral searching can be performed readily.

Infrared spectroscopy can also aid in drawing negative conclusions. For example, if a questioned fibre is thought to be of the modacrylic type, then its spectrum should show the characteristic peak near 2300 cm^{-1} due to the carbon–nitrogen triple bond. If the spectrum does not show this peak, then it can be assumed safely that the fibre is not modacrylic. The concepts discussed so far have been covered in greater detail by many textbooks; a short list of some is presented in the Bibliography.

6.4 FIBRES—THEIR COMPOSITION AND INFRARED SPECTROSCOPY

Fibres are long, thin, tendrils of matter made up from substances called polymers. They can be natural or synthetic in origin and composed of inorganic minerals or organic chemicals. In chemistry, the distinction between organic and inorganic lies in the presence of carbon; organic substances contain carbon, inorganic substances (with a few exceptions) do not.

Polymers are a class of substances that have in common a special structure. They are made up of repeated units of basic structural elements called monomers, just like trains are made up from individual coaches. Many individual chains aggregate together to form polymeric substances. A good example is cellulose, the Earth's most abundant polymer, which is present in such varied articles as plant cell walls, synthetic rayon garments, and paper. Cellulose is a polymer constructed of a monomer called

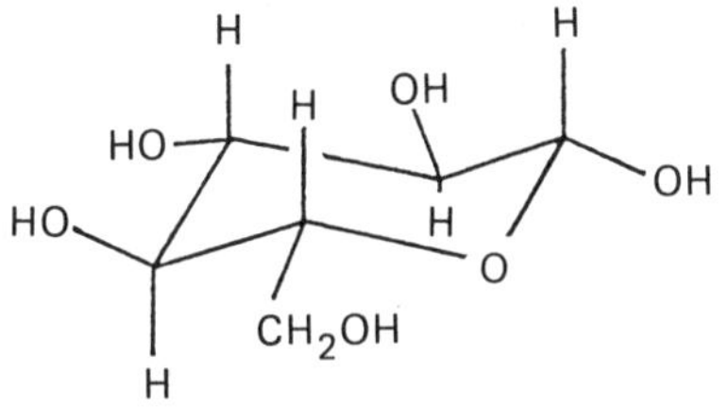

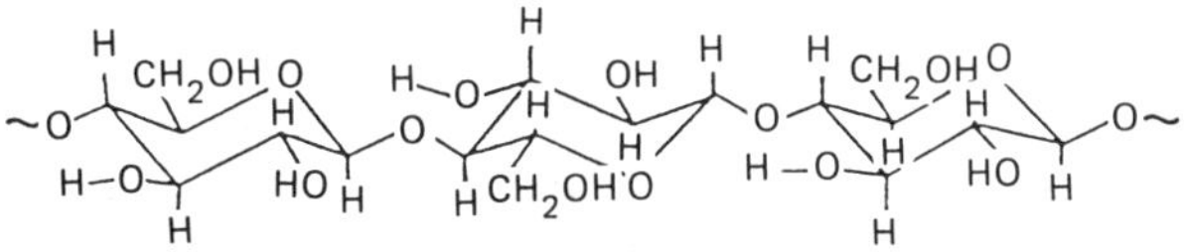

Fig. 6.11. Structure of glucose (A) and a portion of one of its polymers (B). The wavy lines indicate that the structure continues in the same fashion in two directions.

glucose. Many glucose molecules, of structure A in Fig. 6.11, link up as shown in B (Fig. 6.11) to form the polymer.

As a rule in polymer chemistry, the properties of the monomer are completely different from those of the polymer, and the number of monomers used in the structure greatly alters its properties. Glucose is a sweet, highly crystalline substance; none of its polymers is. Glucose is soluble in water, but two of its polymers, starch and cellulose, are not.

It is important to realize that the foregoing treats polymers as though they were composed of discrete strings of monomer which do not interact with one another. If this were actually the case, then polymers would not have the very properties that make them so useful: flexibility, elasticity, toughness etc. Individual chains of polymer strongly interact with each other in a sample by tangling, by lining up in regimented or semi-regimented rows, or by forming aggregates. The degree and nature of interchain interactions can be controlled during manufacture and processing, so that even though two given articles might be composed of identical polymer, their intimate structures might vary.

Undoubtedly, interactions between chains have an effect on some of the bonds within the individual chains, perhaps causing them to weaken or distort. Such a modification must have a concommittant effect on the infrared spectrum of the polymer. Unfortunately it is very difficult to predict exactly what effect the interactions might have on spectral properties.

This limitation means, for identification purposes, that it is wise to compare spectra of unknown *fibres* with spectra derived from standard *fibres*. To compare fibre spectra with those derived from bulk polymer standards, such as films or sheets, might lead to uncertainty because the degree and nature of interchain interactions might vary between fibres and films made from a given polymer.

The simple description of polymers given above might lead one to believe that a bulk sample of polymer is constructed of monomer chains of constant length. This is not the case—within a given sample there is a range of chain lengths which might be narrow or wide. In addition, although two samples might be constructed from the same monomer, the average polymer chain length in the two might be completely different, thereby conferring different physical properties to them.

Using infrared spectroscopy, it is very difficult to differentiate between samples of polymer differing only in the length of their monomer chains. One exception is the instance when the polymers in question have groups of atoms which give rise to a distinctive signal at their chain termini. In this chapter, the term *secondary structure* will be used to describe chain length and interchain associations. Finally, the basic description ignores the fact that many polymers are not composed of a single monomer. A product can be prepared, similar to an alloy, by blending two or more polymers, or by causing two dissimilar monomers to react with each other, and form chains. The latter mixtures are called *copolymers*, and acrylic/modacrylic fibres are good examples. Fibres of these types can differ not only in their composition with respect to the types of monomer used, but also their relative proportions.

For two given fibres there can be many points of comparison such as: their colour, thickness, dye composition, cross-sectional shape, refractive index and birefringence, amount and type of delustrant, melting point, and their polymeric composition.

Despite the limitations described above, infrared spectroscopy is still the method of choice for probing the polymeric construction of fibres. It is very useful in differentiating between fibres of different polymeric groups (nylons from acrylics from polyesters etc.), but at times it might not yield enough information to differentiate between different fibres of the same group; it can rarely differentiate between fibres spun from the same polymer by different manufacturers.

Let us imagine that the spectra in Fig. 6.12 represent two fibres. The spectra are very close in appearance; do they indicate that the fibres are identical? As it turns out, the fibres are nylon-6 and nylon-12, and their structures are given with the spectra. Using the two spectra as an example, it is obvious from inspection (or by automated search) that the fibres are nylons. Our knowledge of polymer chemistry tells us that the nylon class contains many members, most of which differ only in the the number of methylene (—CH2—) groups they contain. Therefore a potential source of useful information in the spectra might be the sizes of the signals due to the methylene groups, measured relative to the sizes of the peaks due to the carbon-oxygen double bonds. The larger relative size of the peak near 3000 cm^{-1} in nylon-12 and the differences in the low frequency ends of the two spectra are significant to the interpretation in this case. Even though this example was an exercise comparing two fibres, it can be seen that the identification of them as nylons was of critical importance. To use the characteristic peaks and troughs of a spectrum merely as a fingerprint, even for comparison purposes, is to leave the full potential of infrared spectroscopy unexploited.

A computer assisted search can be used to provide valuable information in addition to indicating the identity of a fibre. Usually the search routine will provide a list of matches with the questioned polymer and a quality index, which numerically indicates

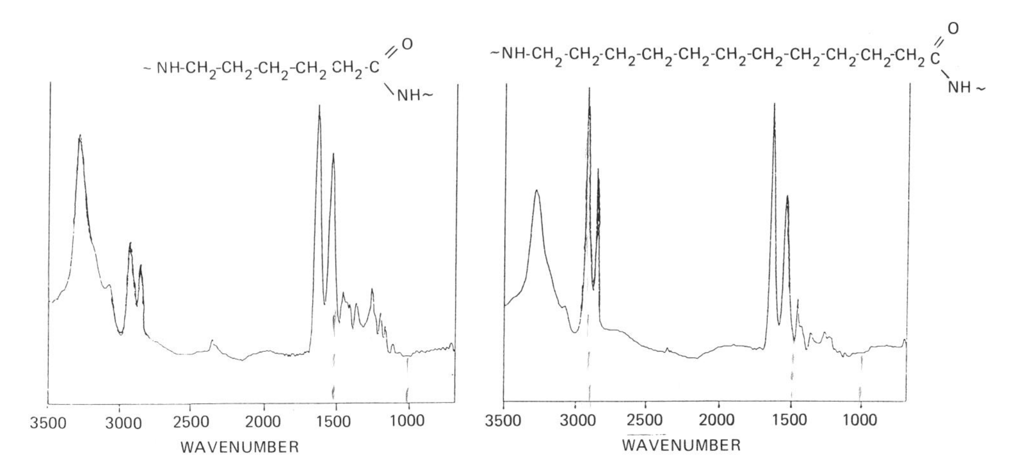

Fig. 6.12. Infrared absorbance spectra of nylon-6 (left) and nylon-12 (right).

how 'good' each match is. If the first ten matches are looked at, and it is noticed that the quality indices are all very close, it indicates that there are many polymers with an infrared spectrum close in appearance to that of the questioned sample. Conversely, it is confidence inspiring if the quality indices indicate one very good match among very poor matches.

The analysis of fibres presents a few difficulties to the infrared spectroscopist, and most of them stem from the fact that the sample contains only a small amount of matter. As the amount of infrared radiation absorbed by a sample is directly proportional to the amount of matter the radiation encounters (Beer's law), then it can be appreciated that a few micrograms of fibre cannot modulate an incident infrared beam to a great extent. This means that any instrumentation must be capable of discerning very small absorbances of radiation, that is to say, must be very sensitive. Some of the techniques used to enable spectrometers to achieve the performance required for single fibre analysis are discussed in section 6.6.

It is hoped that the reader can now appreciate that fibre analysis by infrared spectroscopy is a specialized area, not of forensic science but of polymer science, and it is the small size of the sample that makes it a specialized area. Fortunately the range of polymers used in the manufacture of fibres is fairly limited, compared to other products such as paints, so becoming familiar with their chemistry need not be a daunting task.

In the Bibliography is a list of texts which discuss the application of infrared spectroscopy to polymer analysis in detail beyond that possible in this work. These texts, and the references cited therein, show that the scope of polymer analysis using infrared spectroscopy is wide and that analysis of fibres, particularly in relation to their secondary structure, might still be in its infancy.

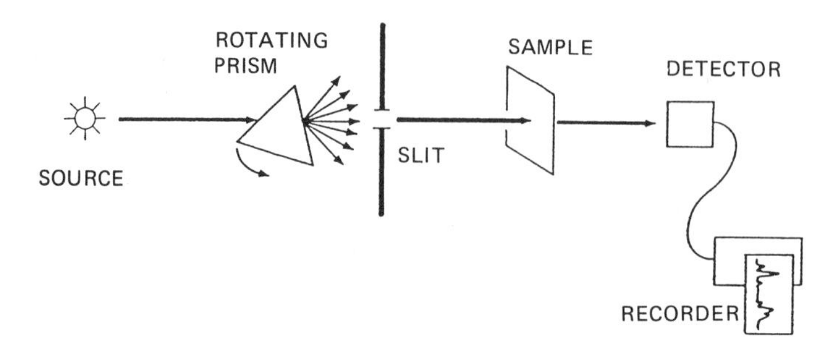

Fig. 6.13. Schematic digram of a conventional spectrometer

6.5 FOURIER TRANSFORM INFRARED SPECTROSCOPY

Fourier transform spectrometers have brought about a renaissance in infrared spectroscopy, and in a few years conventional instruments in forensic science laboratories might be replaced by them entirely. An explanation of Fourier transform techniques firstly requires a more detailed description of conventional spectrometers than was given in section 6.1.

A spectrum is a graph of the intensity of radiation received by the detector, plotted versus frequency. One way of obtaining this plot is to allow the detector to receive only radiation of a specific frequency and allow it to measure the intensity; then allow it to receive radiation of a slightly different frequency and measure its intensity, and so on. The collection of many intensity values, plotted against the frequency at which they were observed, gives the smoothly curved, familiar looking spectrum.

That is how conventional spectrometers operate. A prism, or similar device, is used to disperse the radiation from the source into its component frequencies, and a narrow slit is used to allow only a selected frequency to illuminate the sample at any one time (see Fig. 6.13). If the prism is rotated slowly, then radiation of different frequencies will be allowed to enter the slit, pass through the sample, and be measured by the detector. The output from the detector is then passed on to a recorder arranged to move a chart at the same rate as the prism.

To appreciate the manner in which Fourier transform spectroscopy works, consider the idealized spectrum shown in Fig. 6.14. This unusual spectrum is expected if a filter, which blocks radiation of all frequencies expect $v = 1$, $v = 2$, and $v = 3$, is placed in the beam of a spectrometer. The detector receives only three peaks, of frequency $v = 1$, $v = 2$, and $v = 3$, and intensity $I = 1$, $I = 2$ and $I = 3$. To completely describe this spectrum, all we need to know is that at frequency $v = 1$ there is a peak of intensity $I = 1$, at frequency $v = 2$ there is a peak of intensity $I = 2$, and at $v = 3$ there is a peak of intensity $I = 3$.

A Fourier transform spectrometer contains a device, called an *interferometer*, which irradiates the sample with all the output from the source and operates only on the salient information in the spectrum to produce a coded sum or a *multiplexed signal*.

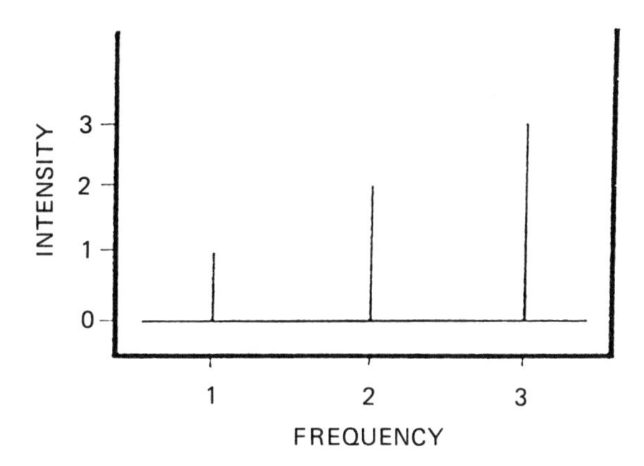

Fig. 6.14. Idealized spectrum composed of three 'peaks'.

It does this by multiplying the waveforms of the spectral frequencies by their respective intensity values and adding the products together. Fig. 6.15 shows such a process operating on the peaks in Fig. 6.14; the multiplexed sum is arrived at by adding together the waveform for frequency $v = 1$ multiplied by 1, frequency $v = 2$ multiplied by 2, and frequency $v = 3$ multiplied by 3.

The multiplexed signal is referred to as an *interferogram* in the terminology of infrared spectroscopy, and only in a very subtle way can spectral information be derived directly from it. However. By a mathematical operation called *Fourier transformation*, the interferogram can be decoded to yield the spectrum.

This might seem a convoluted manner in which to obtain simple spectral information. After all, a real infrared spectrum contains an infinite number of intensity values plotted against frequency. Fourier transform spectrometers, however, enjoy a few advantages over conventional instruments. These advantages enable Fourier transform spectrometers to achieve higher performance in the following areas:

- higher precision and accuracy of peak frequencies;
- higher accuracy of peak intensities;
- higher radiation throughput;
- higher signal to noise ratios.

A thorough discussion of these points, and interferometry in general, is well beyond the scope of this chapter, and the interested reader is referred to the Bibliography where excellent sources of information can be found.

A Fourier transform spectrometer can generate an interferogram and transform it into a spectrum in a few seconds. In practice, however, the spectrometer is usually allowed to acquire consecutively many interferograms (or scans) of the same sample and then add them together to form a 'super' interferogram. Transformation of this yields the same spectral information as that derived from a single interferogram but the noise level in the spectrum is markedly reduced.

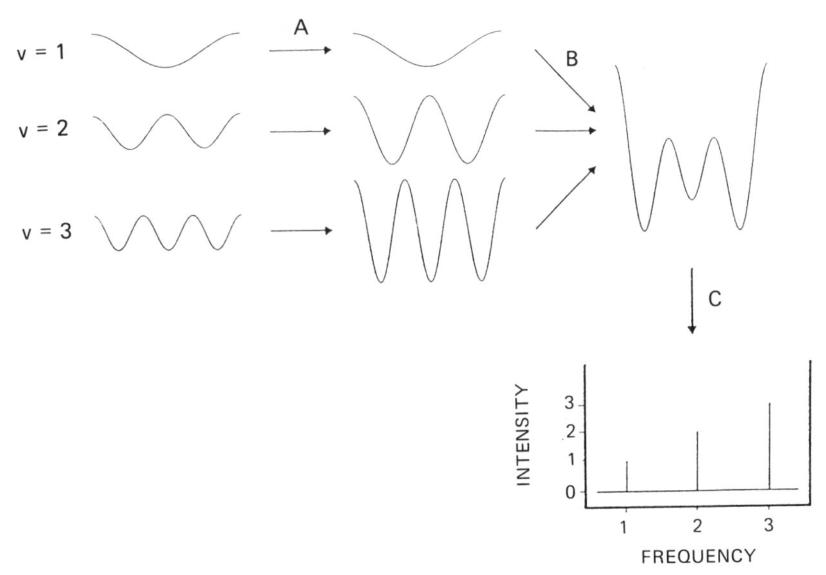

Fig. 6.15. Representation of Fourier transform technique. A—Multiplication of waveforms by corresponding intensities. B—Summation of the resulting products to yield the interferogram. C—Fourier transformation to give the spectrum.

Noise is the stumbling block of many spectroscopic techniques, particularly those which attempt to extract information from tiny samples such as single fibres. To understand the effect of noise, imagine trying to receive a message from a person in a room full of people all talking at once. Even though the person might be speaking at an audible level, the information in the signal is concealed by the voices of the other people in the room. If the signal is to be received and understood it must be intensified—shouted—or the noise reduced to a level where the signal is discernible.

Infrared detectors function in a similar manner, with extraneous radiation being the source of noise. If very tiny signals, such as those arising from a single fibre, are presented to the detector amidst such noise, then the spectrum produced is likely to have some of its information obscured. Fourier transform spectrometers, with their ability to produce spectra of very low noise level, are therefore eminently suited to fibre analysis.

Fig. 6.16 shows the effect of adding together many interferograms in an effort to reduce noise. In this figure, the full-range spectrum was derived from a single polyethyleneterephthalate (PET or polyester) fibre in 256 scans. The three insets show expanded views of spectra of the same sample acquired in 1, 16, and 256 scans. It is important to notice that the absorbance levels of the peaks do not increase with the increasing number of scans—Beer's law must be obeyed. It is the noise level that decreases. In an effort to obtain noise-free spectra, one encounters a situation of diminishing returns. To halve the noise level, the number of scans has to be quadrupled, that is, a spectrum acquired in 64 scans has half the noise level of a spectrum acquired in 16 scans. Spectra are therefore acquired most efficiently in a few tens of scans.

pg 53 - Dye extraction before polarising microscopy?

pg 200 - Are we going to set up the instruments according to what are we going to analyse? Are all techniques available?

Will anyone be with us during the analysis?

If I make a microscope slide w/ my fibre, do I have to remove it to do any further analysis? Is there going to be enough sample?

Suggested ~~module~~ chapters of book?

Since TLC is not confirmatory, we could also perform HPLC. But I cannot extract dye from the same fibre twice. What should I do then?

For Polarised Light Microscopy, do we need to prepare different slides w/ liquids/oils of different RIs?

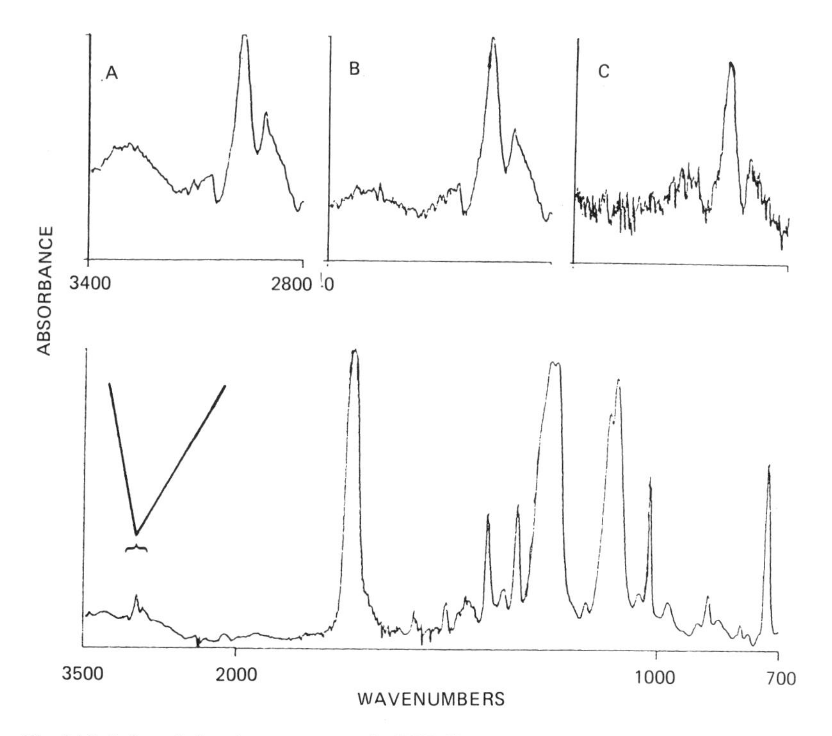

Fig. 6.16. Infrared absorbance spectra of a PET fibre. Bottom—full range spectrum acquired in 256 scans. Top Left—region between 2800 and 3400 cm^{-1} expanded. Centre—same region for a spectrum acquired from the same fibre in 16 scans. Right—same region for a spectrum of the same fibre acquired in 1 scan.

Approximately one-twentieth of a piece of fibre only 5 mm long was used to acquire the spectra shown in Fig. 6.16. It is a testimony to the sensitivity of modern Fourier transform instruments equipped with state-of-the-art sampling accessories that such a large magnification is needed to see the noise, even in a spectrum acquired in one scan.

The infrared radiation used to analyse substances must travel through a few tens of centimetres of atmosphere in addition to the sample as it traverses the optics in a spectrometer. Air contains water and carbon dioxide, both of which absorb infrared radiation. The reason that spectra run on conventional spectrometers do not show peaks due to these two compounds is that these instruments simultaneously use two beams of radiation. One of these beams passes through air and the sample and the other passes only through an equivalent amount of air (the reference beam). Electronics enables the spectrometer to use the reference beam to compensate for the absorbances due to air, as the spectrum is acquired, and produce a 'pure' spectrum of the sample.

Fourier transform instruments use only a single beam. With these spectrometers, two spectra must be acquired in an effort to produce one 'pure' one; these two spectra are called the background (when the beam is allowed to pass only through air) and the foreground (when the beam is allowed to travel through air and the sample). The computer attached to the instrument then removes the contribution due to the air and produces a conventional looking spectrum.

Unfortunately, if a Fourier transform spectrometer is to produce a really 'pure' spectrum, then the atmosphere surrounding the optics must be stable or preferably purified to remove carbon dioxide and water vapour. If the background is acquired in an atmosphere richer in carbon dioxide than the foreground, then the compensated spectrum will have been corrected too far and an inverse peak due to carbon dioxide will be observed. Water and carbon dioxide in air are in the vapour phase, hence their peaks will be broad, complex, and full of rotational fine structure as discussed in section 6.2 and illustrated in Fig. 6.8. Bad compensation leads to artefacts centred at 2300 cm^{-1} (carbon dioxide), and between 1800 and 1600 cm^{-1} and above 3600 (water).

Fourier transform spectrometers have encouraged a renaissance in infrared spectroscopy because, in addition to their low noise capability, they allow large amounts of radiation to be presented to the sample. This is because the optics in Fourier transform instruments do not have the narrow slits present in conventional spectrometers (as shown in Fig. 6.13). Even if the beam is diverted through energy-sapping optics (such as an infrared microscope), or partly blocked by masks or apertures (discussed later), there is still enough radiation getting through to the detector to allow good spectra to be recorded with Fourier transform spectrometers. This means that a wide variety of accessories, difficult to use on conventional spectrometers, can be used with confidence on Fourier transform instruments, thereby allowing a wide variety of samples by be analysed. This versatility is an important consideration in a service laboratory.

6.6 INFRARED ANALYSIS OF FIBRES

In this section, preparation of the sample and its presentation to the spectrometer will be addressed in detail. There are numerous ways in which a large-scale sample can be treated before infrared analysis. A sheet of polymer, for example, can be placed over a hole or aperture in a substrate, such as a sheet of cardboard or metal. If the infrared beam is then allowed to travel through the aperture, any radiation reaching the detector must have travelled through the polymer sheet.

Liquids are usually analysed as a very thin film sandwiched between disks of material transparent to infrared radiation. Unfortunately, glass strongly absorbs infrared radiation and is therefore quite useless as a support medium. The crystalline substances sodium chloride and potassium bromide are transparent to infrared radiation, and great use is made of them in spectroscopy, despite the problem of their high solubility in water.

Powdered potassium bromide has the useful property of forming a transparent solid under pressure. If a finely ground sample is mixed with potassium bromide and then pressed, the sample becomes dispersed in a transparent matrix. The mixture is usually pressed in a die which forms small disks, and these are readily placed in the beam of the spectrometer.

Individual fibres cause problems because they cannot be presented to the spectrometer in the simple ways just outlined. If a single fibre is placed over an aperture or sandwiched between sodium chloride disks and then placed in the spectrometer, most of the beam, which is usually about 11 mm in diameter, will not pass through the fibre and therefore cannot carry any information about it to the detector.

There are two ways to rectify this problem: either to modify the sampling procedure (make the sample occupy more of the beam) or to modify the spectrometer (make the beam smaller). One way of making the sample occupy more of the beam is to prepare it from a number of fibres. For the analysis of standard samples this is acceptable, but it is not for case fibres; to combine them is to assume that they are all identical, and we should not make that assumption. Moreover, the forensic scientist frequently does not have a large number of identical fibres to work with. The remainder of this section is therefore devoted to the analysis of single fibres.

Significant improvements in performance can be achieved, with either conventional or Fourier transform instruments, by placing samples into an infrared beam modified by a beam condenser. This device focuses the beam to a spot much smaller than its normal 11 mm diameter and then enlarges the image of the sample. From the detector's point of view, this is like looking through a magnifying glass—a much smaller field of view is taken in with more detail and less extraneous, unmodulated beam.

Even though a spectrometer equipped with a beam condenser is highly sensitive, attention still needs to be paid to sample preparation. A thin, flat fibre produces a more accurate spectrum than a fibre with a thick, irregular cross-section for reasons which will become apparent later when infrared microscopy is discussed.

There are many ways in which fibres can be flattened. For a fibre that dissolves in solvent, one way is to form a film from it. This can be done be preparing a solution of it and then allowing that solution to evaporate on a smooth, shiny surface. Left behind is a thin, fragile film. This should be treated with boiling water or placed into a vacuum oven to remove traces of residual solvent, which might interfere with the spectrum of the polymer. For a given fibre it might not be possible to find a suitable solvent, and in any event, forming a film from a small, single fibre is difficult to say the least. The process of dissolving the fibre will disrupt any interchain associations and therefore might destroy a potential point of comparison based on secondary structural features.

Flattening a fibre can also be accomplished by placing it on a hard surface, such as a glass microscope slide, and then rubbing it with a hardened steel roller, or by subjecting it to heat and pressure. Surprisingly, placing fibres into a potassium bromide press and subjecting them to a few tonnes force is not a good way to flatten them; it is a very good way of damaging the anvils of the press.

Having obtained a ribbon-like sample by flattening a fibre, or a thin sheet of film by solvent casting, the next step is to present the very fragile sample to the spectrometer. This can best be done as a sandwich between transparent crystals such as potassium bromide or sodium chloride, or merely by supporting the ribbon over an aperture.

A far simpler, more efficient, and universal method is to squash the fibre between two optically finished diamonds and, because diamond is almost completely transparent to infrared radiation, measure the spectrum of it *in situ*. As diamond does absorb a small amount of radiation near 2000 cm^{-1}, cells constructed of very thick

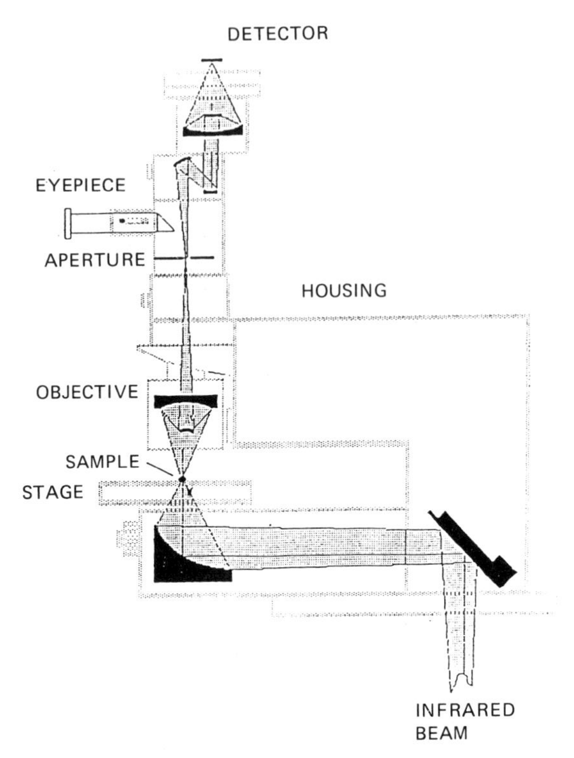

Fig. 6.17. Schematic diagram of an infrared microscope—solid black elements represent mirrors. (Reproduced with permission of Biorad, Aust.)

diamonds can lead to a loss of information near that frequency. Like all of the techniques described, the process of flattening a fibre in a diamond cell can be sufficient to destroy any points of comparison arising from the secondary structure of its polymer.

Cast films, rolled fibres, and flattened fibres captured in diamond cells are all good techniques to combine with a beam condenser; the diamond cell approach would probably win in terms of efficiency and versatility. The only drawback is the exacting task of trying to place the captive fibre in the correct position so that the focused beam falls upon it.

The ultimate beam condenser is an *infrared microscope*, and because its use is the method of choice for fibres analysis, the remainder of this section will be devoted to infrared microscopy

Fig. 6.17 is a schematic diagram of an infrared microscope. The eyepiece is used to look at the sample, enabling it to be brought into focus and positioned with the aid of the movable stage. In this way, only the area of interest is probed by the spectrometer's beam. Microscopes are therefore far easier to use than beam condensers. In the particular microscope shown in Fig. 6.17, the eyepiece is movable; when it is pulled out into the position shown, the infrared beam is free to travel to the detector. When it is pushed in, the image is diverted through the eyepiece. The optics are arranged so that the image viewed through the eyepiece is as close as possible to that presented to the detector.

Fig. 6.17 shows that the optics of an infrared microscope do not contain lenses, and are more reminiscent of a Cassegrainian telescope (hence the use of the term Cassegrain objectives). This is because glass strongly absorbs infrared radiation. It is possible to construct lenses out of material transparent in the infrared, but the most useful substances would deteriorate rapidly. In addition, the human eye has to use visible light to focus the sample and, because infrared radiation does not refract as strongly as visible light, the sample would not be in focus with respect to the spectrometer's beam. Mirrors do not suffer from any of these problems and are therefore used in all infrared microscopes.

Microscopes often come equipped with their own detectors, and this allows them to achieve higher sensitivity than beam condensers. Firstly, the size of the detector can be matched accurately to the optics of the microscope. Usually the element in the detector is about 250 μm square, and this closely matches the size of the field of view of the microscope. This means that the detector is not 'underfilled' by the beam and is therefore not likely to pick up excess noise like a larger detector coupled to a poorly positioned beam condenser. In addition, the detectors for microscopes are usually of a special construction, which is inherently more sensitive than the triglycine sulphate (TGS) detectors fitted as standard to most spectrometers. The high sensitivity detectors, called the *mercury cadmium telluride (MCT) type*, operate at low temperature with the aid of liquid nitrogen, thereby reducing electronic and thermal noise; MCT detectors are about forty times more sensitive than TGS detectors. These features, combined with the accuracy with which a sample can be mounted in the beam, mean that infrared microscopes are able to operate at magnifications much greater than beam condensers.

Even though a fibre is visible under the microscope, and therefore its image must be falling on the detector, sample preparation is very important in infrared microspectroscopy for three reasons. Firstly, the beam from the spectrometer should not be allowed to travel through more than approximately 10–15μm of polymer; to do so runs the risk of some regions of the spectrum reaching 0% transmission. To make matters worse, fibres usually have irregular cross-sections; this means that some portions of the beam pass through a greater thickness of polymer than others. The result is that peak absorbances of untreated fibres can be distorted, thereby making identification difficult and quantitative analysis meaningless. This phenomenon is known as the *wedge cell effect* and is described by Hirschfeld (1979). Secondly, fibres with circular cross-sections act like a short focal length lens, dispersing the infrared beam and ensuring that some of it does not reach the detector (overfilling it). This effect is complex because refraction (the phenomenon occurring in the fibres which causes it to disperse radiation) is frequency dependent; therefore the effect is not constant across the spectrum. Finally, although the field of view of the microscope is small, it is still large compared to the thickness of an unpressed fibre. For a fibre of 20 μm thickness, approximately 90% of the beam is still unmodulated by the sample. This is not a satisfactory state of affairs because stray radiation will be reaching the detector, 'swamping' the modulated radiation. Stray radiation gives the

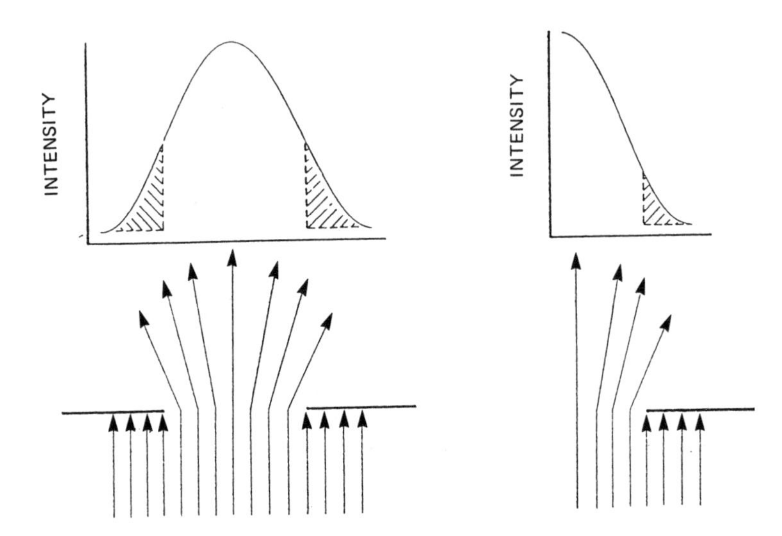

Fig. 6.18. Approximated diffraction patterns through an aperature (left) and past an edge (right).

spectrum the appearance of being vertically compressed, with the baseline being substantially lower than 100% transmittance and strong peaks appearing weaker.

Flattening a fibre will eliminate the problems arising from its irregular cross-section, while apertures can be used to minimize the effects of stray radiation. The aperture referred to in Fig. 6.17 is called a remote aperture, and it is used to mask the image presented to the detector, thereby cutting down stray radiation. Usually the aperture is constructed of four blades, all capable of independent movement, which can be closed down to a slit to view a fibre.

A modification of the system in Fig. 6.17 is called *dual remote aperturing*, and it can offer higher performance. In addition to the aperture above the sample, a dual remote apertured microscope has another below the sample. The lower aperture reduces the beam to a narrow slit, which illuminates only the fibre. The aperture above the sample is closed down on the image of the fibre in the usual fashion.

The biggest problem facing the infrared microscopist, and one which has not been discussed so far in this chapter, is *diffraction*. This phenomenon is observed whenever radiation is forced to pass through a slit, an aperture, or past a high-contrast edge. It is manifested as a complex, spreading pattern. In qualitative terms, diffraction can be visualized as being due to bending of radiation rays through a small angle as they encounter the obstacle.

Fig. 6.18 shows approximated diffraction patterns to be expected when radiation is passed through a slit and past an edge. These are two important models for infrared microspectroscopy because they mimic the effect when radiation passes close to a fibre, or travels through an aperture in a microscope. In Fig 6.18, the curved traces represent the spatial intensity of the radiation that would be received by a detector behind the obstacles. As can be seen, the edges of the shadows are not sharp as our intuition would suggest, but blurred. The shaded areas in the intensity plots graphically

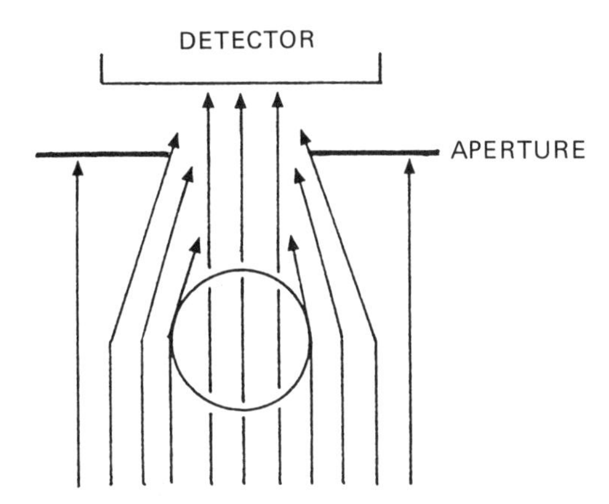

Fig. 6.19. Approximated diffraction past a single, thin fibre in an infrared microscope equipped with a remote aperature. Beam divergence due to the microscope's optics is ignored in an effort to keep the diagram simple.

display the amount of radiation that creeps around the obstacle and illuminates the regions which should be dark shadows. The angle through which radiation is diffracted is a function of frequency, with high frequency radiation being diffracted less than radiation of low frequency. As a consequence, in our macroscopic world where we view everything with high frequency, visible light, the effects of diffraction are very small and our eyes are not able to resolve the blurring.

In an infrared microscope, the detector is receiving infrared radiation, which is severely affected by diffraction. It becomes most noticeable when attempting to observe objects of a dimension approaching the wavelength of the viewing radiation, and for the infrared region, this means objects smaller than about 15 μm. Fig. 6.19 approximates what happens in an infrared microscope when a fibre of about 10 μm is illuminated by infrared radiation and viewed through a single remote aperture closed down to the dimensions of the fibre. As can be seen, the effect of diffraction leads to stray radiation reaching the detector even though masks are in place.

Reflecting optics present in infrared microscope exacerbate the effects of diffraction, therefore, even though the theoretical diffraction limit is approximately 15 μm, stray radiation is still noticeable in spectra of objects up to about 50 μm in diameter. This means that it is imperative to flatten a fibre as much as possible (that is, make it as wide as possible) and to use remote slits closed down to about 70% of the width of the sample if high quality spectra are desired. It is important to realize that diffraction, like the other phenomena discussed, alters only the absorbance values of peaks and not their resolution or frequency.

Although apertures can be used successfully to cut out stray radiation, they also modify the spectrum of an object to some extent. An infrared beam passing through a slit becomes diffracted as shown in Fig. 6.18, with the result that some radiation never reaches the detector. Fig 6.20 shows the performance of a 10 μm focal plane slit, which should, in theory, transmit about 5% of the beam's radiation. As can be

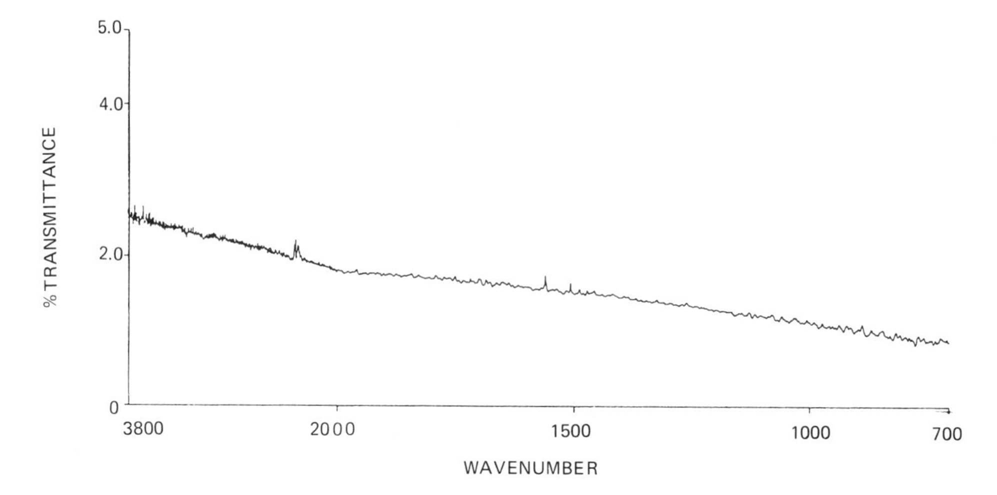

Fig. 6.20. Transmittance of infrared beam through a 10 μm remote aperature.

seen, less reaches the detector; moreover, the actual amount is frequency dependent, with high frequency radiation being least attenuated. Any spectrum acquired through the 10 μm slit would be convoluted with the curve shown in Fig. 6.20. To compensate this effect, either the background spectrum must be run through the same slit or a curve like Figure 6.20 subtracted. The performance curve approaches the ideal as the slit size increases.

Several books devoted to infrared microspectroscopy have been written, and a list of them can be found in the Bibliography. The following method is one we at State Forensic Science find convenient, and it accounts for the bulk of our infrared microspectroscopic analyses of polymers (paints, adhesives, and fibres).

We use a micro-diamond cell manufactured by High Pressure Diamond Optics Inc. in conjunction with an infrared microscope manufactured by Digilab. Although the gems in the diamond cell are very thin and do not seriously attenuate the infrared beam, we usually acquire spectra with the pressed fibre supported by only one of the diamonds. This is accomplished by taking a short length of fibre, pressing it with gentle finger pressure the diamonds, and then carefully separating the two halves of the cell. Usually the flattened fibre will adhere to one of the faces, and it is then presented to the spectrometer.

The use of one half of the cell has a few advantages over using the two halves to sandwich the fibre. Firstly, we find it much easier to view the fibre through the infrared microscope if we do not have to peer at it through one of the diamonds of the sandwich. This means that the microscope can be focused accurately and aperturing accomplished easily. Secondly, the infrared beam suffers only half the attenuation if it is allowed to pass through only half of the cell. As a result, performance in the range between 1900 and 2200 cm^{-1} is better. Finally, the optical interference effects of the cell are minimized if only one half is used. This point will be discussed in greater detail in the next section. The use of half diamond cells has also been described elsewhere (Lin-Vien *et. al.* 1990).

Only gentle pressure is required to flatten fibres, and some, particularly those containing polyacrylonitrile (acrylics or modacrylics), are easily deformed beyond resurrection. We find it is best to err on the side of finesse—press the fibre gently, record its spectrum, press it again, then record a second spectrum, etc.

This approach potentially has two benefits. Firstly, one is less likely to destroy a valuable fibre by reducing it to a smear. Secondly the secondary structure of a fibre can be destroyed by the act of pressing it. If multiple spectra are recorded, then the gradual disappearance of peaks due to secondary structure can be monitored and interpreted. The fibre, pressed gently and adhering to one half of the diamond cell, is then placed under the microscope, brought into sharp focus, and delineated (in our system) by a single focal plane aperture. To minimize the effects of diffraction, we adjust the slits to approximately 70% of the width of the flattened fibre and record the background spectrum by using the same aperture configuration and one half of the diamond cell. With this approach, we routinely obtain spectra of good signal to noise ratio in 16 scans.

One frustrating feature of the micro diamond cell is its ability to take an electrical charge, particularly after it has been rubbed clean with a paper tissue. The charge makes it very difficult to position small samples, such as small chips of paint or short lengths of fibre, because they tend to be repelled by the electric field. However, if the anvil is brushed with a carbon fibre brush, such as the type used to clean gramophone records, the static can be discharged easily.

For identification purposes we use a library of polymer standards (bulk samples) supplied with our spectrometers, but we have also created our own library, using fibre standards and our microscope. Although the compilation of a library is time consuming, the results are well worth it—the spectra are collected, using one's own sampling procedure and apparatus, and are derived from fibres, not bulk polymer samples.

6.7 HOW TO RECOGNIZE A BAD SPECTRUM AND CORRECT THE PROBLEM

Ideally an infrared spectrum should be composed of a collection of smooth, bell-shaped curves (called Lorentzian peaks) emanating from a flat baseline coincident with 100% transmittance (or an absorbance of zero). Except for peaks due to inorganic substances, carbon–oxygen single bonds, oxygen–hydrogen bonds, and nitrogen–hydrogen bonds, peaks in infrared spectra should be narrow. Sometime a spectrum of a fibre will not resemble the ideal, and careful manipulation of sample parameters is necessary to correct it.

Distorted peaks, those that look unnaturally broad or are flat-topped and blunt, are usually the result of a sample which is absorbing too much radiation. Fig. 6.21 shows many peaks distorted by complete absorption (and some stray radiation); in the region between 1000 and 1700 cm^{-1} there should be three major peaks well resolved and similar to those in Fig. 6.12. Subjecting the fibre to a second flattening operation usually cures this problem.

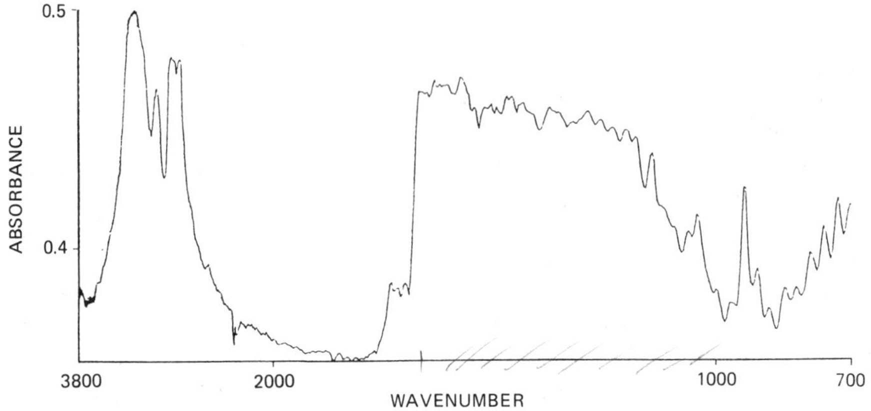

Fig. 6.21. Infrared transmittance spectrum of nylon-6,6 fibre showing distortion due to sample thickness and stray light.

A wayward baseline can be the symptom of different problems. A sloping baseline indicates that the diffraction effects due to the microscope's aperture might not have been correctly compensated for. This arises, for example, when the background is acquired without an aperture and the foreground spectrum is acquired through a narrow aperture.

A phenomenon called *scattering* is also responsible for sloping baselines. Scattering occurs when radiation strikes small opaque particles such as delustrant particles or imperfections in a fibre. Cosmetic action can be taken by instructing a computer to flatten the baseline artificially, and although this action does not cure the problem, it does enable computerized spectral searching to take place more easily. Hirschfeld (1978) has developed a mathematical algorithm which not only corrects a baseline perturbed by scattering but also reduces peak intensity errors caused by it. Another method reported to correct the effects of scattering is to immerse the sample in paraffin, acquire the spectrum, then subtract the spectral contribution due to the oil (Humecki 1988).

Stray radiation, usually the result of radiation diffracting around a fibre, gives a spectrum the appearance of being vertically compressed. The baseline is seen at a high absorbance value (that is, substantially below 100% transmittance) and peaks look short, especially those which should be highly absorbing. The accuracy of the spectrum distorted by stray radiation can be improved by minimizing the effects of diffraction—firstly by increasing the width of the sample and secondly by closing the microscope's aperture down to approximately 70% of the sample width.

If a fibre is so thin that even when flattened out completely it is still only very small, then a noisy spectrum of it is to be expected. A solution to this problem is to allow the spectrometer to acquire a large number of scans, thereby increasing the signal to noise ratio, or to increase the amount of fibre that the spectrometer analyses.

Another method, which allows more of the fibre to be analysed, is to hold it in a pair of fine forceps and move its end close to a small flame, such as that from a match

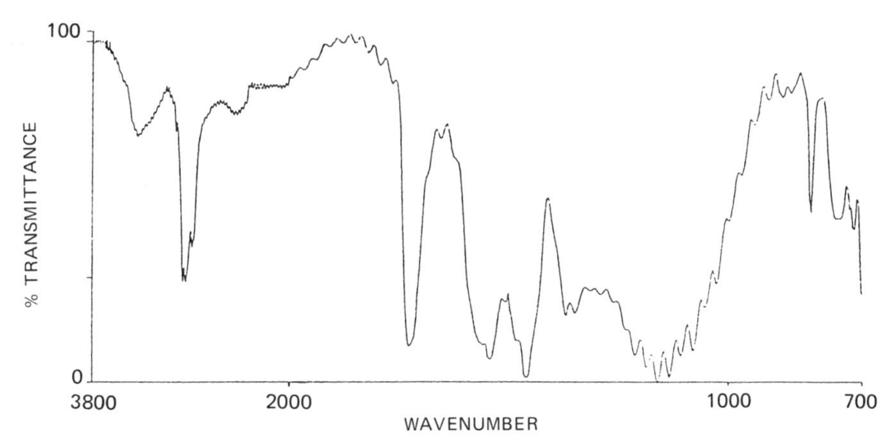

Fig. 6.22. Infrared transmittance spectrum showing interference fringes manifested as a sine curve of frequency 29 cm^{-1} superimposed on the spectrum.

or a microburner. When the fibre becomes hot enough, usually when it is a few millimetres from the flame, the end melts and forms into a small bead—it is important not to place the fibre into the flame because it will burn quickly. In this manner, a few millimetres of fibre becomes concentrated into a small bead, which can be pressed readily in a diamond cell for analysis. This method should be considered a last resort because the chance for error (resulting in burned evidence) is high, and secondary structural features might be created or destroyed.

Interference is an optical phenomenon which causes characteristics patterns to appear in infrared spectra. It is noticed whenever radiation is allowed to travel through thin, highly uniform films and is responsible for the rainbow hues seen in oil films, the iridescent coloration of beetle, butterfly, and bird wings, and sinusoidal modulation of the baseline in infrared spectra. Pressed fibres are films, and many spectra of them exhibit a uniform pattern of small peaks arising solely from interference (see Fig 6.22). The spectrum can be corrected by careful manipulation of its interferogram (see Hirschfeld & Mantz, 1976).

An additional interference effect caused by diamond cells is easy to avoid. If a spectrum has to be acquired by using the entire diamond cell, for example when recording spectra of rubber-like polymers, then the background should be corrected by using only one half of the cell. This is because the entire cell, when it does not contain a sample, encloses a thin film of air, which will give rise to its own interference pattern. As air film interference cannot occur when the sample is sandwiched between the diamonds, it should not be included in any background correction (Lin-Vien *et. al.*, 1990).

The foregoing remedial measures are easy to apply to spectra of common fibres that are familiar to the analysis, but how does one appraise the spectrum of an uncommon fibre? The best approach is to press the fibre cautiously, so that it can be pressed a second or third time. In this way, two or three spectra of the same fibre can be obtained, the effects of pressure, diffraction, and scattering observed, and artefacts recognized. A similar strategy should be adopted when comparing fibres, paying strict attention to sampling procedure so that false exclusions and matches are avoided.

6.8 LATEST DEVELOPMENTS IN VIBRATIONAL SPECTROSCOPY

6.8.1 Infrared dichroism

In section 6.2 it was shown that the resonant interaction between vibrating dipole moments and oscillating electric fields was the mechanism by which matter captures radiation. A beam of radiation is a collection of many rays, and the electric fields associated with them point randomly in a plane perpendicular to the direction in which the beam is travelling.

It is possible to produce radiation which has its electric field pointing in only one direction in this plane—the radiation is then said to be polarized. For radiation to interact with matter, not only must it encounter a dipole moment vibrating at exactly the right frequency but it must also find a dipole moment pointing in the right direction.

On encountering matter that is constructed from randomly oriented molecules, polarized (and unpolarized) radiation will find many dipole moments with which to interact. Fibres, however, are constructed from long chain-like molecules, and it is possible for these molecules to arrange themselves so that some of their atomic groups become oriented in a particular direction; polarized radiation is sensitive to the orientation of these groups. When a favourable alignment occurs between the polarized beam and the groups, a strong infrared absorbance results; if there is a misalignment, then the resulting peak will be weak. This phenomenon is *infrared dichroism*.

The usual method used to probe the chain orientation in fibres is to record two spectra by using polarized radiation, one where the direction of polarization is set parallel to the length of the fibre and one where it is set perpendicular. Groups which do not point in any particular direction will give rise to peaks of equal intensity in the two spectra. Those which are oriented, however, will give a weak peak in one spectrum and a strong peak in the other. The ratios of the peak intensities are called dichroic ratios, and they are potentially useful points of comparison between fibres. Krishnan (1984) has discussed the application of infrared dichroism to polyethyleneterephthalate fibres (polyester, PET), and Jasse (1979) has described the field generally for polymers.

The process of arriving at dichroic ratios for a fibre involves two quantitative analyses being performed on a less than perfect sample. To preserve its secondary structure, the fibre cannot be flattened; therefore its irregular cross-section and narrow diameter mean that stray radiation and the wedge cell effect become problems. It is therefore wise to assume that dichroic ratios obtained from a single fibre are only estimates of the true figure.

6.8.2 Pyrolysis gas chromatography—infrared spectroscopy (Py-GC–IR)

The techniques discussed so far have operated under the paradigm that the integrity of the fibre should not be compromised in any way before or during analysis.

Conversely, Py-GC-IR completely destroys the fibre by a process called pyrolysis—literally meaning to destroy it by heat. The action of heat is to reduce the polymer in a fibre to a characteristic mixture of degradation gases. In Py-GC-IR, the resulting mixture of substances (called the pyrolysate) is presented to a gas chromatograph, which separates the individual components. The separated components of the pyrolysate are then presented, one at a time, to an infrared spectrometer, which records their spectra. From the composition of the pyrolysate can be deduced the polymeric composition of the fibre.

Commercial equipment, which has only recently become readily available, is now approaching the sensitivity required for single fibre analysis.

Py-GC-IR potentially has the ability to differentiate between members of some polymer families more readily than infrared spectroscopy alone, owing to the high resolution capability of modern gas chromatographs. Whereas infrared spectroscopy yields some information regarding the secondary structure of polymers, Py-GC-IR cannot; and it cannot, at present, provide information relating to trace level constituents of a fibre, although it does have scope for improvement. The main limitation with any of the pyrolytic techniques is that they are destructive tests that require a relatively large quantity of fibre.

6.8.3 Quantitative infrared analysis

The computers used to control Fourier transform spectrometers are very useful accessories. Their ability to store and handle large arrays of numbers allows a spectroscopist to measure the sizes of peaks in the spectrum of a mixture and from those values calculate quantitative data related to the mixture.

Analysis of modacrylic/acrylic fibres is one area that could benefit from quantitative techniques. These fibres are formed from a mixture of polyacrylonitrile and another polymer such as polyvinyl acetate. If it is necessary to compare two fibres containing this mixture, it might be critical to determine whether the fibres are formed from the same proportions of polymers.

In theory, all that need be done is compare the size of a peak from the polyacrylonitrile component with the size of a peak from the polyvinyl acetate component for each of the fibres. In practice, the technique is more difficult, but attempts have been made (see, for example, the report by Pandey 1989).

Accurate and precise quantitative data are dependent upon a reproducible sampling technique and the minimization of scattering, diffraction, interference, stray radiation, and wedge cell effects (see Hirschfeld 1979). These requirements might count against the technique being exploited fully in single fibre analysis.

6.8.4 Raman spectroscopy

Raman spectroscopy has been described as the complement of infrared spectroscopy. This is because it produces an analogous spectrum containing vibrational information by the use of a technique entirely different to infrared spectroscopy.

Fourier transform Raman spectroscopy, that most likely to be of use to forensic scientists, uses a red laser beam to probe the sample. Unlike the usual radiation sources for spectroscopy, the laser is an intense beam containing radiation of only one frequency; in Fourier transform Raman spectroscopy, that frequency is usually about 9400 cm^{-1}.

On encountering the sample, the beam is subject to the usual physical constraints, that is it can be reflected, refracted, diffracted, transmitted, or scattered by it. A particular type of scattering, Raman scattering (as distinct from the type discussed in section 6.7, Rayleigh scattering) is the phenomenon associated with Raman spectroscopy. After a molecule has absorbed radiation, it might re-emit it, but not necessarily in the direction in which the radiation was originally travelling. The radiation that is emitted in these other directions is called Raman scattered, and it is this radiation that is measured in Raman spectroscopy.

During the re-emission process, the molecule might retain some energy from the beam, so the radiation it gives out is of a frequency less than that of the laser beam. Therefore, if we find radiation of frequency 9200 cm^{-1} emitted by the molecule, we know that 200 cm^{-1} is the frequency of the radiation kept by it. The energy retained by the molecule is used to increase its vibrational state. The amount of radiation emitted, plotted versus frequency shift from 9400 cm^{-1}, therefore looks like an infrared spectrum and contains exactly the same information.

Although the Raman and infrared spectra contain the same information, the spectra of a given substance are not identical. As mentioned earlier, some peaks in infrared spectra are weak, and some substances, such as oxygen gas, do not give a spectrum at all. It turns out that weak peaks in the infrared spectrum are usually strong in the Raman, and vice versa, so the two techniques really are complementary.

The Raman technique is sensitive enough to be used in conjunction with a microscope, thus single fibres can be analysed readily. The biggest problem with infrared microscopy is diffraction, arising because radiation of long wavelength is used in the analysis. In Raman microspectroscopy, the laser operates at a shorter wavelength; diffraction effects are therefore reduced and the beam is able to probe much smaller areas than infrared radiation. This also means that the optics for Raman microscopes can be of the conventional refracting type. In addition, the current range of high sensitivity infrared detectors can operate only down to about 700 cm^{-1}, Raman spectroscopy allows data to be obtained down to about 200 cm^{-1}. In this region can be found information related primarily to the inorganic components of fibres, such as delustrants. Because Raman spectroscopy is not a transmission technique, fibres do not have to be pretreated before analysis, but there is potential for the surface composition of the fibres to be over-represented in the spectrum.

Raman accessories are now available for commercial Fourier transform infrared spectrometers, therefore a steady growth in the application of Raman spectroscopy to fibre analysis is likely.

6.8.5 Emission spectroscopy

Any article which possesses heat emits infrared radiation, and the chemical bonds present in the substance modulate the emission in much the same way as they modulate radiation transmitted by the substance. The intensity of the radiation emitted by a substance, plotted versus frequency, is called its *infrared emission spectrum*. For a general discussion of this technique, see Bates (1978).

With a Fourier transform spectrometer, it is possible to use a single fibre warmed to a few degrees above room temperature as an infrared source and to record the infrared spectrum emitted by it. The work of Compton (1988) shows that this specialized technique is potentially very useful in identifying surface coats on fibres and might be useful for the analysis of substances which do not readily transmit radiation.

6.9 INFRARED ANALYSIS OF FIBRES—A SUMMARY

Infrared spectroscopy must be included in any scheme directed toward the identification or comparison of fibres because it is the best general method for probing the polymeric construction of them. It is quick, non-destructive, yields a wealth of information to the trained eye, and is capable of the high sensitivity needed to analyse a short piece of single fibre.

An infrared spectrum can be likened to a chemical fingerprint, but the analogy is a bad one. Paradoxically, although the peaks and troughs which make up the spectrum are rich in diagnostic information, an infrared spectrum might not be unique to a given fibre; two fibres made from the same polymer by different factories are likely to be identical so far as their infrared spectral data are concerned. Fingerprints, however, are the opposite, being unique to a person but completely devoid of predictive information.

Every feature of a spectrum, the size, shape, and position of peaks, and even the nature of the baseline, contains information concerning the sample. Not every feature, however, is related to the chemical structure of the sample; some might arise from optical phenomena. Therefore, spectra should be acquired and examined with a view to identifying these spectral artefacts (interference patterns, stray radiation, etc.) so that in comparative work, the possibility of a false negative conclusion being drawn is minimized. Conversely, the chance of overestimating the reliability of a 'match' in comparative tests can be avoided if the theoretical and practical limitations of infrared spectroscopy as applied to polymer analysis are kept in mind.

The most efficient way in which to undertake a comparative analysis of fibres is to perform non-destructive, simple tests first; hopefully, gross, genuine dissimilarities between the fibres will be evident. Tests of this nature are microscopy, visible spectroscopy, and infrared spectroscopy, which probe the morphology, colour, and polymer family of the fibre.

If these tests indicate that the fibres closely match in these attributes, then further work needs to be done. The infrared spectrum should now be examined in greater detail in order to detect any differences between the fibres with respect to their secondary structure. If necessary, the secondary structure can then be examined further by thermal methods such as melting point determination. If applicable, the dyes present in the fibres could be extracted and analysed by a chromatographic technique such as thin layer chromatography or high performance liquid chromatography. As a final point of comparison, the fibres can be subjected to an elemental microanalytical technique, such as X-ray microanalysis or inductively coupled plasma analysis, in an effort to detect any differences between the additives that might have been applied to them (such as metal based dyes, flame retardants, and delustrants).

Once these tests have been performed, the forensic scientist must then offer an opinion as to the interpretation of the scientific data; this must surely be the most difficult aspect of forensic fibre analysis.

BIBLIOGRAPHY

References relating to section 6.2 and 6.3

Avram, M. & Mateescu Gh. D. (1972) *Infrared spectroscopy: applications in organic chemistry*. Wiley, New York, USA.

Bellamy, L. J. (1975) *Infrared spectra of organic compounds*. Chapman & Hall, London, England.

Colthup, N. B., Daly, L. H. & Wiberley, S. B. (1964) *Introduction to infrared and Raman spectroscopy*. Academic Press, New York, USA.

Rao, C. N. R. (1963) *Chemical applications of infrared spectroscopy*. Academic Press, New York, USA.

Walker, S. & Straughn, B. P. (1976) *Spectroscopy*, Vol 2. Chapman Hall, London, England.

Weast, R. C. (ed.) (1976) *Handbook of chemistry and physics*, 56th ed. CRS Press, Cleveland, USA.

References relating to section 6.4

Bolker, H. I. (1974) *Natural and synthetic polymers*. Marcel Dekker, New York, USA.

Bower, D. I. & Maddams, W. F. (1989) *The vibrational spectroscopy of polymers*. Cambridge University Press, Cambridge, England.

Burfield, D. R. & Loi P. S. T. (1988) The use of infrared spectroscopy for determination of polypropylene stereoregularity. *J Appl Pol Sci* **36**, 279.

Cowie, J. M. G. (1973) *Polymers: chemistry and physics of modern materials*. Intertext, Aylesbury, England.

Craver, C. D. (ed.) (1983) *Polymer characterization*. Amer. Chem, Soc., Washington, USA.

Grieve, M. C. & Cabiness L. R. (1985) The recognition and identification of modified acrylic fibres. *For Sci Int* **29**, 129.

Grieve, M. C. (1990) Fibres and their examination in forensic science. In: Maehly, A. & Williams, R. L. (ed.). *Forensic science progress*, Vol 4. Springer Verlag, Berlin, Germany, 44.

Grieve, M. C. & Kotowski, T. M. (1977) The identification of polyester fibres in forensic science. *J For Sci* **22**, 390.

Henniker, J. C. (1967) *Infrared spectrometry of industrial polymers*. Academic Press, London, England.

Hummel, D. O. & Scholl, F. (1988) *Atlas of polymer and plastics analysis*, Vols. 1, 2, and 3, 2nd ed. VCH, New York, USA.

Hummel, D. O. (1966) *Infrared spectra of polymers*. Wiley, Chichester, England.

Nelson, M. L. & O'Connor, R. T. (1964) Relation of certain infrared bands to cellulose crystallinity and crystal lattice type. *J Appl Pol Sci* **8**, 1311.

Smalldon, K. W. (1973) The identification of acrylic fibres by polymer composition as determined by infrared spectroscopy and physical characteristics. *J For Sci* **18**, 69.
Vandeberg, J. T. *et al.* (1980) *An infrared spectroscopy atlas for the coatings industry.* Fed. of Soc. for Coatings Technology, Philadelphia, USA.
Zbinden, R. (1964) *Infrared spectroscopy of high polymers.* Academic Press, New York, USA.

References relating to section 6.5
Bell, R. J. (1972) *Introductory Fourier transform spectroscopy.* Academic Press, New York, USA.
Durig, J. R. (ed.) (1980) *Vibrational spectra and structure, Vol. 8, Infrared interferometric spectrometers.* Elsevier, Amsterdam, Netherlands.
Griffiths, P. R. (1975) *Chemical Fourier transform infrared spectroscopy.* Wiley, New York, USA.
Griffiths, P. R. (ed.) (1978) *Transform techniques in chemistry.* Plenum, New York, USA.
Theophanides, T. (ed.) (1984) *Fourier transform infrared spectroscopy.* Reidel, Dordrecht, Netherlands.

References relating to section 6.6
Bartick, E. G. (1985) Microscopy/infrared spectroscopy for routine sample sizes. *Applied Spectroscopy* **39**, 885.
Cole, A. R. H. & Norman Jones, R. (1952) A reflecting microscope for infrared spectrometry. *J Opt Soc Amer* **42**, 348.
Cournoyer, R., Shearer, J. C. & Anderson, D. H. (1977) Fourier transform infrared analysis below the one-nanogram level. *Anal Chem* **49**, 2275.
Garger, E. F. (1983) An improved technique for preparing solvent cast films from acrylic fibres for the recording of infrared spectra. *J For Sci* **28**, 632.
Hirschfeld, T. (1979) In: Ferraro, J. R. & Basile, L. J. *Fourier transform infrared spectroscopy applications to chemical systems*, Vol. 2. Academic Press, New York, USA, pp 215.
Levy, F. (1988) Energy distribution in the beam of an infrared microscope. *Anal Chem* **60**, 1623.
Lin-Vien, D., Bland, B. J. & Spence, V. J. (1990) An improved method of using the diamond anvil cell for infrared microprobe analysis. *Appl Spec* **44**, 1227.
Messerschmidt, R. G. & Harthcock, M. A. (eds) (1988) *Infrared microspectroscopy theory and applications.* Marcel Dekker, New York, USA.
Roush, P. B. (ed.) (1987) *The design, sample handling and applications of infrared microscopes.* ASTM Special Technical Publication 949, American Society for Testing Materials, Philadelphia, USA.

References relating to section 6.7
Cater, R. O. Carduner, M. C., Paputa Peak, M. C. & Motry D. H. (1989) The infrared analysis of polyethylene terephthalate fibres and of their strength as related to sampled preparation and to particle size. *Appl Spec* **43**, 791.

Hirschfeld, T. & Mantz, A. W. (1976) Elimination of thin film infrared channel spectra in Fourier transform infrared spectroscopy, *Appl. Spec.*, **30**, 552.
Hirschfeld, T. (1978) Scatter eliminated spectrometry by Fourier transform infrared spectrometry, *Anal. Chem.*, **50**, 1023.
Humecki, H. J. (1988) In: Messerschmidt, R. G. & Harthcock, M. A. (eds) *Infrared microspectroscopy theory and applications*, Marcel Dekker, New York, USA. 62.
Lin-Vien, D., Bland, B. J. & Spence, V. J. (1990) An improved method of using the diamond anvil cell for infrared microprobe analysis, *Appl Spec* **44**, 1227.

References relating to section 6.8

Bates, J. B. (1978) In: Ferraro, J. R. & Basile, L. J. (1978) *Fourier transform infrared spectroscopy applications to chemical systems*, Vol. 1. Academic Press, New York, USA. 99.
Bergin, F. J. & Shurvell, H. F. (1989) Applications of Fourier transform Raman spectroscopy in an industrial laboratory, *Appl Spec* **43**, 516.
Chase, B. (1987) Fourier transform Raman spectroscopy, *Anal Chem* **59**, 881A.
Compton, D. A. C. (1988) Novel approaches for analyzing polymeric materials using FT-IR spectroscopy, *Proc Int Sample Tech Conf* September 27–29.
Hirschfeld, T. (1979) In: Ferraro, J. R. & Basile L. J. *Fourier transform infrared spectroscopy applications to chemical systems*, Vol 2. Academic Press, New York, USA, 193.
Jasse, B. & Koenig, J. L. (1979) Orientational measurements in polymers using vibrational spectroscopy, *J Macromol Sci-Rev Macromol Chem.*, **C17**, 61.
Krishnan, K. (1984) Applications of FT-IR microsampling techniques to some polymer systems, *Polymer Preprints*, **25**, 182.
Messerschmidt, R. G. & Chase, D. B. (1989) FT-Raman microscopy: discussion and preliminary results, *Appl Spec* **43**, 11.
Pandey, G. C. (1989) Fourier transform infrared microscopy for the determination of the composition of copolymer fibres: acrylic fibres. *The Analyst*, **114**, 231.

ACKNOWLEDGEMENT

I wish to thank my colleagues at State Forensic Science, in particular Anna Parybyk and Silvana Tridico, for their support, guidance, and encouragement.

7

Fibre identification by pyrolysis techniques

John M. Challinor, BSc
Department of Mines, Chemistry Centre, Forensic Science Laboratory,
125 Hay Street, Perth, Western Australia 6004, Australia

7.1 OUTLINE

This chapter describes the practical aspects of pyrolysis gas chromatography (Py-GC), detection methods including mass spectrometry (MS) and the principles of pyrolysis mass spectrometry (Py-MS) for the identification of fibres. Applications of Py-GC to the identification of synthetic and natural fibre types are illustrated. A brief account of the different thermal degradation mechanisms provides an understanding of the pyrolysis processes. The advantages and disadvantages of the techniques compared to other methods of fibre identification and comparison are outlined. Future developments in pyrolysis techniques are discussed.

7.2 INTRODUCTION

Pyrolysis is the high temperature fragmentation of a substance in an inert atmosphere. This thermal decomposition produces molecular fragments which are usually characteristic of the composition of the macromolecular material. Forensic scientists have used analytical pyrolysis to characterize a wide range of materials found as crime scene evidence. The pyrolysis products have been detected and identified by coupling the pyrolysis unit to a gas chromatograph (GC) or a mass spectrometer (MS) or a combination of both (Py-GC-MS). The infrared spectrometer may also be used as a detector for the GC, or it may be used directly to identify a pyrolysate (pyrolysis-IR).

Pyrolysis gas chromatography (Py-GC) is usually the method chosen because the data give a reliable identification and comparison of analytes and is more cost effective than other methods. Generally Py-GC gives good discrimination, requires minimal sample manipulation, and generates reliable and reproducible results. Low microgram order of sensitivity is often achieved. In spite of these advantages, it is a technique which has been under-used in fibre examinations. Criticisms of poor reproducibility were made in the early days of Py-GC. Contemporary experience has shown that these are unfounded when modern instrumentation and correct techniques are employed. If differences in results exist between laboratories, or even within a laboratory, it is probably because of the differences in the method with which the thermal fragmentation process is conducted. The variables in the pyrolysis process include temperature rise time, pyrolysis temperature, sample mass, the dimensions of the pyrolysis chamber, carrier gas type and flow rate. These factors influence the proportion of primary pyrolysis products, the production of secondary products from recombination processes, and the introduction of catalytic effects. A system which avoids secondary products and minimizes catalytic effects should be favoured (May *et al.* 1977).

Py-GC techniques have progressed from the early days of packed column GC to the use of more refined capillary columns. These have the advantages of improved peak shape and compound separations. They are also capable of chromatographing a wider range of compounds of greater polarity. Recently, pyrolysis derivatization techniques, giving greater structural data, simpler chromatograms, and thus improved sensitivity, have been applied to the identification of a wide range of polymers. (Challinor, 1989).

7.3 PYROLYSER TYPES

Since the introduction of commercial units for analytical pyrolysis applications, essentially three different types have emerged. These are the *pulse mode filament* and *pulse mode Curie point* types and the *continuous mode furnace* types.

The filament type employs a resistively heated platinum coil or ribbon. The coil houses a quartz tube to hold the sample, and the ribbon acts as a surface on which to evaporate the sample. The advantage of this type is that the pyrolysis temperature can be varied continuously from about 200°C to 1000°C.

The *Curie point* system depends on the inductive heating of an alloy wire to its Curie point, using a high frequency oscillating current in a coil surrounding the wire. The wire can be flattened and bent over to hold the sample. The advantages are low dead volume and ease of sample loading.

The *furnace type* involves introducing the sample into an oven unit by a gravity feed mechanism, a magnetic push rod, or a plunger arrangement. Care must be taken to ensure that the pyrolysis zone is free of contaminants from previous experiments before the system is used.

Walker *et al.* (1970) reported a study of the three different pyrolysis systems, using a branch chain alkane as a model compound. The filament pyrolyser gave a higher proportion of lower molecular weight products than the Curie point instrument, suggesting that the 'true' pyrolysis temperature may have been much higher, a result of a faster temperature-rise-time for the particular filament instrument used. Therefore, while any of the systems can be used for the examination of fibres, consideration should be given to true final temperatures, especially when comparing results from different systems. Other factors such as rapid temperature-rise-time, low dead volume, and ease of sample loading should be considered when choosing a pyrolysis system.

7.4 GAS CHROMATOGRAPH CONSIDERATIONS

Modern GCs are equipped for capillary column use, oven temperature programming, and flame ionization detection. By contrast, early Py-GC work was carried out with packed columns for the chromatographic separation. Numerous applications to paint, plastics, adhesives, rubbers, and many other materials have been reported. (Irwin 1982, Wampler 1990). Instrumentation and applications have been described by May *et al.* (1977). Forensic applications of Py-GC using capillary column GC have been described, and the advantages compared to the use of packed columns, were outlined (Challinor 1983).

One criticism of Py-GC as a viable technique is the lack of attention to standardization of analytical conditions with the consequence that interlaboratory comparison of data may not be facilitated. The type of pyrolyser will be the choice of the user, but the GC conditions could be standardized. A common base for fibre examinations could be the adoption of capillary columns, flame ionization detection, and temperature programming from ambient to maximum column temperature at an intermediate rate, for example, 8–10°C min^{-1}. For fibre work, a mid-polarity phase column (for example, OV 17 type or equivalent [cyano-propyl phenyl methyl silicone]) would be favoured. This type of phase gives good peak shape for polar pyrolysis products without baseline drift or the relatively poor phase stability experienced with the more polar polyester (for example Carbowax) phases.

7.4.1 Practical aspects

To achieve optimum performance, attention must be paid to certain practical aspects of the pyrolysis and GC systems. Capillary columns, though now very stable compared to the early types, will become active or adsorbing, with the result that peak tailing or complete loss of product may be experienced. Test mixtures should be used regularly to monitor column performance. Mixtures of a selected range of compounds, recommended by Grob *et al.* (1978), have been found to be satisfactory for this purpose. If column performance deteriorates a coil of the column at the inlet end can be cut off to return the chromatography to an acceptable standard. The flow rate should be maintained by reducing inlet system pressure to compensate for the shorter column. Alternatively a replacement section can be used at the front of the column. Column 'washing' with organic solvents does not usually improve performance to any appreciable extent. The pyrolyser unit should be maintained in a clean condition to avoid carry-over to subsequent samples. Glass/quartz inserts should be regularly washed in a suitable solvent such as dichloromethane. Curie point pyrolysis wires should be replaced regularly when they appear soiled. Resinous pyrolysis products may periodically cause blockage problems at the vent line from the injection system, and this area should be checked periodically by observing deviations from normal behaviour of carrier gas flows and pressures.

7.4.2 Detection systems

The *flame ionization detector* (FID) is the most commonly used detection system. Pyrolysis products are identified by their retention times compared to those of known standard compounds. Other detectors provide more specific responses for compounds eluting from the GC column. The *alkali flame ionisation detector* (AFID) specifically measures nitrogen and phosphorus-containing compounds. The *flame photometric detector* (FPD) is used for monitoring sulphur and phosphorus-containing compounds, and the *electron capture detector* (ECD) is specific for halogenated hydrocarbons and other electron accepting organic compounds. The latter detectors are not normally used for fibre pyrolysis work.

The detection system most favoured is the *mass spectrometer* (MS). MS gives more data about pyrolysis product identity than relative retention times. Electron impact (EI) mass spectral libraries make it possible to rapidly identify many of the pyrolysis products, facilitating interpretation of polymer composition. Chemical ionization mass spectrometry techniques may also be employed. Molecular ions are produced in greater abundance to give molecular weight/molecular formula information of pyrolysis products. The use of the *infrared spectrometer* as a detector and a means of identification of compounds eluting from the GC has developed with the inception of Fourier transform techniques. The two methods of collection of the pyrolysis products are the 'light pipe' and 'matrix isolation' techniques. Wide bore capillary columns are usually required for these operations, and a sensitivity of the order of low nanogram levels is possible.

Further discussion of GC detection systems will not be attempted here. This topic is discussed in some depth by Liebman and Wampler in Liebman & Levy (1985).

7.5 PYROLYSIS MASS SPECTROMETRY (Py-MS)

Py-MS involves the direct introduction of the products of flash pyrolysis into the ionization source of the MS. Optimum reproducibility and sensitivity are achieved by using a custom made interface between the two (Mauzelaar & Kistemaker 1973, Meuzelaar, *et al.* 1973). This dedicates the instrument to a single task. Hughes *et al.* (1978) achieved reasonable results by connecting the pyrolyser to the MS via an empty column in the GC oven. They used this system to examine a range of textile fibres and compared the results with IR. The IR spectroscopic method gave better discrimination, particularly with copolymer fibres. However, Py-MS was capable of characterizing smaller samples, and analysis times were shorter. Poor reproducibility with certain polymer types and difficulties with interpretation of the composition of the polymer and additives were experienced. This problem of reproducibility was studied by Hickman & Jane (1979). The cause was claimed to be the pyrolysis process rather than the electron impact fragmentation process in the mass spectrometer. Irwin (1979) reviewed the progress of Py-MS and discussed the advantages of softer ionization processes. These included *chemical ionization*, Py-CIMS (Saferstein & Manura 1977), *high resolution field ionisation* mass spectrometry, Py-FIMS (Shulton, in Jones & Cramer (1977), and *field desorption* mass spectrometry, FDMS. (Cotter, in Vorhees 1984). The use of probe pyrolysis-mass spectrometry is also useful for characterization and the study of thermal degradation mechanisms. The sample is distilled from the direct insertion probe by using a controlled temperature–time profile, and mass spectra are determined with respect to temperature. Burke *et al.* (1985) carried out a comparison of Py-MS, Py-GC, and IR for the analysis of paint resins, and they emphasized the importance of using more than one technique in forensic comparisons.

Wheals (1980/1, 1985) has reviewed pyrolysis methods including Py-MS. The advantages of Py-MS compared to Py-GC are its speed, ease of data handling, and high sensitivity. The limitations are its higher equipment cost, preferred use of dedicated instrumentation, and difficulty with data interpretation. By comparison, Py-GC profiles facilitate identification of a polymer and assessment of relative concentrations of pyrolysis products which assist in understanding differences in composition of polymers within a class. For these reasons it is likely that Py-GC would be the method of choice for most forensic science laboratories.

7.6 PYROLYSIS INFRARED SPECTROSCOPY

This technique has been employed for the identification of paint, plastics, and rubbers in forensic casework (Smalldon 1969). Milligram samples were pyrolysed in ignition tubes and the pyrolysate transferred to a halide disc which was then analysed in an infrared spectrometer. Comparable results matching the IR spectrum of the polymeric material were claimed. The method is particularly suitable for heavily filled, opaque samples which are difficult to analyse by conventional IR techniques. An IR spectrum of the mixture rather than individual pyrolysis products is obtained, therefore the method has been largely superseded by the techniques previously described. The method is not appropriate for trace fibre examinations because of the relatively large quantity of material required for analysis. A pyrolysis cell using a filament pyrolyser has been recently marketed. This device fits into the optical bench of an FTIR spectrometer and could have potential for the rapid identification of polymers.

7.7 APPLICATIONS

There have been numerous applications of analytical pyrolysis reported in the last two decades. A selected bibliography of applications was reported by Wampler (1989). Wheals (1985) described the forensic applications of the technique, particularly with respect to paint. Wheals (1980) also reviewed analytical pyrolysis techniques applied to fibres and other polymeric materials encountered in crime scene evidence. Perlstein (1983) reported a packed column Py-GC method for identifying fibre types with the aid of identified diagnostic compounds rather than pattern recognition techniques. Challinor (1990) described the use of Py-GC and pyrolysis derivatisation methods in forensic applications.

The most common fibre types occurring in forensic casework include acetate, acrylic, polyamide, polyester, viscose, cotton, and wool. Other fibre types occur less frequently. These include aramid, spandex, and halogenated fibres.

7.7.1 Acetates

Acetates comprise cellulose which has been partly acetylated (approximately 2.3 acetyl groups per cellulose unit). Cellulose triacetate is produced by complete acetylation of cellulose resulting in the formation of three acetylated hydroxyls per glucose monomer unit:

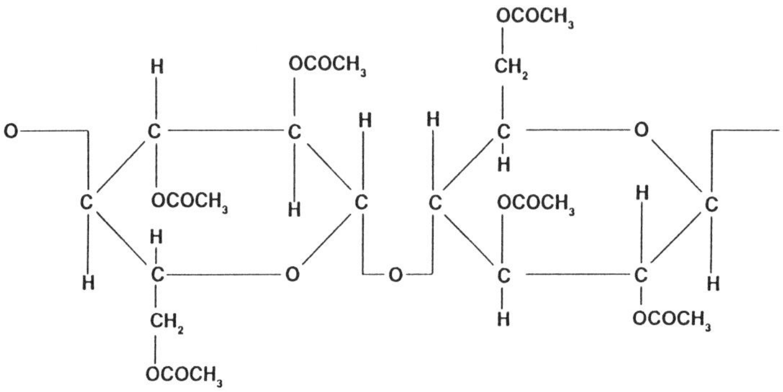

Py-GC profiles of a typical cellulose acetate fibre and a triacetate fibre, Dicel and Arnel, reflect the differences in the degree of acetylation of the two fibre types. (Fig. 7.1). A mid-polarity phase vitreous silica capillary column was used for the separation of the pyrolysis products.

The major pyrolysis product is acetic acid produced by the scission of the acetyl groups from the cellulose molecules. Other pyrolysis products originate from the cellulose. Acetic acid does not chromatograph satisfactorily because of its high polarity. Pyrolysis derivatization with tetrabutyl ammonium hydroxide provides a method for confirming the presence of acetate in a polymer. Acetic acid is converted to butyl acetate by butylation after the group is hydrolysed from the polymer chain (Challinor, 1989). The procedure is also useful for detecting low proportions of acetate in a copolymer.

7.7.2 Acrylics

The acrylics, which contain more than 85% polyacrylonitrile (PAN), undergo pyrolysis at 770°C to give a series of monomer, dimers, trimers, and tetramers. The latter are composed of structural isomers. Homopolymers of acrylonitrile are not common, but they probably include Orlon 81. Chemical properties of the acrylics, including enhancement of dye acceptability, are improved by copolymerizing acrylonitrile (AN) with other monomers. Some of the known comonomers are methyl acrylate (Orlon 42, Courtelle), methyl methacrylate (Crilenka), vinyl pyrollidone (AF100, Zefran),

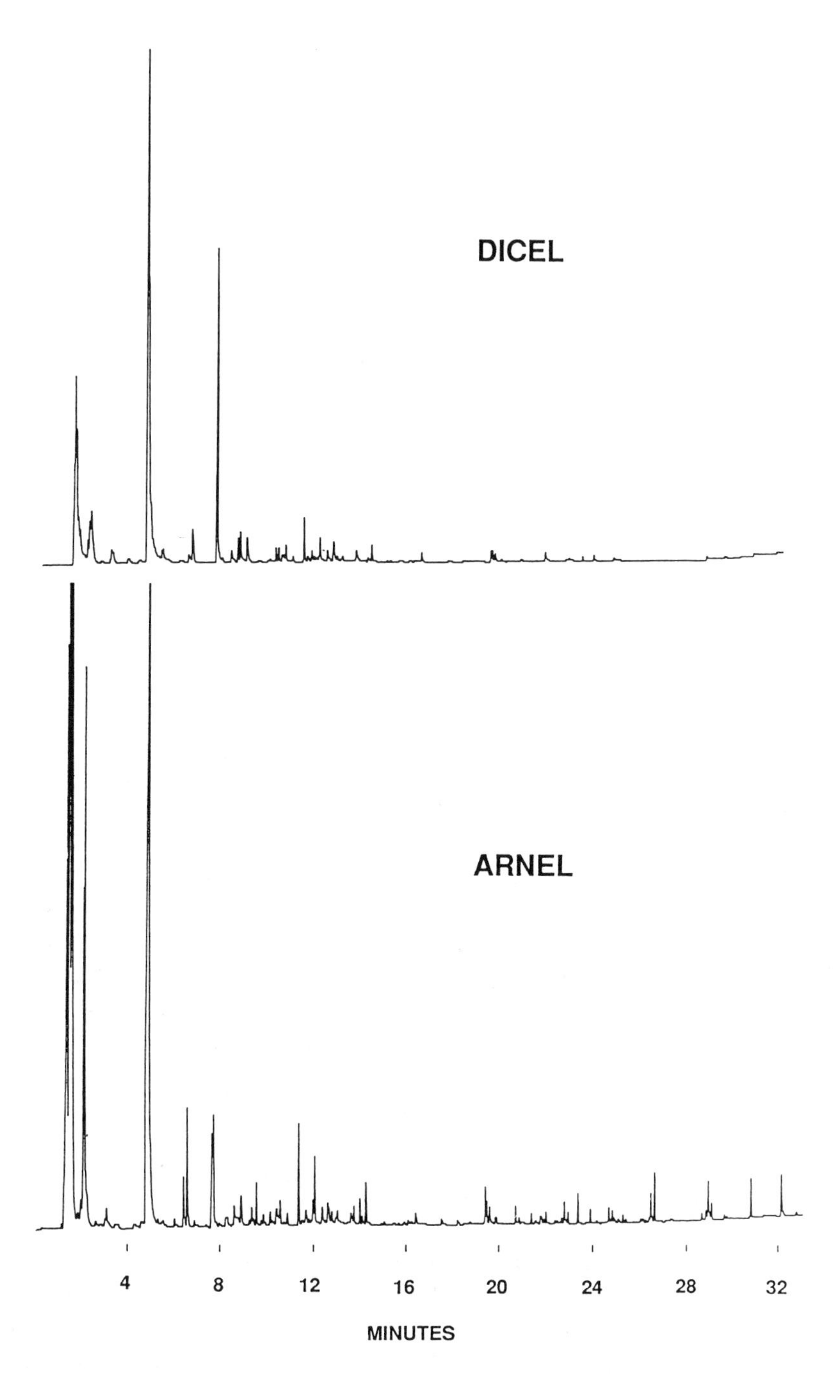

Fig. 7.1. Py-GC profiles of a cellulose acetate fibre (Dicel) and a cellulose triacetate fibre (Arnel).

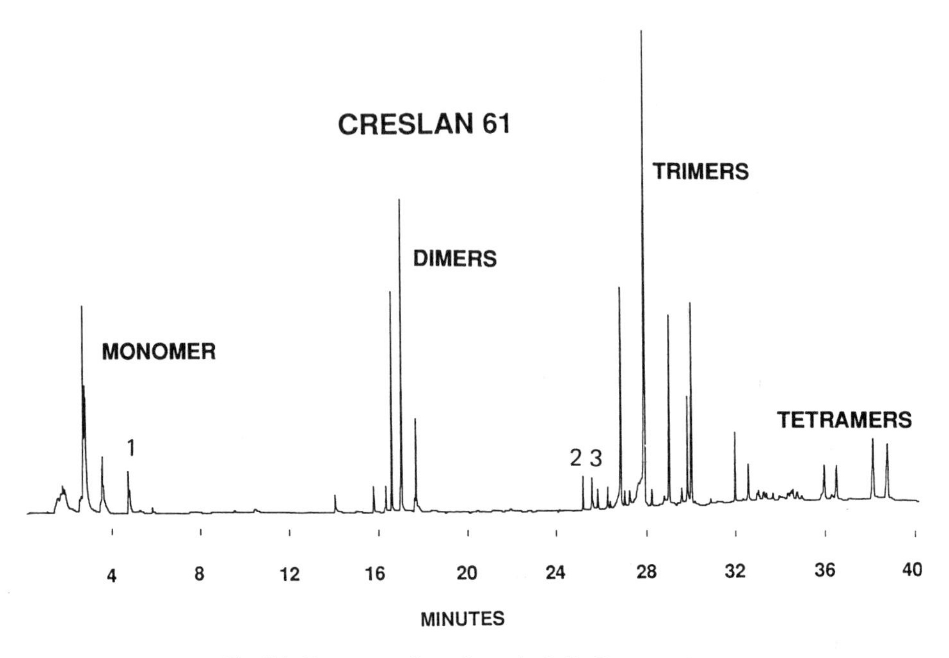

Fig. 7.2. Pyrogram of a polyacrylonitrile fibre (Creslan 61).

vinyl acetate (Acrilan 16, Leacril), and methyl vinyl pyridine (Dolan 86) which also incorporates methyl acrylate. Methyl vinyl pyridine and vinyl acetate are copolymerised with acrylonitrile in Creslan 58. In some cases these comonomers may be detected by Py-GC.

A pyrogram of a typical acrylonitrile copolymer, Creslan 61, is shown in Fig. 7.2. This pyrogram indicates products additional to those found in pyrograms of the homopolymer. Peak 1 may be attributed to acetic acid from vinyl acetate, and peaks 2 and 3 have mass spectra suggesting vinyl pyridine/pyrrolidone origin. Quantification of methyl acrylate and methyl methacrylate comonomers in acrylic fibres has been reported (Saglam 1986). In that study, ten methyl acrylate and five methyl methacrylate copolymers were found to contain between 3.5 and 8% of the comonomers. Coefficient of variation values of +/− 2.5% were achieved. Packed chromatography columns were used.

Thermal degradation of acrylic fibres using Py-GC, nuclear magnetic resonance, and Fourier transform infrared-spectroscopy has been studied (Usami *et al.* 1990). In that study, the pyrolysis products of PAN, including small proportions of comonomers, were identified by mass spectrometry.

In the most recent study of the classification of polyacrylonitrile fibres using Py-GC, Almer (1991) succeeded in subclassifying 85 fibres containing PAN into 9 acrylic groups and 6 modacrylic groups by using pattern recognition techniques. Unfortunately, the comonomers were not identified by mass spectrometry, but it is likely that

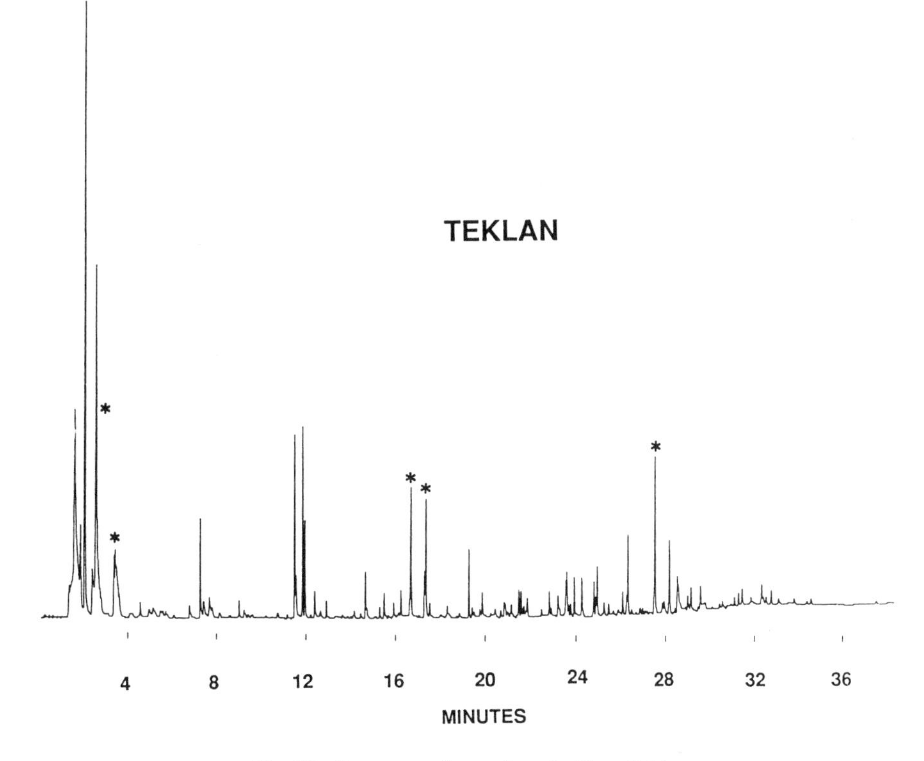

Fig. 7.3. Pyrogram of a modacrylic fibre, Teklan.

it will soon be possible to attribute chemical structures to the comonomers and additives. Bortniak *et al.* (1971) previously used packed column Py-GC to analyse 41 acrylic and modacrylic fibres into 12 acrylic and 3 modacrylic groups.

7.7.3 Modacrylics

Modacrylic fibres include copolymers of acrylonitrile and vinyl chloride (60%) (Dynel) or vinylidene chloride (50%) (Teklan and Verel), (Moncrieff 1975). Pyrolysis profiles of these fibre groups are very different from those of acrylic fibres. Fig. 7.3 shows a pyrogram of Teklan.

The acrylonitrile oligomers are indicated by asterisks. Pyrolysis mass spectrometry, using the softer field ionization technique, has been used to determine sequences in acrylonitrile copolymerized with butadiene and styrene (Plage & Schulten 1991). While this technique is not appropriate to the average forensic laboratory, the work indicates the potential of pyrolysis methods to give more data about polymer composition.

7.7.4 Polyamides

The aliphatic polyamide fibre types most commonly encountered in forensic casework include nylon 6 and nylon 6.6. Nylon 6 is manufactured from caprolactam. Pyrolysis results in the reformation of the monomer in high yield.

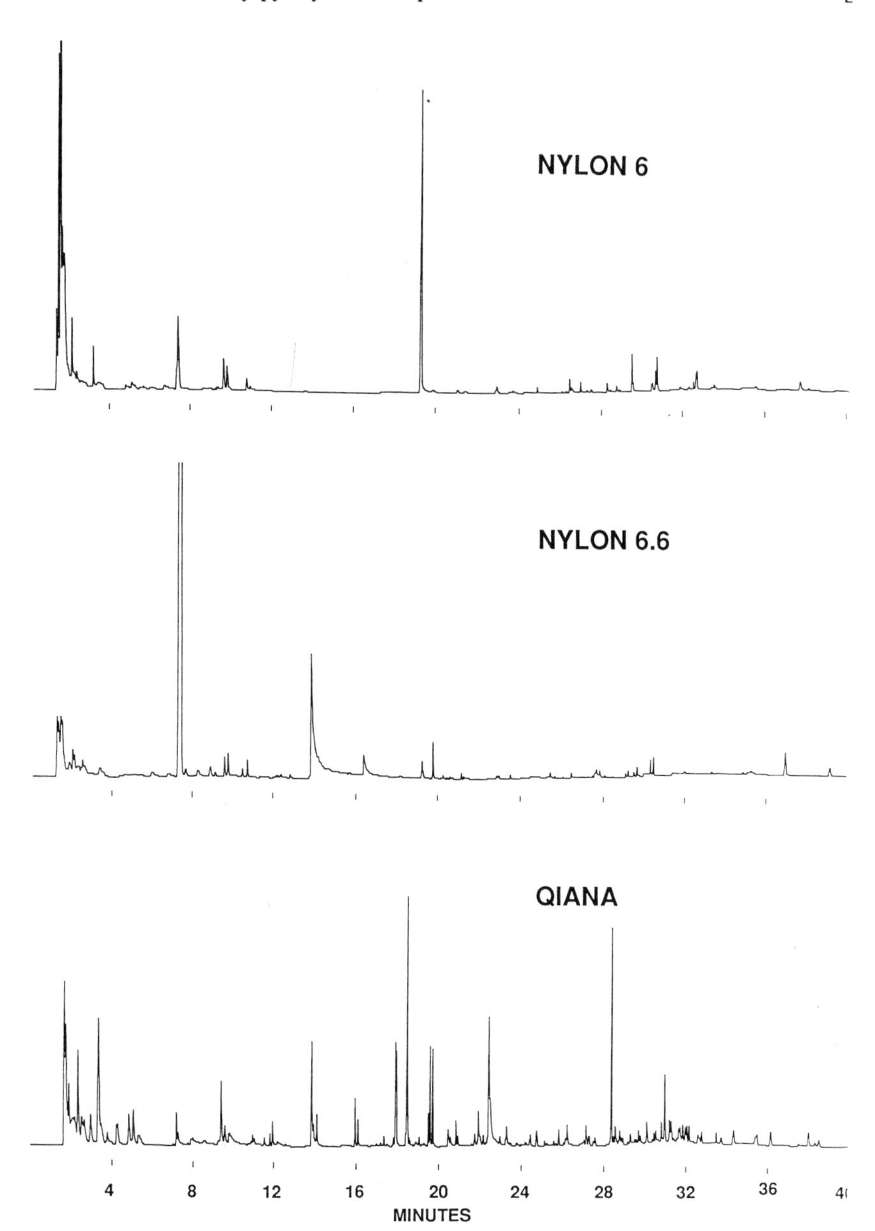

Fig. 7.4. Pyrolysis profiles of nylon 6, nylon 6.6 and Quiana fibres.

Nylon 6.6 is produced from adipic acid (butane dicarboxylic acid) and hexamethylene diamine. Pyrolysis occurs at the CO–NH linkage in the polymer to give adipic acid and hexamethylene diamine monomers. Adipic acid undergoes cyclization after loss of carbon monoxide to give cyclopentanone. Qiana is a polyamide made from *bis*-*para*-aminocyclohexyl methane and dodecanedioic acid. Pyrolysis profiles of the three fibre types are shown in Fig. 7.4.

A polyoxyamide fibre, Vivrelle, recently marketed by Snia Viscosa, Italy, has been differentiated from nylon 6 by the detection of a compound, tentatively identified as cyclopentanediol, by Py-GC-MS. (unpublished work by Grieve and Challinor).

Polyaromatic amides, Kevlar and Nomex, may be readily identified by Py-GC.

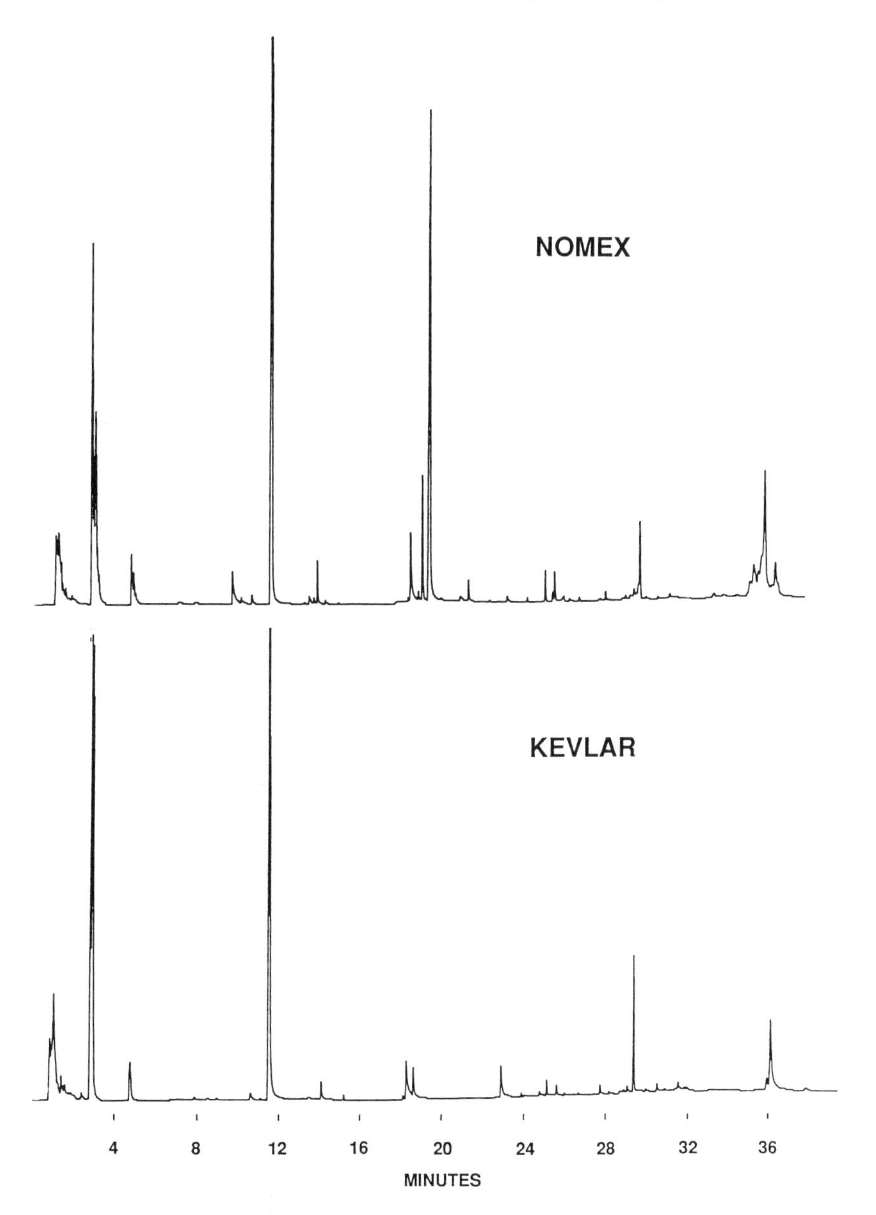

Fig. 7.5. Pyrolysis profiles of two aramid fibres, Nomex and Kevlar.

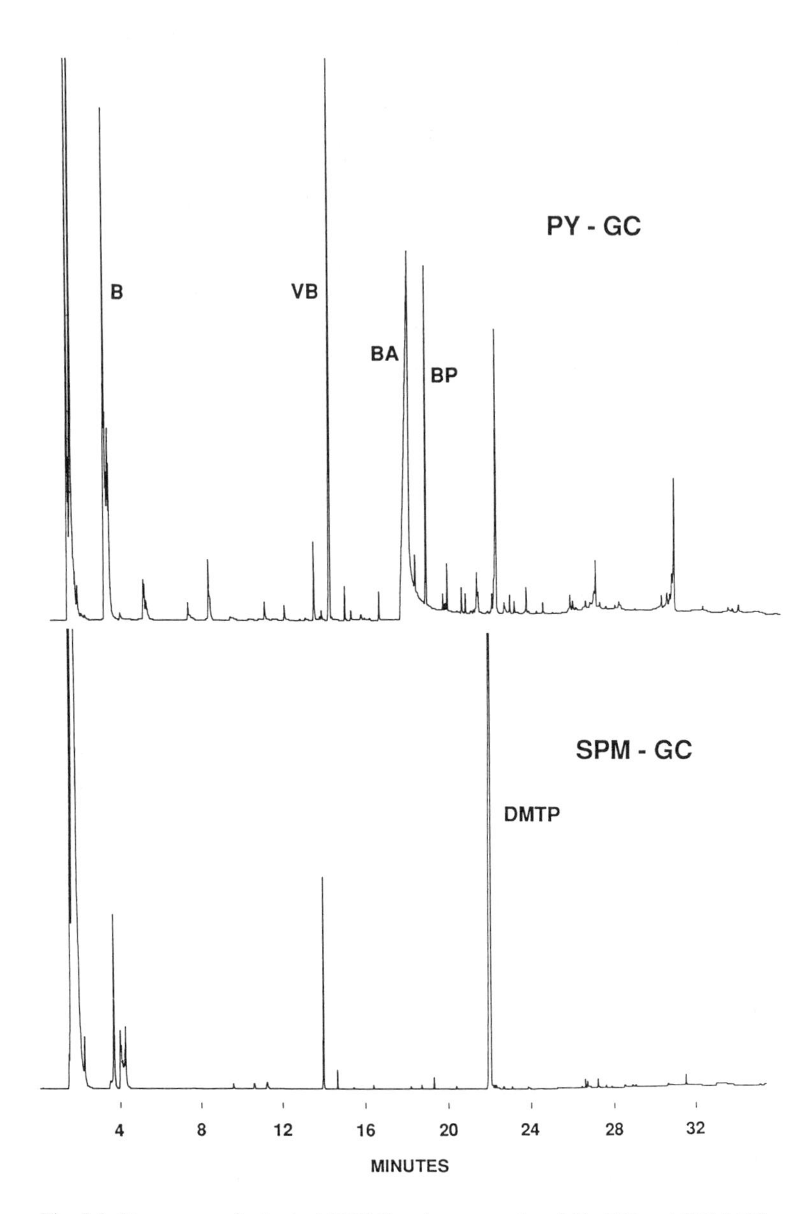

Fig. 7.6. Pyrograms of a typical PET fibre by conventional Py-GC and SPM-GC.

These fibre types require a higher temperature than normally used for regular Py-GC to effect thermal degradation. Pyrolysis at 980°C results in the formation of aromatic compounds. Pyrolysis profiles of Kevlar and Nomex are shown in Fig. 7.5.

Distinctly different pyrograms of closely related polyamide types indicate the discrimination of Py-GC.

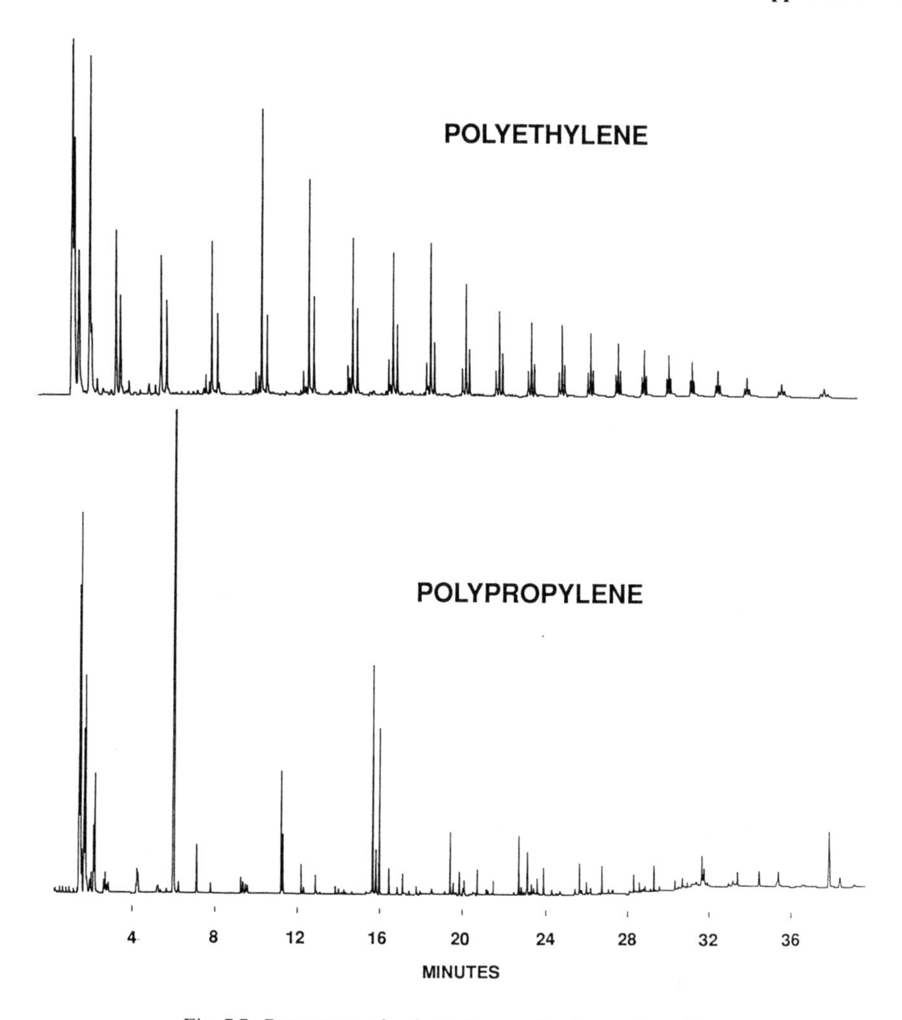

Fig. 7.7. Pyrograms of polyethylene and polypropylene fibres.

7.7.5 Polyesters

Polyesters comprise chains of ethylene glycol and terephthalic acid units linked through ester groups. Pyrolytic scission at these ester linkages results in a range of pyrolysis products which include benzene (B), benzoic acid (BA), biphenyl (BP), and vinyl benzoate (VB). Simultaneous pyrolysis methylation gas chromatography (SPM-GC) of polyethylene terephthalate (PET), in contrast, gives a simpler pyrogram. Pyrograms of PET obtained by the two procedures are shown in Fig. 7.6.

Dimethyl terephthalate (DMTP) is the major pyrolysis product resulting from the high temperature hydrolysis and methylation of the terephthalate polymer. Sensitivity of the order of one microgram is achieved by this method.

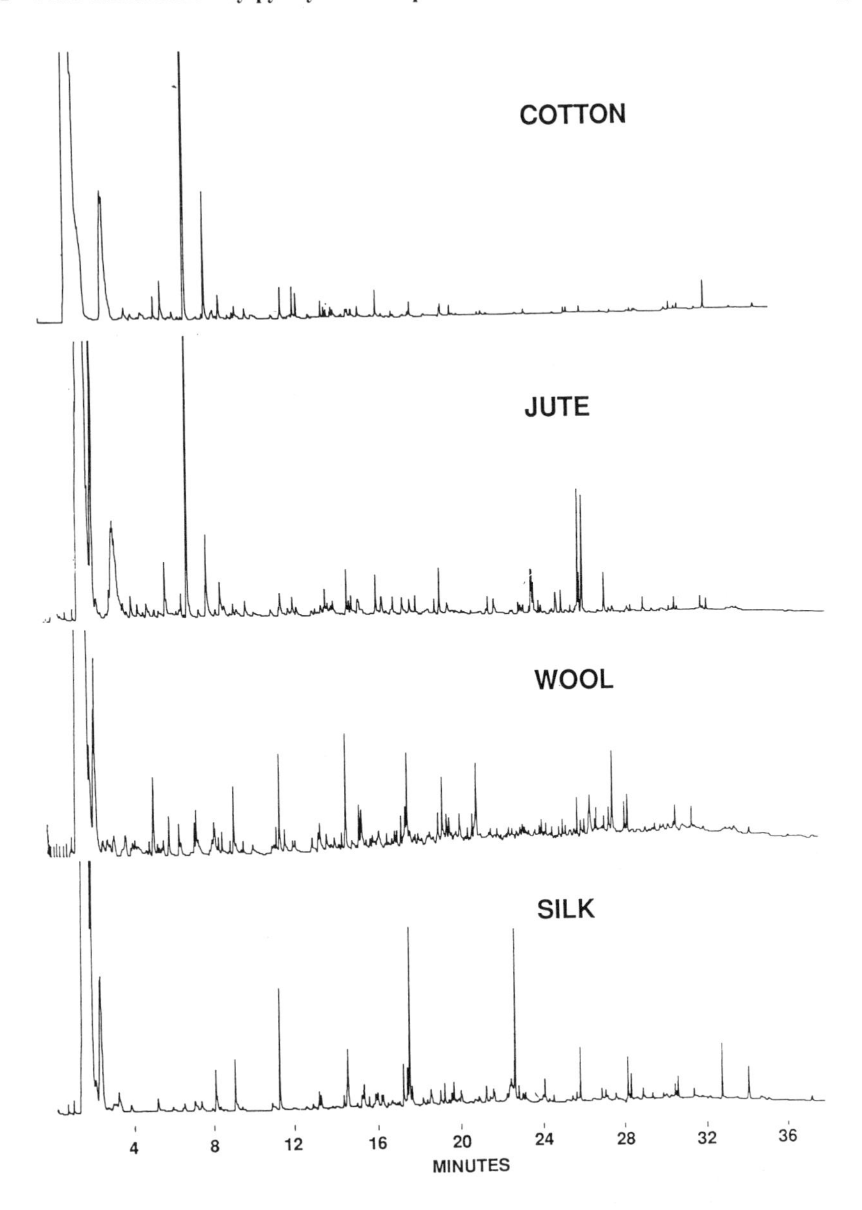

Fig. 7.8. Pyrolysis profiles of mercerized cotton, jute, wool, and silk.

7.7.6 Polyolefines

Polyethylene fibres are used in ropes and carpet. Polyethylene thermally degrades to a homologous series of alkanes, alkenes, and diolefines which result from the random scission of the polymethylene chains. Separation of these nonpolar pyrolysis products is best achieved by using a nonpolar methyl silicone phase column of the OVI or equivalent type.

Polypropylene fibres, in contrast, give rise to branched chain oligomers on pyrolysis, with the trimer, dimethylheptene, predominating. Pyrograms of the two olefinic fibres are shown in Fig. 7.7.

It is possible to determine the stereoregularity of polypropylene by the ratio of pyrolysis products in the trimer, tetramer, and pentamer region (Tsuge *et al.* in Vorhees 1984).

7.7.7 Natural fibres

The natural fibres include cellulosic and proteinaceous fibres. Morphological detail, determined by optical microscopy, is usually sufficient for identification. However, where these fibres are degraded by adverse environmental conditions or are otherwise unrecognizable, Py-GC may facilitate their identification. Fig. 7.8 shows typical pyrograms of cotton, jute, wool, and silk.

7.8 PYROLYSIS MECHANISMS

There are four main pathways for thermal degradation of polymers. In some cases, more than one mechanism occurs during pyrolysis of the polymer.

(*a*) *Random chain cleavage* occurs in olefinic and vinyl polymers which have a polymethylene 'backbone' structure. Usually a series of oligomers is produced by random fragmentation at sites along the polymer chain. In the simplest case, polyethylene fragments to give a homologous series of alkanes, mono-olefines, and di-olefines.

In some vinyl polymers, for example, polyacrylonitrile, the side chain does not

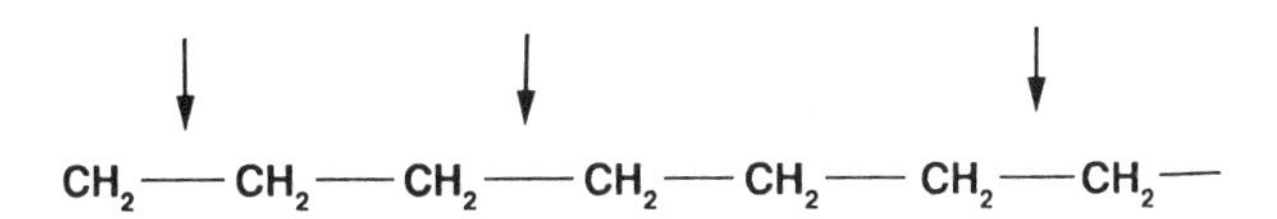

fragment but remains attached to the polymethylene chain, and a series of aliphatic oligomers are produced. However, dehydrogenation and loss of HCN, followed by chain scission and cyclization, gives some aromatic oligomers. (Usami *et al.* 1990).

(*b*) *Directed chain cleavage* occurs when thermolysis takes place at sites of comparatively weak bond strength. Polyamides and polyesters undergo this type of cleavage at CO–NH and CO–O bonds, respectively.

$$— (CH_2)_x — CO \overset{\downarrow}{—} O — (CH_2)_y —$$

$$— (CH_2)_x — CO \overset{\downarrow}{—} NH — (CH_2)_y —$$

The pyrolysis products of polyamides are usually dibasic acids and diamines, although nylon 6 undergoes cyclization of aminohexanoic acid to give a monomer, caprolactam. The dibasic acid in nylon 6.6, adipic acid, suffers a loss of carbon monoxide to give cyclopentanone (Senoo *et al.* 1971). Polyethylene terephthalate cleaves at the CO–O bond in addition to other sites in the polymer chain to give decarboxylation and recombination products, (Sugimura *et al.* 1979).

(*c*) *Side chain scission* takes place in some vinyl polymers where a pendant group to the polymethylene chain breaks off. The backbone chain then fragments and cyclization takes place to yield aromatic compounds.

$$\left[-CH_2-\underset{}{\overset{X}{\overset{|}{C}H}}- \right]_n \longrightarrow -CH_2-\dot{C}H-CH_2-\dot{C}H- + HX$$

Chlorofibres, polyvinyl and polyvinylidene chloride, undergo this type of thermolysis mechanism. Chlorine radicals are split off from the polymethylene chain, and combine with hydrogen radicals generated by thermolysis of the backbone chain to produce hydrogen chloride. Fragments of the backbone chain cyclize to produce aromatic compounds including benzene, toluene, and polyaromatics.

(*d*) *Chain depropagation* takes place in polymers having polymethylene backbone chains. 'Unzipping' takes place to give high yields of monomer.

$$\left[-CH_2-\underset{\underset{CO_2CH_3}{|}}{\overset{\overset{CH_3}{|}}{C}}- \right]_n \longrightarrow CH_2{=}\underset{\underset{CO_2CH_3}{|}}{\overset{\overset{CH_2}{|}}{C}}$$

One example of this is polymethylmethacrylate (Perspex), found as a comonomer in some acrylic fibres.

7.9 ADVANTAGES AND DISADVANTAGES

Experience has shown that Py-GC is a powerful technique for the identification and discrimination of polymers. Differentiation of most fibre types is a major advantage of Py-GC. Compositional data of fibres within a class are clearly indicated in pyrograms. For example, polyamides are readily distinguished. The method also has the potential for the identification of additives.

Some of the more negative aspects may be resolved by developments in instrumental design and pyrolysis technique. Although the method is destructive, for many forensic investigations there is often sufficient material for subsequent examination where necessary. The removal of polymeric matrix from material may, indeed, facilitate the examination of more intractable components. For example, some pigments and dyes may be recovered unchanged from the pyrolysis zone, which can then be subsequently identified by other means.

Sensitivity is limited to the low microgram level compared to a lower order of magnitude for FTIR. For example, with fibres giving simple pyrograms (Nylon 6) a 10 mm single fibre weighing approximately 3 micrograms can be readily identified. This limit of detection is adequate for many forensic examinations. Pyrolyser design, derivatization methods, and the use of more sophisticated MS techniques may improve sensitivity levels. Standardized conditions have not yet been adopted, although UK Home Office forensic laboratories standardized on Carbowax packed columns in the 1970s. Many regular users of Py-GC have adopted one phase type as their standard to facilitate profile comparisons. Analysis times are of the order of 30 minutes. The time taken to handle samples and interpret data from pyrograms is, however, short. Therefore, the actual time required to perform the examination is comparable to other techniques. Sample automation, available on some commercial pyrolysers, would improve throughput.

7.10 FUTURE DEVELOPMENTS

Py-GC has developed in the past two decades from the pioneering work of de Forest (de Forest 1974) and others in the 1970s to the present where researchers, such as Tsuge, carry out sophisticated work on pyrolysis mechanisms. Although pattern recognition techniques have a place in the identification of and comparison of polymers, the identification of pyrolysis products and the study of pyrolysis mechanisms will lead to a more thorough interpretation of polymer composition. With the development and greater use of MS and other identification methods, it is expected that there will be advances in this direction.

Sensitivity for trace quantities of fibres is a problem that needs to be addressed. Improvements in sensitivity can probably be achieved by improvements in instrumental design and modifications to the pyrolysis process. Developments in the design of the pyrolysis chamber, GC injector system, capillary columns, and detectors will give improvements in sensitivity. Selected ion monitoring mode in mass spectrometry for diagnostic pyrolysis products may also be used to identify trace quantities of polymeric material.

Modifications to the pyrolysis process are being developed. An example of this is the use of pyrolysis derivatisation techniques (Challinor 1990, 1991). Pyrolysis alkylation tends to produce simpler pyrograms, but at the same time giving more compositional data. Direct results of simpler pyrolysis profiles are greater sensitivity and ease of interpretation of pyrolysis profiles. The procedure is particularly pertinent to polymers which can be hydrolysed and derivatized, and it includes the polyesters and polyurethanes. Variations on this theme may be expected.

In summary, pyrolysis in conjunction with gas chromatography and/or mass spectrometry is a currently useful technique offering greater and easier discrimination of some polymers than alternative techniques such as infra-red spectroscopy. Reproducibility can be achieved within a laboratory, and this should not be seen as a problem to be associated with the pyrolysis technique. Whilst sample size for fibres is still a limiting factor in some instances the future should lead to improvements in sensitivity resulting in a reduction in sample size.

Pyrolysis techniques should not be seen as an either/or with infrared spectroscopy but rather they should be viewed as complementary techniques.

BIBLIOGRAPHY

Almer, J. (1991) Subclassification of polyacrylonitrile fibres by pyrolysis capillary gas chromatography. *Can Soc For Sci J* **24**, 51–64.

Bortniak, J. P., Brown, B. S. A. & Sild, E. H. (1971) Differentiation of microgram quantities of acrylic and modacrylic fibres using pyrolysis gas liquid chromatography. *J For Sci* **16**, 380–392.

Burke, P., Curry, C. J., Davies, L. M. & Cousins, D. R. (1985) A comparison of pyrolysis mass spectrometry, pyrolysis gas chromatography and infrared spectroscopy for the analysis of paint resins. *For Sci Int* **28**, 201–219.

Challinor, J. M. (1983) Forensic applications of pyrolysis capillary gas chromatography. *For Sci Int* **21**, 269–285.

Challinor, J. M. (1989) A pyrolysis derivatisation gas chromatography technique for the structural elucidation of some polymers. *J Anal Appl Pyrolysis* **16**, 323–333.

Challinor, J. M. (1990) Pyrolysis gas chromatography—some forensic applications. *Chemistry in Australia*, April 1990, 90–92.

de Forest, P. R. (1974) The potential of pyrolysis gas chromatography for the pattern individualisation of macromolecular materials. *J For Sci* **19**, 113–120.

Grob, K., Grob, G. & Grob, K. (1978) Comprehensive standardised quality test for glass capillary columns. *J Chrom* **156**, 1–20.

Hickman, D. A. & Jane, I. (1979) Reproducibility of pyrolysis-mass spectrometry using three different pyrolysis systems. *Analyst* **104**, 334–347.

Hughes, J. C., Wheals, B. B. & Whitehouse, M. J. (1978) Pyrolysis—mass spectrometry of textile fibres. *Analyst* **103**, 482–491.

Irwin, W. J. (1982) *Analytical pyrolysis—a comprehensive guide*. Marcel Dekker, New York.

Jones, C. E. R & Cramers, C. A. (1977) *Analytical pyrolysis*. Elsevier.

Liebmann, S. A. & Levy, E. J. (1985) *Pyrolysis and GC in polymer analysis*. Marcel Dekker. New York.

May, R. W., Pearson, E. F. & Scothern, D. (1977) *Pyrolysis gas chromatography*. Analytical Science Monographs No. 3. The Chemical Society.

Meuzelaar, H. L. C. & Kistemaker, P. G. (1973) A technique for fast and reproducible fingerprinting of bacteria by pyrolysis mass spectrometry. *Anal Chem* **45**, 587–588.

Meuzelaar, H. L. C., Posthumus, M. A., Kistemaker, P. G. & Kistemaker, J. (1973) Curie point pyrolysis in low voltage electron impact ionisation mass spectrometry. *Anal Chem* **45**, 1546–1549.

Moncrieff, R. W. (1975) *Man-made fibres.* Newnes Butterworth.

Perlstein, P. (1983) Identification of fibres and fibre blends by pyrolysis gas chromatography. *Analytica Chimica Acta* **155**, 173–181.

Plage, B. & Schulton, H. R. (1991) Sequence in copolymers studied by high mass oligomers in pyrolysis—field ionisation mass spectrometry. *Die Angewandte Makromolecular Chemie* **184**, 133–146.

Saferstein, R. & Manura, B. S. (1977) Pyrolysis mass spectrometry—a new forensic science technique. *J For Sci* **22**, 748–756.

Saglam, M. (1986) Qualitative and quantitative analysis of methyl acrylate and methyl methacrylate of acrylonitrile fibres by pyrolysis gas chromatography. *J Applied Polymer Science* **32**, 5719–5726.

Senoo, H., Tsuge, S. & Takeuchi, T. (1971) Pyrolysis gas chromatography analysis of 6-6.6 nylon copolymers. *J Chrom Sci* **9**, 315–318.

Smalldon, K. W. (1969) The identification of paint resins and other polymeric materials from the infrared spectra of their pyrolysis products. *For Sci Soc J* **9**, 135–140.

Sugimura, Y. & Tsuge, S. (1979) Studies on the thermal degradation of aromatic polyesters by pyrolysis gas chromatography. *J Chrom Sci* **17**, 34–37.

Tsuge, S. & Ohtani, H. (1984) Pyrolysis gas chromatographic studies on the microstructures of stereospecific polypropylenes. In *Analytical pyrolysis, techniques, and applications.* K. J. Vorhees, (ed.), Butterworths.

Usami, T., Itoh, T., Ohtami, H. & Tsuge, S. (1990) Structural study of polyacrylonitrile fibres during oxidative thermal degradation by pyrolysis gas chromatography, solid state C-13 nuclear magnetic resonance and Fourier transform infrared spectroscopy. *Macromolecules* **23**, 2460–2465.

Vorhees, V. J. (1984) *Analytical pyrolysis, Techniques and applications.* Butterworths.

Walker, J. Q. & Wolf, C. J. (1970) Pyrolysis gas chromatography: A comparison of different pyrolysers. *J Chrom Sci* **8**, 513–518.

Wampler, T. (1989) A selected bibliography of analytical pyrolysis applications 1980–1989. *J Anal & Applied Pyrolysis* **16**, 291–322.

Wheals, B. B. (1980/1981) Analytical pyrolysis techniques in forensic science. *J Anal Applied Pyrolysis* **2**, 277–292.

Wheals, B. B. (1985) The practical application of pyrolytic methods in forensic science during the last decade. *J Anal Applied Pyrolysis* **8**, 503–514.

8

Information content: the interpretation of fibres evidence

Michael Grieve, BSc
Bundeskriminalamt KT 34, Postfach 1820, Wiesbaden, Germany

8.1 INTRODUCTION

On completion of a case involving fibre transfer, the analyst is faced with one of his most difficult tasks—interpreting the value and significance of his findings and expressing it concisely so that it can be clearly understood, without the risk of ambiguity, by scientists, lawyers, and laymen alike.

In this chapter the factors influencing interpretation of fibre transfer evidence will be systematically considered. Reference will be made to experimental work that can be used to help the analyst in making his decisions. Some examples of casework findings will be discussed.

Previously published papers of a general nature concerning fibres evidence include those by Grieve (1983), Deadman (1984a), Bressee (1987), and Fong (1989). More detailed discussions can be found in chapters by Gaudette (1988) and Grieve (1990) in *Forensic Science Handbook*, vol. 2 and *Forensic Science Progress*, vol. 4 respectively.

The value of fibre evidence depends on many factors which are often complex and interacting. They are summarized below.

Factors that are known (or that are usually determinable):

(1) the circumstances of the case;
(2) the time that has elapsed before collection of the evidence;
(3) the suitability of the fibre types for recovery and comparison;

(4) the extent of the comparative information derived from the samples;
(5) the number of types of matching fibres;
(6) whether or not there has been an apparent cross transfer of fibres;
(7) the quantity of matching fibres recovered;
(8) the location of the recovered fibres;
(9) the methods used to conduct the examinations.

Unknown factors;

(1) the degree of contact that has occurred and the pressure undergone;
(2) the degree of certainty that specific items were definitely in contact;
(3) donor fibre shed potential;
(4) frequency of occurrence of the matching fibre types.

A list of potential aids to interpretation is shown below. Most of the information available has been derived from research projects. These are so labour intensive, however, that they must necessarily be limited in number.

(1) Transfer studies.
(2) 'Target fibre' studies.
(3) Studies on differential shedding.
(4) Experiments on the alteration of fibre characteristics in a manner consistent with localized conditions specific to a case.
(5) Manufacturers enquiries.
(6) Use of a data collection.
(7) Use of a Bayesian approach.

8.2 THE INFLUENCE OF CASE CIRCUMSTANCES

It has been advocated (Starrs 1990) that in the interests of complete impartiality it is preferable that analysts should be in total ignorance of the circumstances of the case that they are dealing with. I cannot subscribe to this viewpoint.

Fibre cases are very time consuming and labour intensive. Knowledge of the alleged circumstances can help to focus the proposed examination in directions that are likely to produce findings yielding the highest information content. There is absolutely no point in carrying out a lengthy examination to show that fibres are present in locations where they may reasonably be expected to be after a legitimate transfer, or to show that there is no transfer of fibres to an item that turns out not to have been present at the crime scene.

In my opinion, it is necessary, with cooperation from the case agent, to make sure that the following information is provided.

- A detailed account of what is alleged to have happened, what is known to have happened, who was involved, and to what extent?
- Where did the offence occur? If it was in a dwelling or in a car, who is the owner/occupier?

- Did a rape or assault take place on the floor, or on a bed or couch? Have the appropriate fibre standards been provided?
- At what time did the offence occur? When was the evidence collected?
- Did any person involved have legitimate access to the crime scene, if so, when?
- What exactly was being worn by the participants? Were any items of clothing removed?

The answers should give the analyst an insight into the likelihood of the possibility of transfer or persistence of fibres from certain items. This is important if a Bayesian approach toward interpretation of the findings is to be used (see section 8.10). It may be possible to restrict the examination to underclothes only. Information may be gained that will indicate whether transfers are likely to have been by primary or by secondary means.

It is vital that an assurance be given that after the offence, handling and packaging of the evidence has been properly carried out without creating an opportunity for cross contamination. It is of prime importance that garments from all individuals involved be strictly segregated and individually packed. Underclothes must never be packed together with items of outer clothing. Items from suspect and victim should be clearly labelled in a manner that reflects that they have been dealt with by different agents. Submissions by the defence that transfer of seemingly incriminating fibres took place subsequent to the offence will render even the best fibre transfer evidence worthless if they cannot be confidently rejected.

The following factors will tend to lower the evidential value of apparent fibre transfers.

(1) Any doubt, no matter how slight, that contamination may have occurred.
(2) Indications of previous legitimate contact between suspect and victim, e.g. that they were drinking, dining, or dancing together, or were together socially in a particular environment. Time may play an important role here.
(3) Offences which occur in a family environment, for example incest. The number of transferred fibres recovered may assume extreme importance in such instances.

Conversely, certain circumstances may tend to enhance the evidential value.

(1) When apparent transfer can be demonstrated between individuals totally unknown to one another, particularly a cross transfer involving several colours/types of fibre.
(2) When the delay between the offence and collection of the evidence has been minimal. This reduces the likelihood of matching fibres, present in quantity, having originated from other putative sources.
(3) When transfer of incriminating fibres has occurred onto, or better still, between items of underwear in cases of a sexual nature.
(4) When the results of transfer analysis follow in detail a pattern that is totally consistent with the reported circumstances.

8.3 THE GENERAL INFLUENCE OF FIBRE TYPES

After some experience of dealing with fibre transfer cases it soon becomes apparent that certain generic types of fibres are much more frequently encountered than others. This can be confirmed by data collection. A perusal of the HOCRSE fibre data collection (Laing *et al.* 1987), which currently contains some 18 000 samples, shows that the most frequently encountered type is cotton, followed by acrylic, polyester, polyamide, viscose, and wool.

This corresponds reasonably well with a study of approximately 3000 'foreign' fibres found by Grieve & Dunlop (1990) on the surface of casework undergarments, where the order is the same except that polyamide is in last place. Such rank orders are inevitably subject to regional variations. Wool fibres are more common in clothing manufactured in Great Britain than in America; in Australia, the percentage of cotton garments in relation to those containing synthetic fibres is very high.

It follows that cotton fibres, being so frequently encountered, and being inherently more difficult to characterize by virtue of having few morphological characteristics, do not represent as good value, evidentially, as synthetic fibres. It also follows that the evidential value will increase in cases involving recovery of the less frequently encountered types of synthetic fibre. Some examples might be acetates, olefins, chlorofibres, modacrylic fibres, or even acrylic fibres of unusual polymer composition. The HOCRSE data collection reflects that most acrylics are of the acrylonitrile/methyl acrylate variety (63%), followed by those co-polymerised with vinyl acetate (32%) whereas those containing methylmethacrylate are represented by only 4%.

Within a generic type, colour also plays a role in relation to frequency, even when referred to in a subjective sense. Black cotton fibres are, for example, much more frequent than purple or turquoise ones. From the 'foreign' fibre study of Grieve and Dunlop referred to earlier, cotton fibres represented 47% of the total; of the cotton fibres 45% were black, which had an overwhelming majority over red (20%) and blue (13%).

Previous work by Grieve *et al.* (1988a) reinforces the opinion that black cotton fibres are of relatively low evidential value. The frequency of matching black cottons was found to be 1 in 58, as opposed to 1 in 148 for red and 1 in 259 for blue. These figures were obtained from a total of 1035 comparisons for each colour after examinations with a comparison microscope equipped for transmitted and fluorescent light microscopy and by microspectrophotometry. If transmitted light comparison microscopy is the only comparative technique used, the frequency of matches (1 in 4, 1 in 3, and 1 in 7 respectively) rises to such an extent that it renders them evidentially worthless.

Deadman (1984b) divided fibres into three categories. Firstly, those that are indisputably common and which are used in the production of a massive number of articles. The best example is white cotton. Because the sources of these fibres are so numerous, and those from different sources are impossible to distinguish, they can have little value as evidence.

The second category are those fibres which for some reason may be defined as uncommon. They may possess a very unusual morphological characteristic, may have been manufactured in a small quantity for a limited period, or may be obsolete. Another possibility is that the fibres have been produced for a very specific end use, perhaps for an item with which members of the general public would not be expected to have contact. A particularly good, and now classical example of an unusual fibre type, is the Wellman 181B nylon carpet fibres involved in the Wayne Williams case in Atlanta, almost ten years ago.

All other fibres types lie somewhere between these extremes, and guides to their frequency can be provided only by the use of databases, by direct enquiry to fibre producers and dyers, or by research into the probability that fibres of a specific type will be present on randomly selected items of clothing.

At the beginning of any transfer examination, the items involved will have to be scrutinized to decide whether they contain suitable target fibres. These are the fibres that the analyst will look for to see if he can show that a transfer may have taken place. Optimally, they need to be brightly coloured, or highly fluorescent, and to shed well. The search will be easier if the fibres being sought are of a contrasting colour to the background.

If most of the garments in a case are made of fibres types inherently unsuitable as 'targets', this will reduce the general value of the fibres evidence, as the potential to verify the allegations will be reduced.

Unsuitability may be the result of the fibre types being extremely common, such as those originating from white cotton items, from blue jeans, or from some types of uniform (where many potential sources of the same fibre type are concentrated in a limited area).

Garments made from smooth shiny fabrics will not shed or retain fibres easily. Commonly encountered items of this type are anoraks, sport wear, and lingerie. These fabrics are very often made from polyamide or fine polyester fibres. Of course, the recovery of a low number of target fibres from (or on) a garment with these surface characteristics will assume much greater importance than the recovery of only a few fibres believed to originate from a good donor.

Fibres may be unsuitable as 'targets' for technical reasons. Colour is a prime characteristic necessary for fibre recovery and comparison. Fibres with a low dye concentration will present difficulties on both accounts. Even if their recovery is possible, spectral comparisons are more difficult to interpret, and dye comparison by thin layer chromatography will not be possible. Their evidential value may fall accordingly.

In the case of one very common fibre type, blue cotton from denim articles, the fact that the colour of fibres from diverse sources is indistinguishable (with the colour lying in a small area of the chromaticity diagram), has been established by research and documented by Laing *et al.* (1986) and by Grieve *et al.* (1988a).

There are other types of fibres where experience may suggest to the fibre examiner that they are common, and therefore of limited evidential value, but, as suggested by Gaudette (1988) and Grieve (1990), these suppositions need to be backed up by research which results in hard facts.

A good example is the black, pigmented viscose rayon fibres which are a low percentage component of pale grey sweat suits and running shorts of American manufacture. The other components are normally microscopically colourless cotton and polyester fibres, and are therefore unsuitable 'targets'.

The black viscose fibres occasionally differ in diameter and in delustrant concentration. Attempts at spectral colour comparisons result in the production of featureless spectra. It would be useful to know whether the products of one manufacturer can be differentiated from those of another. Enquiries to date suggest that these fibres are all produced by the same manufacturer and are distributed by many of the more than 40 athletic suit and shirt fabric suppliers within the United States.

8.4 THE INFLUENCE OF THE EXTENT OF FIBRE CHARACTERIZATION WITHIN A GENERIC TYPE

8.4.1 The individuality of synthetic fibres

To the uninformed, there is a tendency to believe that all fibres fitting a simple descriptive term, such as blue polyester, are indistinguishable. This is of course far from the truth. Polyester fibres produced by one company may differ in several respects; differences in cross-sectional shape, diameter, concentration of delustrant particles, and in alterations caused by processing. All of these features and their combinations will be apparent to the microscopist.

By the time this range of production variations has been subjected to dyeing, the result will be an extensive range of differentiable fibres within a generic type–colour combination. Even allowing for the fact (Pailthorpe 1990) that dye manufacturers are considerably reducing their vast range of dye mixtures, a small number of dyes can still be used to produce a large number of colour combinations by computation, using the CIELAB system. For example 12 Serilene dyes, used to dye polyester staple, can be used to give 220 possible colour combinations.

If one takes a particular synthetic fibre, and attempts to find exactly matching fibres from a different source, it is very difficult. So-called 'target fibre studies' have effectively demonstrated this.

In 1986 Cook & Wilson searched 335 items of clothing for the presence of frequently encountered fibres indistinguishable from those used in the manufacture of four popular brands of garment. A total of only 11 matching fibres, of which nine were of one type (blue wool), were found on ten garments, with no more than two fibres on any item.

In the same year Jackson & Cook (1986) published the results of searching the front seats of 108 vehicles for two common target fibre types, red wool (ladies pullover) and brown polyester (mens trousers). From 8436 recovered fibres, 37 matched the red wool and 8 matched the brown polyester in all respects.

The methods of comparison were the same as those in the previous study: comparison microscopy (bright field and fluorescence), microspectrophotometry, and thin layer chromatography.

The maximum number of matching fibres (red wool) found on any seat was 13; in any one vehicle 20. Fibres matching both types of target fibre were found in only one vehicle. A likely source for these fibres was forthcoming from amongst items of clothing belonging to the owner of the vehicle. The study demonstrated that when large numbers of more than one type or colour of fibre are recovered from a car seat, the evidence for primary contact appears to be highly significant.

Further research, carried out by the Northern Ireland Forensic Science Laboratory, was referred to by Cook in 1989. It concerned the presence of fibres indistinguishable from those found in four types of mask in 99 samples of hair combings . Three masks were made from acrylic fibres, the other from a particular type of blue wool knit fabric. The highest number of matching fibres of any type was 13 (blue wool) with a maximum of 5 fibres per hair sample. One of the targets was an unusual fibre, a grey 'tiger tail' acrylic. No matching fibres of this type were recovered. The evidential value will increase considerably when matching fibres of an unusual variety *are* recovered.

This study, which was obviously designed to enhance the evidential value of fibres recovered from suspects hair after the wearing of masks for terrorist acts, was a good example of research being applied to assist with the interpretation of an oft encountered but localized situation.

The value of fibre evidence can be enhanced considerably by demonstrating that fibres of the same generic type and colour as those apparently incriminating a suspect could not have originated from other putative sources which might be put forward by the defence. In a South Wales murder case, black acrylic fibres recovered from tapings of the victim's body were found to match those in the pullover worn by the accused. 31 other samples of black acrylic fibres, some of which could be distinguished microscopically from those in question, were examined by thin layer chromatography. Their dye composition was compared with that of the fibres from the suspect's pullover. None of them matched. This is an effective rebuttal of the proposal that 'of course, black acrylic pullovers are very common'.

8.4.2 Possession of common features

The more characteristics that two fibres can be shown to have in common, the greater the chance that they originated from a single source. The extent of the comparative information that can be obtained from any fibre sample is dependent on the techniques and the equipment that are available to examine it. The size of a single fibre is often a limiting factor. Failure to use state-of-the-art comparative techniques may result in severe reduction of the evidential value of the findings, as with limited comparison the number of potential alternative sources will rise considerably. According to the present state of the art, optimal comparison of coloured fibres should include comparison microscopy, using bright field and fluorescent light, followed by microspectrophotometry, and finally, if sample size permits, extraction of the dye and subsequent examination by chromatography.

Determination of polymer composition normally requires polarized light microscopy. More detailed information, which will allow sub-typing, can be obtained after using infrared spectroscopy and/or melting point determination.

The following features may enhance the evidential value of matching fibre types:

(1) when they are of an infrequently seen polymer composition;
(2) if they are characteristically soiled or partly burnt;
(3) if they have been overdyed;

(4) if they display a form of coloration that is unusual with respect to a particular generic type, for example pigmentation in acrylic or polyester fibres;
(5) if they are dyed with a class of dye not commonly used for that type of fibre, for example an acrylic dyed with an acid dye;
(6) if they display manufacturing faults, voids, inclusions, cross markings, or porosity marks;
(7) if they possess channels or antistatic inclusions;
(8) if they are of an unusual cross-sectional shape;
(9) if they have been treated to make them flame retardant;
(10) if they have undergone unusual physical processing;
(11) if their appearance has been modified as a result of subjection to localized conditions.

A good example of how a detailed study of one particular characteristic can help to provide a wealth of comparative and investigative information can be found in the work of Palenik & Fitzsimmons (1990) on cross-sections. Cross-sections can provide information on the fibre manufacturer, the spinning process used, the end use, physical processing, fibre quality, and the dyeing method. A particular feature of trilobal carpet fibres, the *modification ratio*, can be used to suggest a manufacturer and thus open the way to production enquiries. The Wayne Williams carpet fibre, Wellman 181B nylon 6, has a very unusual modification ratio of 2.7. Full details of this case and the subsequent enquiries are given in the publications by Deadman, (1984a,b).

Fibres with especially unusual cross-sections are used by manufacturers to 'tag' their bulked continuous filament carpet yarns. Certain cross-sections may be produced for exclusive use, for example bullet-shaped polypropylene fibres produced by Phillips for use by General Motors.

Great care must be taken not to make false exclusions after fluorescence examinations where unexpected differences are noted between control 'target' fibres and recovered fibres which match in all other respects. More detailed investigations, and the taking of further control samples, will normally show that these differences arise owing to local conditions. Variations may be apparent when:

(1) surface fibres fluoresce differently from those deeper within the garment;
(2) alterations have been caused by excessive wear, for example in the central area of a well worn car seat;
(3) the fabric has been washed or dry cleaned recently or has some localized staining;
(4) alterations have been caused by exposure of the recovered fibres to the effects of the elements, after their deposition. The fading of fibres after their immersion in a river was a factor playing a role in the Wayne Williams case. Control experiments on target fibres taken from a bedspread showed that after immersion in river water, they faded in an identical manner to fibres recovered from bodies subjected to similar treatment.

Once an apparent explanation for inconsistencies is forthcoming, additional tests may result in consistent findings, after which the evidential value of the matching fibres is likely to be enhanced.

Sometimes, a particular dyeing process may result in fibres gaining a very characteristic appearance (Cook 1989). An example is the Neochrome dyeing process used by Courtaulds to dye Courtelle acrylic fibre. After spinning, the fibres are steam stretched before passing continuously through an open dye bath. Differential dye penetration of the less crystalline areas of the fibre will result in the appearance of so called 'tiger tails'. The overdyeing of 'shoddy' fibres will also produce fibres where traces of the original colour are still visible.

Certain manufacturers use trace metals in their fibres as a means of coding their own products. Magnesium, cobalt, lithium, and manganese have been used by polyester producers in the United States. These may be present in a quantity as low as 2–5 ppm, so that this much discussed phenomenon is not likely to help forensic scientists to characterize a few millimetres of single fibre.

8.5 NUMBER OF MATCHING FIBRES RECOVERED

Interpretation of the results of a fibre transfer examination is strongly influenced by the number of matching fibres recovered. The problem is that the number of fibres originally transferred will always be unknown. Does the number recovered represent a high percentage or a low percentage of this number? For example, 20 matching fibres from an obviously high shedding donor, could, depending on other circumstances, be attributable to a secondary transfer. If on the other hand, the donor is a very poor shedder, the recovery of 20 fibres may be a strong indication of contact.

I believe that to maximise information helpful to interpretation, *all* matching fibres should be recovered from the fibre tapings made from an item, unless the numbers are obviously very substantial. Information may be lost when searching is terminated after the recovery of perhaps 15–20 matching fibres. (If the fibres have a very characteristic appearance on the tapes it may be possible to estimate the number remaining without actually removing them, which can certainly save time.) If the tapings are not all searched, because large numbers are being recovered, I think this fact should be made evident.

It is important to have some way of estimating the shedding potential of a garment. One way of doing this is to use a short length of adhesive tape applied to the garment surface, but this has been shown by Coxon *et al*, (1990) to be somewhat unreliable. These authors propose the use of a simple mechanical method to transfer fibres from the actual donor to the appropriate type of recipient fabric surface.

There are a number of reasons that could be used to account for the recovery of only a low number of matching fibres.

(1) There may have been a considerable delay between the offence and the recovery of the items. Fibre persistence has been discussed by Pounds & Smalldon, (1975) and by Robertson *et al.* (1982).
(2) If there has been a delay, fibres will be lost by continued wearing of the recipient garment, by other garments being worn over it, or by the fibres being situated on an area of the garment particularly prone to contact with other surfaces. Redistribution of fibres after their original deposition has been discussed by Robertson & Lloyd (1984).

(3) It could be due to the donor being a poor shedder, or to a very low contact pressure, or to contact having been only fleeting and involving a limited area.

(4) It could be due to the after effects of garments having been washed or dry cleaned. The effects of these treatments on fibre persistence have been the subject of studies by Robertson & Olaniyan (1986) and by Grieve *et al.* (1988b). Both studies indicate that some fibres may be expected to persist. Fibres may also survive several secondary contacts and remain undisturbed on seats for a long period (Grieve *et al.* 1988*b*).

(5) When a garment is composed of a fibre mixture, the various fibre types will not necessarily be transferred in the same ratio as that in which they are present. The transfer of some types may be expected to be substantially less than others. Transfer will depend on how easily the fibres fragment. The process is called 'differential shedding', and it has been documented in papers by Parybyk & Lokan (1986) and Salter *et al.* (1987). An interesting case report illustrating this phenomenon was reported by Mitchell & Holland, (1979).

A recent case in this laboratory entailed the searching of car seats for fibres which could have originated from a skirt and halter top made from a mixture of acetate fibres and nylon fibres. Many acetate fibres were recovered, but not a single nylon fibre was found. Tests revealed that this was consistent with the shedding behaviour of this fabric.

(6) The presence of only a few matching fibres may mean that they have been deposited by means of a secondary or tertiary transfer as opposed to a primary one. Secondary transfer may be defined as the first indirect transfer after primary transfer, taking place via an intermediary object. In trials carried out by Jackson & Lowrie (1987) involving items of clothing, secondary transfer of at least one fibre took place at least 50% of the time. However, in only 3% of 120 trials were more than 5 fibres transferred. With trials involving seating, not only was secondary transfer more common, but the maximum number of fibres recovered in one trial was 74. Secondary transfer in these trials was made immediately after the primary contact.

(7) Owing to the possible inefficiency of using adhesive tape to recover the transferred fibres. This was found by Jackson and Lowrie in the above study to vary between 32 and 100%. Further information on the efficiency of methods of fibre recovery was provided by Pounds (1975).

8.6 INTERPRETATION AND METHODOLOGY

It is important within the context of fibre examinations to ask the question 'to what extent does the choice of methodology affect the amount of information gained from the overall examination?'

It is undoubtedly important to reassure both the courts and the public that standard procedures are being used and that recognized steps are taken to apply quality control procedures. The American Society of Crime Laboratory Directors (ASCLD) have a set of guidelines in which they outline requirements for laboratories to meet their Accreditation scheme. Some of these requirements are as follows:

> 'the laboratory must maintain written copies of appropriate technical procedures for each discipline in which the laboratory performs analysis'
> 'procedures and instruments used must be those generally accepted in the field or be supported by data gathered and recorded in a scientific manner'
> 'new technical procedures must be thoroughly tested to prove efficacy in examining evidence material before being introduced into casework'
> 'conclusions of an examiner must fall within the consensus of opinions of experienced, able, and knowledgeable individuals in the field of forensic science or be supported by sufficient scientific data'

Some comments with respect to fibre examinations may be appropriate. Opponents of standardizing procedures feel that this results in an inflexible 'cook book recipe' approach which denies flexibility, a feature frequently necessary in fibre examinations. It is, however, fundamental that analysts working in one laboratory or laboratory system, follow a common set of procedures. New employees may bring fresh ideas, which, provided that they are demonstrably superior, may replace existing techniques.

Some aspects of fibre examination lend themselves to standardization much more readily than others. Within the classical microscopical methods for measuring fibre elongation, refractive indices, and birefringence there is little scope for variation. On the other hand, during recent years (1986–1990) at least six variations of method for obtaining fibre cross-sections have been proposed. Considerable variation is still found in the choice of mounting medium used by different laboratories for fibre slide preparations (Collaborative Testing Services, Inc., Herndon, VA. – Crime Laboratory Proficiency Testing Program).

In both the choice of sectioning method and of mounting medium, strict criteria have to be met (Grieve & Kotowski 1986, Cook & Norton 1982). Provided that these are adhered to, there is no reason why different methods should not be considered acceptable, and thus the flexibility of choice remains.

A positive aspect of standardization is that it facilitates quality control. Again some techniques lend themselves to this more than others. The identification of fibre polymers by infrared spectroscopy is easily verifiable, provided that the instrument has been properly set up and calibrated and the spectrum is produced under defined operating conditions. The record obtained can be verified against both in-house reference collections and library spectra. The importance of comparisons against manufacturers' standards run on the same instrument cannot be overemphasized.

The disadvantage of another popular method of identifying fibre polymers, capillary column pyrolysis-GC, is that comparisons with spectra recorded under a different set of operating conditions are not possible, even though the same instrument may have been used.

Analytical techniques should preferably provide a permanent record, allowing visual comparison of the results of analysis of material from different sources. If necessary, the results can be verified by an independent party, either by reference to published literature or by repeating the analysis, not necessarily on the same instrument.

It is worth mentioning that the methods used to identify and compare fibres were not specifically designed for that sole purpose. They are fundamental analytical and microscopical techniques that have been tried and tested in countless scientific fields for many years.

One hopes that standardization of methodology used for fibre examinations will lead to a more unified basis from which to draw conclusions on completion of the examination. This should help to eliminate the situation where, after examination of the same samples by different laboratories, different opinions have been forthcoming (Collaborative Testing Services Inc., Herndon, VA. – Crime Laboratory Proficiency Testing Program).

With fibre transfer examinations it is particularly difficult to monitor a part of the examination crucial to success, and that is the process of searching for and recovering fibres from adhesive tape lifts. Interpretation of the findings will depend on how well this has been carried out. The number of matching fibres recovered will be affected by the degree of conscientiousness and skill employed in the searching process. Decisions on the possibilities of primary versus secondary transfer and whether the fibres may have been deposited as the result of a short or long-term transfer will be influenced. This problem illustrates the need for the proficiency of trainees in this area to be thoroughly investigated.

The standardization of methodology is difficult. Factors of availability, cost of equipment and materials, experience and expertise of staff, use of non-destructive methods, and problems caused by sample size all apply. The best that can be achieved perhaps, is to suggest the application of 'minimum standards'—a consensus of opinion based on common sense and practical experience, and to hope that they will be adopted as successful guidelines.

The promotion of minimum standards is a subject that will certainly arouse controversy and provoke considerable discussion. The following suggestions may provide a basis for argument. All of them represent established practices, widely practised among fibre examiners.

(1) All possible anti-contamination procedures must be followed and documented, both during collection and examination of the evidence. If suspicions that this situation has been compromised cannot be dispelled, a fibre transfer examination should not be made.
(2) Fibres should be collected from fabric surfaces by using adhesive tape, the fibre bearing surface of which must remain sealed during the recovery process.
(3) A mountant which fulfils recognized criteria (Cook & Norton 1982) should be used to prepare the fibres for microscopic examination.
(4) Recovered fibres should be mounted individually, so that the examination of any specific fibre can be precisely documented.
(5) Before making comparisons, fibres must be identified. The generic type of synthetic fibres should be determined.
(6) In making microscopical comparisons, the standard and the recovered fibre samples must be viewed simultaneously.
(7) A sufficient number of standards should be taken so that any in-garment variation can be accounted for. As an extension of this, examination of standards by microspectrophotometry and thin layer chromatography should include suffi-

cient replicate runs to constitute a representative sample. It is particularly important to consider variations due to dye concentration or to dye batching.

(8) Whenever there is sufficient material available, thin layer chromatography should be used to complement examinations by microspectrophotometry.

(9) All instrumental analyses must be carried out under prescribed, documented operating conditions. Identifications based on instrumental analysis should be confirmed by reference to samples of known origin, run on the same instrument.

(10) Thin layer chromatography should be carried out under standard conditions as described by Laing & Boughey (1990). The use of reference dye solutions (Wiggins *et al.*, 1988) is recommended.

(11) Unless exceptional circumstances dictate otherwise, fibre comparisons should be made only with reference to a sample of known origin (Grieve 1990).

8.7 DATA COLLECTIONS

With the establishment of a databank, a great deal of useful information can be accumulated which, if used intelligently, can assist in making decisions on fibre frequency. The Home Office Forensic Science Service in Great Britain has the benefit of such a collection, which currently contains details on some 18 000 fibres. The collection has been described by Laing *et al.* (1987). A similar system is at present being established in Germany (Adolf 1989). The Royal Canadian Mounted Police in Canada have established a different form of database for storing details of fibre standards obtained from manufacturers. Details have been given by Carroll *et al.* (1988).

The biggest disadvantage of data collections from casework garments is that, because of the time it takes to assemble them, they cannot keep up with fashion trends. A seasonal colour, not hitherto found amongst the data, may suddenly spring into prominence in leading retail outlets. Also, although the collection may contain thousands of records, most of these fibres will automatically differ in colour and generic type from any particular questioned fibre. The actual number of relevant comparisons of fibres with very similar characteristics will inevitably be much smaller. Differences in dye concentration and composition cannot be taken into account, because of difficulties in storing this information.

Where a collection of this type really comes into its own is in providing information on the frequency of occurrence of certain morphological characteristics within various generic types. Some illustrations taken from the HOCRSE data collection are as follows.

The most common type of acrylic fibres are copolymerized with methylacrylate, and of these approximately 70% have round cross-sections. From those copolymerized with vinyl acetate, over 90% are bean shaped.

95% of all round polyesters are delustred. This figure rises to 100% for all the octalobal shaped polyesters encountered. Bright polyamide fibres with a round section are much less frequent than delustred ones, although with trilobal polyamides, 81% of nylon 6.6 has been observed to be delustred, but only 18% of nylon 6.

The characteristics which lend themselves best to this treatment are polymer sub-types, cross-sectional shape, delustrant, and fibre diameter (certain fibre diameters tend to have specific end uses). Combinations are possible, and depending on how low the resultant percentage is, so the evidential value of a recovered fibre with similar characteristics will rise accordingly.

It is not normally feasible for a small laboratory to model a data collection on the HOCRSE one, because of the time required to make the necessary replicate colour measurements on the Nanospec microspectrophotometer system described by Laing *et al.* (1986).

This does not totally preclude the saving of data from all target fibre samples passing through any laboratory. Many of their features must be recorded in any event during the course of normal note taking. Colour may be treated subjectively. All features can be noted in tabular form, which is then computerized and saved under the appropriate generic type in order to speed up searching time. The information generated can be most helpful in evaluating the frequency of some of the characteristics of synthetic fibres but it does not substantially assist for wool and cotton with their limited descriptive features.

8.8 TRADE ENQUIRIES

It may be possible to derive information from an exhibit that will allow persual of a chain of enquires that, if successful, will provide accurate information concerning the production of import figures for that particular item. This approach has been used frequently and very successfully in the United Kingdom.

The starting point in the chain will normally be from a label in the garment. If the garment is of foreign origin, it may be possible, through the Association of Importers, to gain access to the address of the manufacturer. Alternatively, a label may lead back to a retail outlet and from there to a chain buying office, garment producer, and yarn producer.

Garments often bear more than one label. The information to be gained from them may include some or all of the following: retailer, country of origin, composition, cleaning instructions, printer's reference, manufacturing company, garment style number, and date of production.

Certain garments or items may particularly lend themselves to trade enquiries, for example fashion wear with unusual fabric construction or a range of fibre types and colours in a striking design. Cases involving items containing bicomponent fibres, items trimmed with artificial fur, tights (pantie-hose) and ropes with an unusual construction have also been the subject of successful enquiries. Naturally, the making of trade enquiries is highly dependent on building up a network of contacts within the textile trade.

Reports of trade enquiries in the forensic press are scarce. Some examples can be found in papers by Mitchell & Holland (1979) Deadman (1984a,b). Wiggins & Allard (1987) conducted a survey of the major manufacturers of loose car seat covers and of integral coverings of car seats. Fabric suppliers were subsequently contacted, and personal visits to a cloth producer and a dye-house were made. The samples were compared with casework seat covers received during 1982.

This survey contains a wealth of information on the production, dyeing, and supply of fabric used for car seats and seat covers. Two examples are given showing how the acquisition of information from production sources can be used to enhance the evidential value of case work findings.

8.9 REPORT WRITING

When his analytical work is completed, the examiner is faced with the task of summarizing his findings. I believe that, provided due caution is exercised, the examiner is obliged to give some indication of how he considers the value of his/her evidence. The expert witness is under a moral obligation to be as helpful as possible when explaining the complexities of scientific evidence. Needlessly complicated and badly structured reports full of complex scientific terminology which end with either vague or no conclusions are not what is required. The end product should be succinct, and as informative as possible to the reader. Over-caution and reluctance to express any opinions will certainly not assist those seeking guidance.

The opinions expressed in a fibres report many depend on whether or not the examiner has a database at his/her disposal, as so much weight is placed on assessment of the frequency of occurrence of particular fibre types. It was evident from the survey carried out by Lawton *et al.* (1988) that it is common for fibre examiners to make clear, either in writing or verbally, that owing to the mass production of textile items, recovered fibres cannot be stated categorically to have originated from one particular textile to the exclusion of all other textiles made from fibres with exactly the same characteristics.

Even without the aid of a databank, there are certain circumstances which may allow the examiner to give more weight to his findings. Recovery of high numbers of matching fibres of several types and colours, cross transfer, unusual fibres, a pattern of recovery that is highly consistent with alleged events will all help in this respect. In these instances phrases like '*the findings are strongly indicative of contact having occurred between....*' or '*it is highly probable that these fibres originated from*'.... are, in my opinion, fully justified.

It should be possible for the examiner to decide whether his findings are consistent with a recent direct contact, a casual or non-recent direct contact, or a secondary contact. Often, alternatives may be feasible.

I believe that any report on a fibre transfer examination should contain the following information, which is important to the reader in explaining why certain conclusions have been reached.

It should clearly state the objectives of the examination and, as not all fibres are suitable for recovery, state which target fibres have been chosen. If some items are not examined, or some fibre types not chosen, this should be stated, together with an explanation. The number of recovered matching fibres of different varieties should be given; and, if a search was carried out for fibres of a particular type and none of these were recovered, that must be made clear. The techniques used for fibre comparisons must be clearly stated.

A conclusion should be given at the end, summarizing the findings by referral to previous paragraphs. The following points should be taken into consideration and mentioned if they appear to strengthen or weaken the evidence.

(1) The number of matching fibres recovered in relation to the sheddability of the donor.
(2) The possibility of primary versus secondary transfer.
(3) The number of types of fibre examined, and whether cross transfer was apparent.
(4) Whether transfer to any particular garments, for example underclothes, is specially significant.
(5) Any possibility of contamination.
(6) It may be possible to offer a hypothetical explanation for results that are apparently not consistent with the alleged circumstances.

A report which states that no evidence of fibre transfer was apparent, *does not of course mean that no contact took place.* Recovery techniques are not 100% efficient, fibres which had been transferred may have been already lost, and the nature of donor and recipient surfaces may play a substantial role. Garments which are apparently good donors may have been taken off, or have been covered by another item, effectively preventing substantial contact.

Factors which may result in a low recovery of fibres from a fabric which appears to be a good donor have already been outlined in section 8.5.

8.10 THEORIES PERTAINING TO THE INTERPRETATION OF FIBRE EVIDENCE

There are various theoretical concepts that have been applied to estimating the significance of fibre evidence. Gaudette (1988) discussed the application of error probabilities to associative evidence in general, and to fibre evidence in particular. The main source of association errors with fibres is due to their mass production; they cannot be stated to originate from one specific source to the exclusion of all others.

Gaudette states that provided that a fibre examination has been conclusive and there has been no technical error on the part of the examiner, two types of error may occur:

Type 1 error: the fibre originated from the known sample, but the examiner concludes that it is not consistent with having done so (incorrect exclusion).

Type 2 error: the fibre did not originate from the known sample, but the examiner concludes that it is consistent with having done so (incorrect association).

Gaudette concludes that the value of fibre evidence will be inversely proportional to the probability of a type 2 error. A type 2 error could result in the presentation of incorrectly incriminating evidence. The probability of type 2 errors occurring in fibre comparisons is unknown. This leaves the examiner dependent on the results of target fibre studies, which are limited because they are so labour intensive, and certain basic facts. The facts listed below reduce the probability of associative fibre evidence being presented as the result of a coincidence.

(1) When all the parameters that can be used to characterize fibres are considered there are a very large number of distinguishable types in existence. The probability of finding a specific type (unless it is very common) is correspondingly very low.
(2) Even with a common fibre type there is no certainty that this will be present on any randomly chosen garment.
(3) Fibres adhering to the surface of textiles are likely to be the result of a recent contact, particularly when a large number of matching fibres are recovered.

Stoney (1987a) discussed interpretation issues in multiple fibre cases. He outlined factors that make fibres present at a crime scene relevant to an offender. He stated that '*one major way that relevance is shown is when we can infer timely, direct contact from some fibre source*'.

This statement strongly supports the need to compare recovered fibres with a sample of known origin (Saferstein 1981, Deadman 1984a,b, Grieve 1990). Comparison of fibres removed from different sources, without reference to a known standard, is a practice to be discouraged, except for the purpose of providing investigative leads.

Stoney (1987b) also suggested a set of 15 fundamental principles which describe the complexity of associative evidence evaluation. His aims in doing this and defining terminology were to focus theoretical discussions and to identify fundamental differences in viewpoint. Among the fundamental principles described was the concept that the value of the associative evidence is assessed by using a likelihood ratio, that is, by taking a Bayesian approach.

The interpretation of fibre evidence by a statistical approach is fraught with difficulty, because of the great problems of obtaining information on production figures for certain fibre types, their end uses, and the distribution data for the finished products. This may be possible for specific cases through manufacturers/importers enquiries, but on a case by case basis it is impracticable, especially for small laboratories with limited resources.

Attempts to use a statistical approach without proper foundation can lead to highly speculative and dangerously misleading conclusions; some examples are quoted in the textbook by Gianelli & Imwinklreid (1986).

The use of a non-statistical approach may well be suitable for the handling of fibre transfer evidence. A comprehensive general explanation of the Bayesian approach, with specific examples of its use, can be found in Evett (1990). Non-statistical probabilities are those that are vague by their very nature, unlike statistical probabilities which can be precisely estimated. For example, we can work out the odds of throwing two dice and getting a certain total.

Using a Bayesian approach will result in the production of an end value which is called the likelihood ratio (LR).

$$\mathrm{LR} = \frac{\text{the probability of the evidence if C is true}}{\text{the probability of the evidence if C}' \text{ is true.}}$$

C is the event that the suspect is the person who committed the crime.
C′ is the event that the suspect is not the person who committed the crime.

The evidence in a fibre case will be the matching fibre types.

Buckleton & Evett (1989) have written a paper on aspects of Bayesian interpretation of fibre evidence which highlights various problems. To calculate the likelihood ratio, it is necessary to know not only the number of matching fibre types recovered on a particular garment, but also the number of non-matching fibre types present on that item. This number is not usually known and is generally considered irrelevant. The problem arises, because by using an average, non-case specific number for this term of the equation can affect the likelihood ratio very considerably. Buckleton & Evett (1989) used data supplied by Fong & Inami (1986) in relation to the number of types, or groups, of foreign fibres found on clothing.

Grieve *et al.* (1990) researched the possibility of arriving at this number in a way that would be case specific, but would be practicable and economical in terms of time taken. Using fibre tapings prepared from colourless items of underwear (the most significant garments in sexual assault cases), they found it perfectly feasible to make a rapid estimation of the number of 'foreign' fibre types on the tapings by using a stereomicroscope.

The average time taken to do this was only 12.5 minutes. A 20% loss of accuracy was suffered in comparison to removing the fibres, mounting them, and using high power microscopy, but the average time required for that was considerable— an average of 3 hours 10 minutes. In spite of the loss of accuracy, acceptance of the number arrived at is much preferable to the use of a non-specific average number.

Evett (1990) points out that a numerical statement for the Likelihood Ratio is unlikely to convey much to a court, unless it is very small, or very large. He suggests that it would be possible to assign a verbal convention to the likelihood ratio. For example this could range from values of the LR falling between 1–10 which could be interpreted as 'the evidence weakly supports C' to the other end of the scale where the LR value was > 1000, which would be interpreted as 'the evidence very strongly supports C'. Values of less than 1 would correspondingly translate into sentences supporting C′.

Bayesian treatment can help to formulate the most relevant questions in an individual case in the form of an 'expert system'. The answer, although subjective, can be combined with objective information to provide the most accurate guide to interpretation currently possible. Study is required to appreciate the relevant concepts, and complicated mathematics may be necessary, factors which tend to deter the operational scientist. With some application, however, and with the conviction that the improvement in interpretation justifies the effort, these difficulties can be overcome.

8.11 EXAMPLES OF CASEWORK FINDINGS

In all examples quoted comparisons were made using bright field, polarized and fluorescent light microscopy, microspectrophotometry and, whenever possible, by infrared spectroscopy and thin layer chromatography.

Example 1

The subject attempted to murder his girl friend by attaching an explosive device to her car. It failed to detonate, and was submitted to the laboratory for examination. The timing device and batteries were held together by adhesive tape.

Numerous fibres were recovered from the sticky surface of the adhesive tape, including animal hairs, blue denim cottons, a pale green polyester, and two yellow polypropylene fibres. As a result of this examination, the subject's apartment was searched for possible sources of these fibres.

The search revealed a folding polypropylene camp seat, a green polyester/cotton sports shirt, a pair of blue jeans, and a black cat! All of the fibre types matched those from their putative sources.

Factors which increase the value of the findings are:

(1) the protected location of the fibres under the tape surface indicates that they were deposited at the time the device was being assembled;
(2) a combination of four types of matching fibre, even though two of them are in themselves common;
(3) the fact that polypropylene fibres are not very common, and infrared spectroscopy and melting point examination showed them to be of the same type as those in the camp chair;
(4) information which revealed that only 12 chairs of this type had been sold from the local PX, one of them to the suspect.

Example 2

The partly clothed body of a young woman was found lying on a blood soaked bed sheet in her apartment. The items submitted to the laboratory bore numerous red acrylic fibres. These fibres led to the apprehension of a suspect as they matched those in a red cardigan that he had worn, and had been photographed in, while dancing with the victim at a party.

This man was known to have been in the victim's apartment on previous occasions, the number of transferred fibres found on various items was therefore to be a crucial interpretive issue.

The recovery of many long fibres ensured that the polymer and dye compositions of the recovered fibres could be compared with those in the cardigan. A check with the HOCRSE database revealed that these fibres were not of a particularly common variety. Other potential donors, including the victim's own clothes, were examined but none contained similar fibres.

Extensive transfer tests were conducted in the laboratory, using the actual cardigan as a donor. These have been reported by Grieve *et al.* (1988b). The findings successfully refuted possible suggestions that the large numbers of fibres found on the bedsheet, nightdress, and wrist binding of the deceased had originated from a casual or secondary transfer during several hypothetical situations which could have occurred during an innocent visit to the apartment. When confronted with this evidence, the suspect admitted his guilt.

Example 3

The subject approached the victim from the rear, placed a hand over her mouth, dragged her into a nearby field, and raped her. The man was arrested, and his jeans and T-shirt seized as evidence within four hours.

The victim was wearing a dress made of very characteristic turquoise viscose rayon fibres. Over 1200 of these fibres were recovered from the suspect's T-shirt, mainly from the chest area. In addition, numerous examples of the same type were recovered from his jeans. Over 80 blue cotton fibres matching those in the suspects T-shirt were recovered from the victim's dress.

The exceptionally high number of recovered fibres supports the interpretation of a recent, considerable, primary contact. The presence of so many fibres on the chest area of the T-shirt is consistent with the victim's back resting and rubbing against the subject as he dragged his victim into the field. Again, the turquoise viscose fibres are not of a particularly common variety, and this, together with a demonstrated cross transfer involving all items submitted, helped to strengthen the evidential value of the findings.

Example 4

The suspect broke into the bedroom of the victim, who was wearing panties and a nightshirt and attempted to rape her. He was unknown to the victim and had not been on the premises before. He was wearing a fashion pullover, shirt, jeans, and underpants. The only items containing suitable target fibres were the pullover and the victim's magenta-coloured panties, which had been ripped.

The pullover contained two varieties of acrylic fibres (75%), grey wool fibres (10%), and coarse, projecting ribbon viscose fibres in three colours (15%). Examples of six of these fibres types were found on the victim's nightdress in proportions approximating to those contained in the pullover. Both types of acrylic fibre (7 and 5 fibres respectively) and one matching grey wool fibre were recovered from the panties. Subsequent in-lab. transfer tests showed that ribbon viscose fibres of this variety have a very low shed potential.

The victim's panties were made of magenta cotton, and the torn waist band contained pink and magenta nylon fibres. Examples of these three fibre types were found on the suspect's shirt, and a matching pink nylon fibre was found on his underpants. The evidential value is strengthened by an apparent cross transfer involving eight fibre types. The relatively low number of matching fibres transferred from, and adhering to, the panties can be accounted for by their limited surface area as they were of the micro-bikini variety. The apparent transfer from the pullover was consistent with both its composition and its shedding potential.

Examples 5 & 6

(Recovery of only one matching fibre)

- The suspect had invited a friend to his apartment and subsequently had intercourse with her, against her will, on top of his bed. He denied this, and claimed that at no time had she been in the bedroom.

  As the victim was naked at the time of the offence and as a demonstration of contact between the outer clothing would of course be of no value, a fibre transfer examination was not requested.

However, in the combings from the victim's pubic hair was a long, dark golden yellow polyester fibre. This fibre was not delustred and was pentalobal in cross-section, making it very characteristic. It matched the fibres in the suspect's bedspread.

While not helping with the question of consent, the unusual fibre type, together with its recovery from a seemingly incriminating location, did lend support to the victim's statement.

- After a breaking and entering, involving the smashing of a wooden storeroom door, a single long curly acrylic fibre was recovered, caught on splintered wood. The fibre was of an unusual polymer composition and was dyed yellow-brown, yielding an interesting chromatogram. It matched the fibres in a brown pullover taken from a suspect.

  As in the previous example, the value of the fibre lies very much in the unusual characteristics together with its point of recovery, indicating that the wearer of a garment containing these fibres had contact with the door during or shortly after it was splintered. Overemphasis should not be placed on single fibres, but I believe these two examples illustrate that, provided that they are highly characteristic, they should not be dismissed out of hand, on the grounds of quantity alone.

Example 7

(Low evidential value).

During the course of a party, the suspect is alleged to have removed the clothing of a six year old child in the bathroom and indecently assaulted her. He was wearing casual jeans made from black cotton and a wool pullover, mainly dark blue, but also containing small areas of red, green, and pale blue wool fibres. The child was wearing pyjamas and a pair of white cotton underpants. The pyjamas did not contain suitable target fibres, being made of very pale green (microscopically colourless) cotton.

Recovered from the pyjamas were 11 dark blue and two red fibres matching those in the pullover and 14 grey cottons matching those in the suspect's jeans. Two matching dark blue wools and seven grey cottons were recovered from the underpants.

In his defence, the suspect stated that the child had sat on his knee earlier in the evening, thus explaining the presence of the matching fibres on the pyjamas. The two dark blue wool fibres found on the underpants could be attributable to a secondary transfer. The grey cotton fibres were of a type with a totally flat, featureless spectrum (visible range), which research by Grieve *et al.* (1988a, 1990) have shown to be common, and to be present on many undergarments.

8.12 CONCLUSION

It should be evident from this chapter that there are no simple answers or solutions to the interpretation of fibre evidence. The basis of any meaningful analysis lies firstly in a full, detailed, and scientifically sound examination. Only with this sound foundation is it safe to take the next step of attempting to analyse what significance, if any, should be attached to the finding of fibres. Many relevant questions have been posed in this chapter. It will not always be possible to answer these, and it is important that lawyers in particular have realistic expectations. It is equally important that the forensic scientist presents evidence in an open, even handed manner and recognises legitimate alternative scenarios to explain his or her findings. It is hoped that the information contained in this chapter will assist both the providers and the users of fibres evidence with a sensible framework and set of standards with which to operate.

BIBLIOGRAPHY

Adolf, F. P. (1989) Faserdatei. *Proceedings, 5th Kriminaltechnisches Symposium - Textilkunde, Berlin.* 18–20.

Bressee, R. R. (1987) Evaluation of textile fibre evidence: a review. *J For Sci* **32**, 510–521.

Buckleton, J. S. & Evett, I. W. (1989) Personal communication.

Carroll, G. R., Lalonde, W. C., Gaudette, B. G., Hawley, S. L. & Hubert, R. S. (1988) A computerized data base for forensic textile fibres. *Can Soc For Sci J* **21**, 1–10.

Cook, R. (1989) Personal communication.

Cook, R. & Norton, D. (1982) An evaluation of mounting media for use in forensic textile fibre examination. *J For Sci Soc* **22**, 57–64.

Cook, R. & Wilson, C. (1986) The significance of finding extraneous fibres in contact cases. *For Sci Int* **32**, 267–273.

Coxon, A., Grieve, M. C. & Dunlop, J. (1991) Development of a method to determine fibre sheddability of casework garments and a further look at fibre transfer. *J For Sci Soc* (in press).

Deadman, H. (1984a) Fibre evidence and the Wayne Williams trial, Part 1. *FBI Law Enf Bull,* March, 13–20.

Deadman, H. (1984b) Fibre evidence and the Wayne Williams trial, Part 2. *FBI Law Enf Bull,* **53**, 10–19.

Evett, I. W. (1990) The theory of interpreting scientific transfer evidence. In: Machly, A. & Williams, R. L. (eds) *Forensic Science Progress,* Vol. 4. Springer Verlag, Berlin, Heidelberg, 141–180.

Fong, W. (1989) Analytical methods for developing fibres as forensic science proof: a review with comments. *J For Sci* **34**, 295–311.

Fong, W. & Inami, S. H. (1986) Results of a study to determine the probability of chance match occurrences between fibres known to be from different sources. *J For Sci* **31**, 65–72.

Gaudette. B. G. (1988) The forensic aspects of textile fibre examination. In: Saferstein, R. (ed.), *Forensic science handbook,* Vol. 2. Prentice Hall, Englewood Cliffs, NJ, 209–272.

Gianelli, P. C. & Imwinkelreid, E. J. (1986) *Fibres—scientific evidence.* The Mitchie Co., Charlottesville, VA, 1040–1060.

Grieve, M. C. (1983) The role of fibres in forensic science examinations. *J For Sci* **28**, 877–887.

Grieve, M. C. (1990) Fibres and their examination in forensic science. In: Maehly, A. & Williams, R. L. (eds) *Forensic Science Progress*, Vol. 4. Springer Verlag, Berlin, Heidelberg, 44–125.

Grieve, M. C. & Dunlop, J. (1990) A study of foreign fibres found on garments and its applicability to a Bayesian interpretation of fibre evidence. Presented at 12th International Association of Forensic Scientists meeting, Adelaide, Australia.

Grieve, M. C., Dunlop, J. & Haddock, P. S. (1988a) An assessment of the value of blue, red and black cotton fibres as 'target' fibres in forensic science investigations. *J For Sci* **33**, 1332–1344.

Grieve, M. C., Dunlop, J. & Haddock, P. S. (1988b) Transfer experiments involving acrylic fibres. *For Sci Int* **40**, 267-277.

Grieve, M. C. & Kotowski, T. M. (1986) An improved method for preparing fibre cross sections. *J For Sci Soc* **26**, 29–34.

Jackson, G. & Lowrie, C. N. (1987) Secondary transfer of fibres. Presented at 11th International Association of Forensic Scientists meeting, Vancouver, Canada.

Jackson, G. & Cook, R. (1986) The significance of fibres found on car seats. *For Sci Int* **32**, 275–281.

Laing, D. K., Boughey, L. & Hartshorne, A. W. (1990) The standardisation of thin layer chromatographic systems for comparisons of fibre dyes. *J For Sci Soc* **30**, 299–308.

Laing, D. K., Hartshorne, A. W. & Harwood, R. J. (1986) Colour measurements on single textile fibres. *For Sci Int* **30**, 65–77.

Laing, D. K., Hartshorne, A. W., Cook, R. & Robinson, G. (1987) A fibre data collection for forensic scientists—collection and examination methods. *J For Sci* **32**, 364–369.

Lawton, M. E., Buckleton, J. S. & Walsh, K. A. J. (1988) An International survey of the reporting of hypothetical cases. *J For Sci Soc* **28**, 243–252.

Mitchell, E. J. & Holland, D. (1979) An unusual case of identification of transferred fibres. *J For Sci Soc* **19**, 23–26.

Pailthorpe, M. (1990) Recent developments in the colouration of fibres encountered in forensic examinations. Presented at 12th International Association of Forensic Scientists meeting, Adelaide, Australia.

Palenik, S. & Fitzsimmons, C. (1990) A simple method for the sectioning of single fibres, Parts 1 & 2. *The Microscope* **38**, 187–195; 313–320.

Parybyk, A. & Lokan, R. J. (1986) A study of the numerical distribution of fibres transferred from blended fabrics. *J For Sci Soc* **26**, 61–68.

Pounds, C. A. (1975) The recovery of fibres from the surface of clothing for forensic examinations. *J For Sci Soc* **15**, 127–132.

Pounds, C. A. & Smalldon, K. W. (1975) The transfer of fibres between clothing materials during simulated contacts and their persistence during wear. Part 2—Fibre persistence. *J For Sci Soc* **15**, 29–37.

Robertson, J., Kidd, C. B. M. & Parkinson, H. M. P. (1982) The persistence of textile fibres transferred during simulated contacts. *J For Sci Soc* **22**, 353–360.

Robertson, J. & Lloyd, A. K. (1984) Observations on the redistribution of textile fibres. *J For Sci Soc* **24**, 3–7.

Robertson, J. & Olaniyan, D. (1986) The effect of garment cleaning on the recovery and redistribution of transferred fibres. *J For Sci* **31**, 73–78.

Saferstein, R. (1981) *Criminalistics—An introduction to forensic science*. Prentice Hall, NJ, 172.

Salter, M., Cook, R. & Jackson A. R. (1987) Differential shedding from blended fabrics. *For Sci Int* **33**, 155–164.

Starrs, J. (1990) But query: forensic scientist? Plenary lecture, presented at 12th International Association of Forensic Scientists meeting, Adelaide Australia.

Stoney, D. (1987a) Interpretation issues on multiple fibre cases. Presented at 11th International Association of Forensic Scientists meeting, Vancouver, Canada.

Stoney, D. (1987b) Fundamental principles in the evaluation of associative evidence. Presented at 11th International Association of Forensic Scientists meeting, Vancouver, Canada.

Wiggins, K. G. & Allard, J. E. (1987) The evidential value of fabric car seats and car seat covers. *J For Sci Soc* **27**, 93–101.

Wiggins, K. G., Cook, R. & Turner, Y. J. (1988) Dye batch variation in textile fibres. *J For Sci* **33**, 998–1007.

Index